Probes of Structure and Function of Macromolecules and Membranes

VOLUME II

Probes of Enzymes
and Hemoproteins

Johnson Research Foundation Colloquia

Energy-Linked Functions of Mitochondria
Edited by Britton Chance
1963

Rapid Mixing and Sampling Techniques in Biochemistry
Edited by Britton Chance, Quentin H. Gibson, Rudolph H. Eisenhardt, K. Karl Lonberg-Holm
1964

Control of Energy Metabolism
Edited by Britton Chance, Ronald W. Estabrook, John R. Williamson
1965

Hemes and Hemoproteins
Edited by Britton Chance, Ronald W. Estabrook, Takashi Yonetani
1966

Probes of Structure and Function of Macromolecules and Membranes
Volume I   Probes and Membrane Function
Edited by Britton Chance, Chuan-pu Lee, J. Kent Blasie
1971

Probes of Structure and Function of Macromolecules and Membranes
Volume II   Probes of Enzymes and Hemoproteins
Edited by Britton Chance, Takashi Yonetani, Albert S. Mildvan
1971

Proceedings of the Fifth Colloquium of the
Johnson Research Foundation, School of Medicine,
University of Pennsylvania, Philadelphia, Pennsylvania,
April 19-21, 1969

Probes of Structure
and Function of
Macromolecules and Membranes

VOLUME II
Probes of Enzymes and Hemoproteins

Edited by

Britton Chance
Takashi Yonetani

Johnson Research Foundation, School of Medicine
University of Pennsylvania, Philadelphia, Pennsylvania

Albert S. Mildvan

Institute for Cancer Research
Fox Chase, Philadelphia, Pennsylvania

Academic Press New York and London **1971**

CHEMISTRY

ACADEMIC PRESS, INC.
111 Fifth Avenue, New York, New York 10003

United Kingdom Edition published by
ACADEMIC PRESS, INC. (LONDON) LTD.
Berkeley Square House, London W1X 6BA

LIBRARY OF CONGRESS CATALOG CARD NUMBER: 77-158814

PRINTED IN THE UNITED STATES OF AMERICA

CONTENTS

PART 1. APPLICATION OF PROBES TO THE MECHANISM OF OLIGOMERIC ENZYMES

v

CONTENTS

vi

PART 2. STRUCTURAL INTERACTIONS IN
LIGAND BINDING IN HEMOPROTEINS

CONTENTS

CONTENTS

CONTENTS

CONTRIBUTORS

Adams, M. J., Department of Biological Sciences, Purdue University, Lafayette, Indiana

Antonini, Ernaldo, Centro di Biologia Molecolare, Consiglio Nazionale delle Ricerche, Istituto di Chimica Biologica, Citta Universitaria, Rome, Italy

Argos, Patrick, Department of Physiology and Biophysics, Washington University School of Medicine, St. Louis, Missouri

Asakura, Toshio, Johnson Research Foundation, School of Medicine, University of Pennsylvania, Philadelphia, Pennsylvania

Banaszak, L. J., Department of Physiology and Biophysics, Washington University School of Medicine, St. Louis, Missouri

Bayne, Ronald A., Department of Chemistry, University of South Florida, Tampa, Florida

Beetlestone, J. G., Department of Chemistry, University of Ibadan, Ibadan, Nigeria

Beinert, H., Institute for Enzyme Research, University of Wisconsin, Madison, Wisconsin

Bernhard, S., Institute of Molecular Biology, University of Oregon, Eugene, Oregon

Berger, R. L., Laboratory of Technical Development, National Institutes of Health, Bethesda, Maryland

Blokzijl, M. F. Y., Laboratory of Biochemistry, B. C. P. Jansen Institute, University of Amsterdam, Amsterdam, The Netherlands

Blumberg, W. E., Bell Telephone Laboratories, Inc., Murray Hill, New Jersey

Boers, W., Laboratory of Biochemistry, B. C. P. Jansen Institute, University of Amsterdam, Amsterdam, The Netherlands

Brand, K., Max-Planck Institut für Ernährungsphysiologie, Dortmund, Germany

Brand, Ludwig, Department of Biology, McCollum-Pratt Institute, Johns Hopkins University, Baltimore, Maryland

Bright, Harold J., Department of Biochemistry, School of Medicine, University of Pennsylvania, Philadelphia, Pennsylvania

Brill, Arthur S., Department of Materials Science, University of Virginia, Charlottesville, Virginia

Brocklehurst, J. R., Department of Biochemistry, Oxford University, Oxford, England

Brunori, Maurizio, Centro di Biologia Molecolare, Consiglio Nazionale delle Ricerche, Istituto di Chimica Biologica, Citta Universitaria, Rome, Italy

Bunkenburg, J., Johnson Research Foundation, School of Medicine, University of Pennsylvania, Philadelphia, Pennsylvania

Caughey, Winslow S., Department of Chemistry, University of South Florida, Tampa, Florida

Chance, Britton, Johnson Research Foundation, School of Medicine, University of Pennsylvania, Philadelphia, Pennsylvania

Chiancone, Emilia, Centro di Biologia Molecolare, Consiglio Nazionale delle Ricerche, Istituto di Chimica Biologica, Citta Universitaria, Rome, Italy

Conrad, Richard H., McCollum-Pratt Institute, Johns Hopkins University, Baltimore, Maryland

Czerlinski, G. H., The Medical School, Northwestern University, Chicago, Illinois

De Vijlder, J. J. M., Laboratory of Biochemistry, B. C. P. Jansen Institute, University of Amsterdam, Amsterdam, The Netherlands

Drabkin, D., Department of Biochemistry, School of Dental Medicine, University of Pennsylvania, Philadelphia, Pennsylvania

Drott, Henry R., Johnson Research Foundation, School of Medicine, University of Pennsylvania, Philadelphia, Pennsylvania

Edelstein, Stuart J., Section of Biochemistry and Molecular Biology, Cornell University, Ithaca, New York

Ehrenberg, Anders, Biofysiska Institutionen, Stockholms Universitet, Karolinska Institutet, Stockholm, Sweden

Eigen, M., Max-Planck-Institut für Physikalische Chemie, Göttingen, Germany

Englander, S. W., Department of Biochemistry, School of Medicine, University of Pennsylvania, Philadelphia, Pennsylvania

Fabry, T. L., IBM Watson Laboratory, Columbia University, New York, New York

Gibson, Quentin H., Section of Biochemistry and Molecular Biology, Cornell University, Ithaca, New York

Gilmour, M., Department of Chemistry, Yale University, New Haven, Connecticut

Gohlke, James R., McCollum-Pratt Institute, Johns Hopkins University, Baltimore, Maryland

Gutfreund, H., Molecular Enzymology Laboratory, Department of Biochemistry, University of Bristol, Bristol, England

Hagman, Lars-Ove, Institute of Inorganic and Physical Chemistry, University of Stockholm, Stockholm, Sweden

Hartzell, C. R., Institute for Enzyme Research, University of Wisconsin, Madison, Wisconsin

Heimbürger, Gunilla, Biofysiska Institutionen, Stockholms Universitet, Karolinska Institutet, Stockholm, Sweden

Heitz, James R., McCollum-Pratt Institute, Johns Hopkins University, Baltimore, Maryland

Hochstrasser, R., Department of Chemistry, University of Pennsylvania, Philadelphia, Pennsylvania

Horecker, B. L., Department of Molecular Biology, Albert Einstein College of Medicine, New York, New York

Hunsley, J. R., Department of Biology, Harvard University, Cambridge, Massachusetts

Huntley, Thomas E., Department of Biochemistry, School of Medicine, University of Connecticut Health Center, Farmington, Connecticut

Ilgenfritz, Georg, Department of Biology, State University of New York at Buffalo, Buffalo, New York

Irvine, D. H., Department of Chemistry, University of Ibadan, Ibadan, Nigeria

Kayne, F. J., Johnson Research Foundation, School of Medicine, University of Pennsylvania, Philadelphia, Pennsylvania

Keyes, Melvin, Laboratory for Biophysical Chemistry, Chemistry Department, University of Minnesota, Minneapolis, Minnesota

Kierkegaard, Peder, Institute of Inorganic and Physical Chemistry, University of Stockholm, Stockholm, Sweden

Kim, J., IBM Watson Laboratory, Columbia University, New York, New York

King, Tsoo E., Department of Chemistry, State University of New York at Albany, Albany, New York

Koenig, S. H., IBM Watson Laboratory, Columbia University, New York, New York

Kowalsky, A., Department of Biophysics, Yeshiva University, Bronx, New York

Kretsinger, R. H., Laboratoires de Biophysique et de Biochemie Genetique, Universite de Geneve, Geneve, Switzerland

Larsson, Lars-Olor, Institute of Inorganic and Physical Chemistry, University of Stockholm, Stockholm, Sweden

Leigh, J. S., Jr., Johnson Research Foundation, School of Medicine, University of Pennsylvania, Philadelphia, Pennsylvania

Levine, Michael, Department of Physiology and Biophysics, Washington University School of Medicine, St. Louis, Missouri

Love, Warner E., Thomas C. Jenkins Department of Biophysics, Johns Hopkins University, Baltimore, Maryland

Lumry, Rufus, Laboratory for Biophysical Chemistry, Chemistry Department, University of Minnesota, Minneapolis, Minnesota

Marmsen, B. J. M., Laboratory of Biochemistry, B. C. P. Jansen Institute, University of Amsterdam, Amsterdam, The Netherlands

Margoliash, E., Protein Section, Department of Molecular Biology, Abbott Laboratories, North Chicago, Illinois

Mason, H. S., Department of Biochemistry, University of Oregon Medical School, Portland, Oregon

Mathews, F. Scott, Department of Physiology and Biophysics, Washington University School of Medicine, St. Louis, Missouri

McConnell, H. M., Stauffer Laboratory for Physical Chemistry, Stanford University, Stanford, California

McCoy, Sue, Department of Orthopedic Surgery, University of Virginia School of Medicine, Charlottesville, Virginia

McMurray, C. H., Molecular Enzymology Laboratory, University of Bristol, Bristol, England

McPherson, A., Jr., Department of Biological Sciences, Purdue University, Lafayette, Indiana

Mildvan, Albert S., Institute for Cancer Research, Fox Chase, Philadelphia, Pennsylvania

Moss, Thomas, IBM Watson Laboratory, Columbia University, New York, New York

Muijsers, A. O., Laboratory of Biochemistry, B. C. P. Jansen Institute, University of Amsterdam, Amsterdam, The Netherlands

Myer, Y. P., Department of Chemistry, State University of New York at Albany, Albany, New York

Nicholls, P., Department of Biochemistry, State University of New York at Buffalo, Buffalo, New York

Ogata, R. T., Stauffer Laboratory for Physical Chemistry, Stanford, California

Orme-Johnson, W. H., Institute for Enzyme Research, University of Wisconsin, Madison, Wisconsin

Ozols, Juris, Department of Biochemistry, School of Medicine, University of Connecticut Health Center, Farmington, Connecticut

Padlan, Eduardo A., Physics Department, University of the Philippines, Quezon City, Philippines

Park, Jane Harting, Department of Physiology, Vanderbilt University Medical School, Nashville, Tennessee

Peisach, J., Departments of Pharmacology and Molecular Biology, Albert Einstein College of Medicine, Yeshiva University, Bronx, New York

Perham, Richard N., Department of Biochemistry, Cambridge University, Cambridge, England

Phillips, W. D., Central Research Department, E. I. Du Pont de Nemours & Co. (Inc.), Wilmington, Delaware

Radda, G. K., Department of Biochemistry, Oxford University, Oxford, England

Reed, T., Department of Physiological Chemistry, School of Medicine, University of Würzbürg, Würzbürg, Germany

Riggs, A. F., Department of Zoology, University of Texas, Austin, Texas

Rossmann, M. G., Department of Biological Sciences, Purdue University, Lafayette, Indiana

Rumen, N. M., Department of Dermatology, Medical College of Georgia, Augusta, Georgia

Sandberg, Howard E., Molecular Biophysics Laboratory, Washington State University, Pullman, Washington

Schejter, Abel, Department of Biochemistry, Tel Aviv University, Tel Aviv, Israel

Schevitz, R. W., Department of Biological Sciences, Purdue University, Lafayette, Indiana

Schillinger, W. E., IBM Watson Laboratory, Columbia University, New York, New York

Schoenborn, B. P., Department of Biology, Brookhaven National Laboratory, Upton, New York

Schonbaum, Gregory R., Department of Biochemistry, University of Alberta, Edmonton, Canada

Schuster, T. M., Biological Sciences Group, University of Connecticut, Storrs, Connecticut

Seamonds, Bette, Williams Pepper Laboratory, Department of Pathology, University of Pennsylvania, Philadelphia, Pennsylvania

Shulman, R. G., Bell Telephone Laboratories, Murray Hill, New Jersey

Simon, Sanford R., State University of New York at Stony Brook, Stony Brook, New York

Slater, E. C., Laboratory of Biochemistry, B. C. P. Jansen Institute, University of Amsterdam, Amsterdam, The Netherlands

Smiley, I. E., Department of Biological Sciences, Purdue University, Lafayette, Indiana

Smillie, Lawrence B., Department of Biochemistry, University of Alberta, Edmonton, Canada

Smythe, George A., Department of Chemistry, Arizona State University, Tempe, Arizona

Strittmatter, Philipp, Department of Biochemistry, School of Medicine, University of Connecticut Health Center, Farmington, Connecticut

Stryer, L., Department of Molecular Biophysics and Biochemistry, Yale University, New Haven, Connecticut

Suelter, C. H., Department of Biochemistry, Michigan State University, East Lansing, Michigan

Tasaki, Akira, Faculty of Engineering Science, Osaka University, Osaka, Japan

Theorell, Hugo, Department of Biochemistry, Nobel Medical Institute, Stockholm, Sweden

Thomas, J. A., Department of Biochemistry, School of Medicine, University of South Dakota, Vermillion, South Dakota

Tiesjema, R. H., Laboratory of Biochemistry, B. C. P. Jansen Institute, University of Amsterdam, Amsterdam, The Netherlands

Trentham, D. R., Molecular Enzymology Laboratory, Department of Biochemistry, The Medical School, University of Bristol, Bristol, England

Tsernoglou, D., Department of Physiology and Biophysics, Washington University School of Medicine, St. Louis, Missouri

Turley, Patricia A., Department of Chemistry, Yale University, New Haven, Connecticut

Turner, D. C., Department of Biology, McCollum-Pratt Institute, Johns Hopkins University, Baltimore, Maryland

VanGelder, B. F., Laboratory of Biochemistry, B. C. P. Jansen Institute, University of Amsterdam, Amsterdam, The Netherlands

Velick, S. F., Department of Biochemistry, University of Utah Medical School, Salt Lake City, Utah

Wade, M., Department of Physiology and Biophysics, Washington University School of Medicine, St. Louis, Missouri

Weber, G., Division of Biochemistry, Department of Chemistry, University of Illinois, Urbana, Illinois

Weiner, Henry, Department of Biochemistry, Purdue University, Lafayette, Indiana

Welinder, Karen, Department of Biochemistry, University of Alberta, Edmonton, Canada

Wilson, D. F., Johnson Research Foundation, School of Medicine, University of Pennsylvania, Philadelphia, Pennsylvania

Wonacott, A. J., Department of Biological Sciences, Purdue University, Lafayette, Indiana

Wyman, Jeffries, The "Regina Elena" Institute for Cancer Research, Rome, Italy

Wüthrich, Kurt, Institut für Molekularbiologie and Biophysik, Eidgenössisch Technische Hochschule, Zürich, Switzerland

Yonetani, Takashi, Johnson Research Foundation, School of Medicine, University of Pennsylvania, Philadelphia, Pennsylvania

Yong, F. C., Department of Chemistry, State University of New York at Albany, Albany, New York

PREFACE

A chemical description of the mechanism of action of an enzyme would constitute the epitome of a correlation between structure and function. The use of various probes and physical techniques with enzymes permits an approach to this goal.

Hence, as a second part of the colloquium, the soluble enzymes are considered with a special emphasis upon the dehydrogenases and the hemoproteins. Whether this made a logical combination of enzymes, the organizers found it difficult to decide; it seems, nevertheless, a tradition initiated by Hugo Theorell that the work on dehydrogenases nicely complements work on the hemoproteins, and, indeed, it has occurred similarly in this colloquium, not only with respect to outlines of reaction mechanisms at the kinetic level, but particularly with respect to the types of novel techniques employed in the study of the two classes of enzymes, fluorescence probes, spin labels, and x-ray crystallography.

Since the preceding Johnson Foundation Colloquium on "Hemes and Hemoproteins" held in 1966, spectacular advances have been made in the theory and application of various physical methods to hemoproteins. These advances, coupled with further refinement of the three-dimensional molecular structures of myoglobin and hemoglobin, have made possible correlations of physical data from multiple techniques on functionally significant structural changes upon ligand binding in myoglobin and hemoglobin. The structural changes in the monomeric unit and their possible role in cooperative phenomena in dimers and tetramers can be now discussed in detail. X-ray crystallography, ultraviolet and infrared spectroscopy, high-resolution NMR, nuclear relaxation, EPR, magnetic susceptibility, field- and temperature-jump relaxation, flash photolysis, and equilibrium methods have been applied to myoglobin and hemoglobin to examine ligand-induced conformation changes and the mechanism of cooperativity and information transfer between subunits.

Finally, the cytochromes themselves, peroxidases and catalases, electron-transporting and redox catalysts of the hemoprotein family (including cytochrome oxidase) are also treated with particular emphasis on the functions of their sixth coordination positions.

The organizers of the various sessions were: Session I: J. K. Blasie (Chairman), A. Azzi, H. R. Drott, M. Cohn, H. Schleyer, A. Kowalsky; Session II: M. Cohn (Chairman), A. S. Mildvan, A. Kowalsky; Session III: T. Yonetani (Chairman), H. Schleyer, H. R. Drott, G. R. Schonbaum, C. P. Lee; Session IV: C. P. Lee (Chairman), B. T. Storey, A. Azzi, L. Mela, J. K. Blasie, T. Ohnishi; Session V: D. DeVault (Chairman), B. Chance, D. Mayer, J. A. McCray, J. K. Blasie, H. Schleyer, A. Kowalsky.

Volume I contains Sessions I, IV, and V; Volume II contains Sessions II and III.

Britton Chance
Takashi Yonetani
Albert S. Mildvan

CONTENTS OF VOLUME I

Probes of Enzymes and Hemoproteins

PART 1

APPLICATION OF PROBES TO THE MECHANISM OF OLIGOMERIC ENZYMES

COMPLEXES OF DOGFISH MUSCLE LACTATE DEHYDROGENASE WITH COENZYME AND SUBSTRATE: THEIR RELATION TO THE APOENZYME STRUCTURE AS SEEN AT 5 Å RESOLUTION

M.J. Adams, A. McPherson, Jr., M.G. Rossmann,
I.E. Smiley, R.W. Schevitz, and A.J. Wonacott

Department of Biological Sciences
Purdue University, Lafayette, Indiana 47907

A 5 Å resolution electron density map of dogfish M_4 LDH, which crystallizes in spacegroup F422 with a = 146.9 Å, c = 155.1 Å and a single polypeptide chain (MW 35000) in the asymmetric unit (1) has been obtained using heavy atom derivatives binding to one or more of three sites in the subunit (2). The molecule can be seen to have 222 symmetry with the molecular center at z = 1/4.

Crystals have been obtained in which the coenzyme (NAD), or parts of it, have been diffused into the apoenzyme. Adenosine binds without change of molecular symmetry; NAD, however, and larger fragments of the coenzyme, e.g., AMP and ADP, cause a reduction in molecular symmetry from 222 to 2.

Co-crystallization of LDH, NAD, and pyruvate (the product of oxidation) has resulted in the formation of crystals with spacegroup P422, with a = 95.4 Å, c = 86.1 Å. The asymmetric unit is again a single polypeptide chain and the molecule has 222 symmetry. Packing of the tetramers is different from that in the apoenzyme.

The 5 Å resolution model of LDH, with some subsequent improvements resulting from incorporating a further derivative into the phasing, will be used in considering the changes which occur on binding coenzyme or its parts or in forming the abortive ternary complex.

The Apoenzyme Model at 5 Å

In the 5 Å electron density distribution, the spaces between molecules and the surfaces between subunits could be recognized. The individual subunit is irregularly shaped and seems to have two halves joined by a "neck", a region wide enough to accomodate four or five polypeptide chains (Figure 1). The complete molecule is seen in Figure 2. The packing can be seen schematically in Figure 3 (from ref. 3).

Figure 1. Single subunit of 5 Å model of dogfish M_4 LDH.

Figure 2. Model of complete molecule at 5 Å resolution.

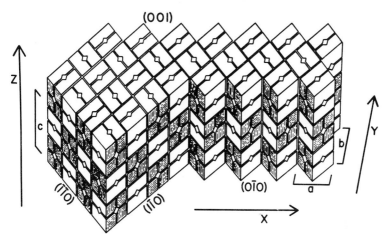

Figure 3. Schematic model of packing (from ref. 3).

Two of the three heavy atom sites used in phase determination were obtained by reaction of mercurials hydroxymercuribenzoate (HMB), dimercuryacetate (DMA), Baker dimercurial (BDM), and mercuric chloride (in the later map) with the apoenzyme. In solution, on addition of four equivalents of HMB to dogfish LDH (4,5), no activity is lost. In the crystal, HMB at the same molar ratio substitutes to site A which lies between subunits (see Figure 4, a view of one half of the molecule looking down the 4_2 axis). At higher molar

Figure 4. View of 1/2 molecule viewed down 4_2 axis.

7

ratios, HMB and mercuric chloride bind to a second site (site B), which is more buried and is close to the "neck" of the subunit. Substitution at this "essential" site results in loss of activity.

Platinum (II) containing compounds (e.g., platinum ethylenediamine dichloride (PED), which has been used for phase determination) substitute a third heavy atom site. The sites of substitution of PED are shown in an (h0ℓ) difference Fourier projection (Figure 5). Site A is the non-essential

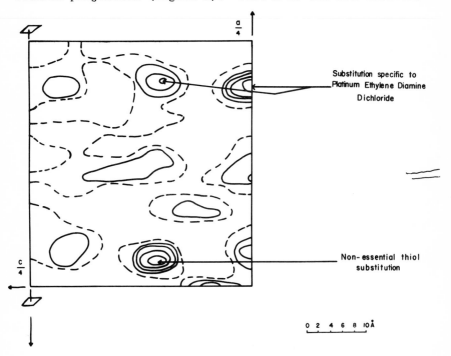

Figure 5. (h0ℓ) difference Fourier projection of PED derivative.

thiol site found in, for example, HMB. Site C is close to a subunit boundary and its position on the molecule can be seen in Figure 2. Higher concentrations of platinum (II) compounds [(Pt^{II}(NH$_3$)$_4$)$^{++}$ and (Pt^{II}Cl$_4$)$^{--}$] induce a conformation change of the same kind as does NAD.

Binding of Coenzyme Fragments

Diffusion of various coenzyme fragments into LDH has resulted in intensity changes or a spacegroup change. Some of these results are shown in Table I. We have shown by a kinetic analysis that adenosine, AMP and ADP are competitive inhibitors for dogfish M_4 LDH while nicotinamide is a non-competitive inhibitor (Figures 6a and 6b).

TABLE I

Compound	Concen-tration	Photograph	Inhibition
Adenosine	1 mM	Intensity changes	Dogfish M_4, competitive inhibitor
AMP	10 mM	Difference scarcely detectable	Dogfish M_4, competitive inhibitor
	25 mM	Space group changes	
ADP	1-4 mM	No differences	Dogfish M_4, competitive inhibitor
	10 mM	Space group changes	
Adenosine diphosphate ribose	1 mM	Space group change	Bovine heart, competitive inhibitor
NAD	1.4 mM	Space group change	Active with substrate
NADH	.05 mM	Space group change	Active with substrate
Nicotinamide	10 mM	Intensity changes	Dogfish M_4, non-competitive inhibition
Methyl nicotinamide	50 mM	No changes	Dogfish M_4, no inhibition

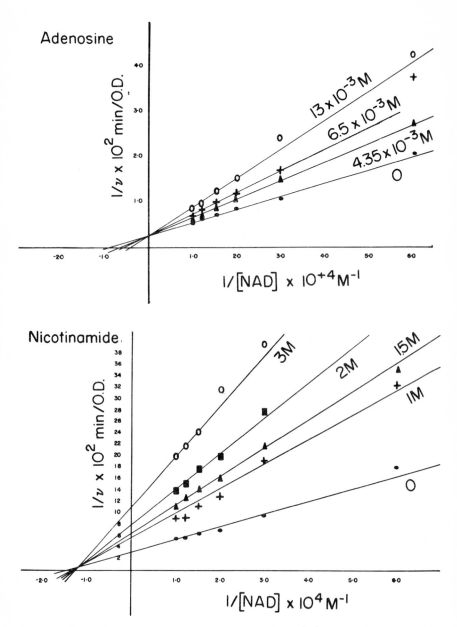

Figure 6. Lineweaver Burk plot showing (A) inhibition of LDH by adenosine and (B) inhibition of LDH by nicotinamide.

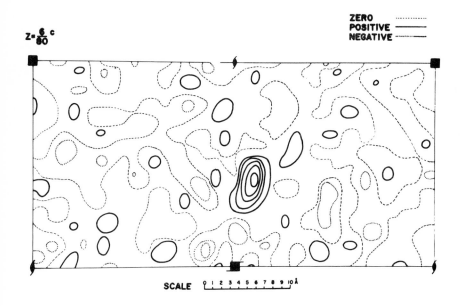

Figure 7. Section z = 6/80 of adenosine difference map at 5 Å resolution.

Of those compounds which do not change the spacegroup three dimensional X-ray data to 5 Å resolution have been obtained for the adenosine complex. Using these data and phases obtained from the most recent 5 Å refinement (including HMB, DMA, BDM, PED, and $HgCl_2$ rather than $NaAuCl_4$) a difference map has been obtained (Figure 7) with one peak three times the background. This peak is at z = 6/80,* and is between the subunits in a cavity which is readily accessible from the edge of the molecule and is about 12 Å from the C

*Use of a previous set of data and poorer phases more dependent on A site substitution, resulted in a peak of similar size close to the A site. Unless both "old" data and "old" phases are used, this site cannot be reproduced. It must be assumed that this previous peak (1) must be the result of systematic phasing errors and errors in the data.

11

site. The closeness of the adenosine site to the C site and
the fact that only platinum (II) derivatives give spacegroup
changes similar to NAD, suggests that adenosine might bind
close to the NAD binding site in the binary complex. This
result is supported by the observation of competitive inhibition for adenosine.

The Binary Complex with NAD

A photograph of the h0ℓ zone showing the change of spacegroup to $C4_22_12$, arising on diffusion of NAD into the apoenzyme is seen in Figure 8. The same spacegroup change results
when NADH and LDH are co-crystallized. The new spacegroup requires the loss of one 2-fold axis perpendicular to c. The
new asymmetric unit contains two polypeptide chains and the
required molecular symmetry is only 2-fold instead of 222.

The relationship between the molecules in the two spacegroups may be investigated either by making use of the

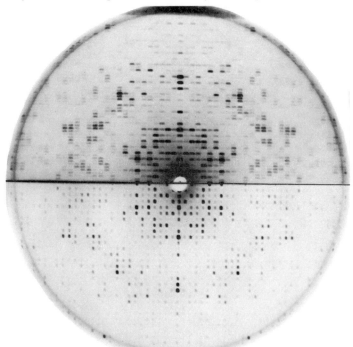

Figure 8. Photograph of (h0ℓ) zone of dogfish M_4
LDH and of NAD complex.

structure amplitudes and phases of the apoenzyme or indepen-
dently of these. A difference Patterson $(|F_{APO}| - |F_{NAD}|)^2$
of the (h0ℓ) zone of the spots common to the apoenzyme and
the binary complex is shown (Figure 9). The use of this function assumes the same signs for these terms in the apoenzyme
and the NAD complex. An (h0ℓ) Patterson of the new spots of
the coenzyme complex is seen in Figure 10. This does not make
use of the apoenzyme data. Both functions show as their major
feature a large negative peak about 5 Å from the origin.

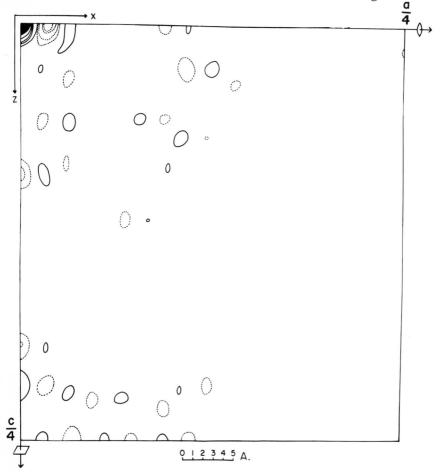

Figure 9. (h0ℓ) difference Patterson between NAD-LDH complex and LDH. ℓ even coefficients only.

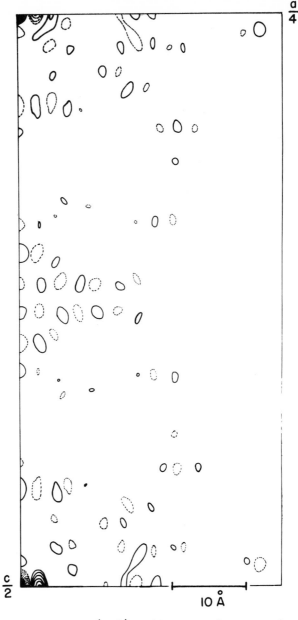

Figure 10. (h0ℓ) Patterson of new spots of NAD–LDH complex.

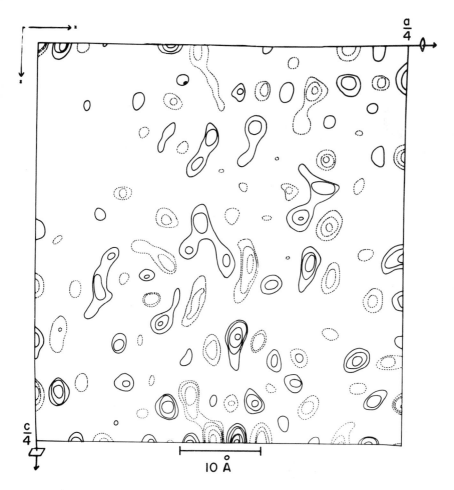

Figure 11. (h0ℓ) difference Fourier between NAD–LDH complex and LDH. Resolution, 2.8Å.

Since the difference Patterson, assuming similar signs for the (h0ℓ) reflections of both spacegroups seems valid, the signs may be used explicitly. An (h0ℓ) difference Fourier projection (Figure 11) shows pairs of positive and negative peaks about 5 Å apart in regions distant from the 4-fold axis. The region close to the 4-fold axis, which is the model of the apoenzyme is mainly solvent, is empty. At least

15

one subunit must therefore have moved by about 5 Å. The
direction of the shift is along (110) since the (h0ℓ) Patterson projection (containing 2 molecules rotated by 90°) shows
only one vector.

If NAD is diffused into HMB containing LDH crystals, the
same spacegroup change results. A set of signs for the centric zones which are independent of the apoenzyme and which
include "old" and "new" type reflections may, therefore, be
obtained if the mercury positions are determined. Further,
these heavy atom positions in the binary complex give an indication of the minimum movement of the subunits. A difference Fourier of the (h0ℓ) zone ($F_{HMB+NAD} - F_{NAD}$), phased by
the two independent mercury positions was calculated (Figure
12). One of the two mercury atoms is in the same position
as in the apoenzyme. The second, with smaller substitution,
has moved away from the 4_2 axis in the (110) direction. This
shift of the A sites of half the subunits of 6.5 Å is consistent with the vector of about 5 Å obtained from the difference Pattersons and the Fourier based on the apoenzyme
phases. It suggests a movement of two of the subunits away
from each other on diffusion of NAD.

The Abortive Ternary Complex

A ternary complex LDH-NAD-pyruvate has been reported
both for heart and muscle isoenzymes of LDH (6,7,8,9,10) and
has been shown to be associated with an absorption band at
340 μ. Crystals of a ternary complex of dogfish M_4 LDH, NAD,
and pyruvate have been obtained from 2 M ammonium sulphate at
pH 8.0. Both the solution before crystallization and re-
dissolved crystals show the 340 μ absorption band. Co-crystallization of LDH, NAD, and oxalate lead to crystals which
are isomorphous with those of the abortive ternary complex.

The spacegroup of these crystals is $P4_2 1 2$ with one polypeptide chain as the asymmetric unit and molecules with 222
symmetry. A photograph of the (hk0) zone of the abortive
ternary complex is compared with that for the apoenzyme in
Figure 13.

16

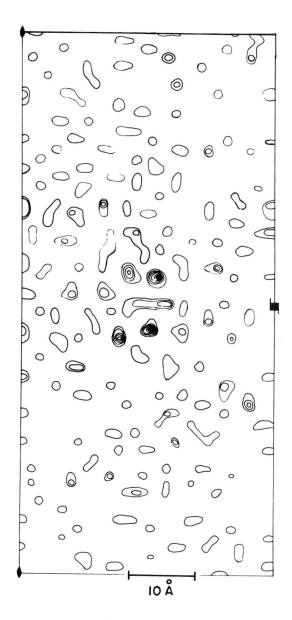

10 Å

Figure 12. (h0ℓ) difference Fourier
between HMB-NAD-LDH and NAD-LDH.

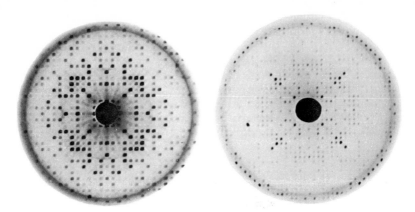

Figure 13. Photographs of (hk0) zones of apoenzyme and ternary complex.

Specific intensity differences have been observed on diffusion of HMB, $HgCl_2$ and $PtCl_6^{--}$ solutions. A difference Patterson $(|F_{HgCl_2}| - |F_{ternary}|)^2$ for the (h0ℓ) zone is shown (Figure 14). A three dimensional study of these derivatives is being carried out.

Addendum*

Recently the structure of dogfish M_4LDH has been obtained at 2.8 Å resolution (10), 5 Å resolution electron density and difference electron density maps of the abortive ternary complex (11), and the binary complex (12).

References

1. Rossman, M.G., B.A. Jeffery, P. Main, and S. Warren. Proc. Nat. Acad. Sci., 57, 515 (1967).

2. Adams, M.J., D.J. Haas, B.A. Jeffery, A. McPherson, Jr., H.L. Mermall, M.G. Rossman, R.W. Schevitz, and A. Wonacott. J. Mol. Biol.,

3. Labaw, L.W., and M.G. Rossman. J. Ultrastruc. Res.,

*Note added June 12, 1970.

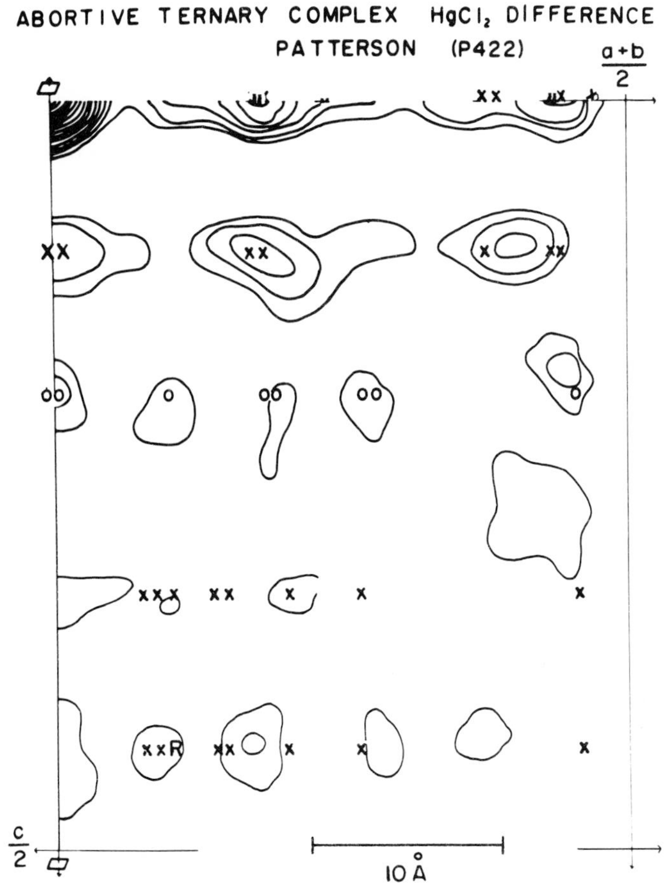

Figure 14. Difference Patterson for mercury chloride derivative of ternary complex. X,O vectors for possible interpretation.

4. DiSabato, G., and N.O. Kaplan. Biochemistry, 2, 776 (1963).

5. Fondy, T.P., J. Everse, G.A. Driscoll, F. Castillo, F.E. Stollzenbach, and N.O. Kaplan. J. Biol. Chem., 240, 4219 (1965).

6. Fromm, H.J. Biochim. Biophys. Acta, 52, 199 (1961).

7. Fromm, H.J. J. Biol. Chem., 238, 2938 (1963).

8. Gutfreund, H., R. Cantwell, C.H. McMurray, R.S. Criddle, and G. Hathaway. Biochem. J., 106, 638 (1968).

9. Kaplan, N.O., J. Everse, and J. Admiraal. Ann. N.Y. Acad. Sci., 151, 400 (1968).

10. Adams, M.J., G.C. Ford, R. Koekoek, P.J. Lentz, Jr., A. McPherson, Jr., M.G. Rossmann, I.E. Smiley, R.W. Schevitz, A.J. Wonacott. Nature, in preparation.

11. Koekoek, R., M.G. Rossmann, I.E. Smiley. J. Mol. Biol., in preparation.

12. Adams, M.J., A. McPherson, Jr., M.G. Rossmann, R.W. Schevitz, A.J. Wonacott. J. Mol. Biol., in press.

DISCUSSION

Gutfreund: I would like to make a comment in connection with the structural change involved in the ternary complex formation. When lactic dehydrogenase is dissolved at pH 6, the sedimentation constant is slightly lowered to 6.8 from the value of 7.2 in neutrality. If you add NAD, the sedimentation constant remains unchanged. If you add NAD and pyruvate (they are both in the ternary complex that you use for crystallography), the sedimentation constant at pH 6 goes up to the neutral value of 7.2. These changes in sedimentation constant do not involve a change. Activity sedimentation in the presence of substrates always gives s = 7.2.

The other point about structural change that I want to make is that kinetically one only sees one step in the combination of the nucleotides and the enzyme, but there are two steps after the second substrate adds and one should look for evidence whether NAD alone gives any structural change in solution. The crystallographic change you get with NAD alone may be something which specifically applies to the crystal.

Finally, I have heard that you have found that NAD is, in fact, bound across two subunits. Is this correct?

Adams: The NAD is bound across two subunits (1).

We can co-crystallize the LDH and NADH and get the same space group as we get when we soak NAD into LDH, so that if it is a crystallographic phenomenon, at least this is certainly a low-energy state of the binary complex, and it does have lower symmetry than does the ternary complex.

Gutfreund: Is the binary complex in the low energy state in the crystal?

Adams: In the crystal, yes, certainly. But it isn't just a case of wondering, as we would have had NAD^+ only been soaked in, whether it was the start of a movement which couldn't go any further because of restrictions of crystal forces.

Bernhard: If you have 2-2-2 symmetry in the crystal before you add the effector (coenzyme), and adding the coenzyme gives a conformational change, how close to the original center of symmetry are the various coenzyme molecules that are coming in?

Adams: All the information that we have about the coenzyme is based, really, on the adenosine sites in the apoenzyme, which is about here and is probably 20-30 Å away from the 2-2-2 point, which is the center of the molecule. On the other hand, a shift is defined in terms of the most active -SH position, which is not the coenzyme position, but it is the only position that we can define with the heavy atom used. We know where it is in the apoenzyme, we know where it is in the binary complex. This is probably about 15-20 Å from the center of the tetramer, so that we know that a point 15 Å away moves by about 6 Å. We don't really know what happens closer to the NAD site which we are assuming for now is where the adenosine binds.

Bernhard: The point of my question is that if one binds across two subunits, depending on how close to the symmetry axis this is taking place, one will wind up in the product with (to some extent) non-equivalent binding sites. Since part of this binding site is presumably involved in the catalytic site, if the catalytic site on one subunit and the "A" binding site on another subunit are fairly close, one should not expect four equivalent catalytic sites in a tetramer composed of covalently identical subunits.

Theorell: Does the binding with the coenzyme cause any change in the conformation within the subunits?

Adams: From the evidence which we have at the moment from the difference Fourier, using the apoenzyme phases, it did appear that there was a consistent 5-6 Å shift all over the molecule; there were positive and negative peaks 5 Å apart; it looked as if the shift of the subunit is larger than the movements within the peptide chain of the unit, but until we get further information, I would not like to be sure.

Theorell: I ask, because in the alcohol dehydrogenase there is a gross conformational change following the hooking-on of the co-enzyme, but in your case there is a small change?

Adams: I think this is a smaller change, and probably not
very much of the enzyme moved by more than the 6-7 Å shift
which occurred at the site of the enzyme that we have defined.

Leigh: I should like to ask about the activity of the enzyme in the crystalline state.

Adams: We have shown that the enzyme is active in the crystalline state.

Reference

1. Adams, M.J., A. McPherson, Jr., M.G. Rossmann, R.W. Sche-
    vitz, A.J. Wonacott. J. Mol. Biol., in press.

SOME RECENT RESEARCH ON ALCOHOL DEHYDROGENASES

Hugo Theorell

Department of Biochemistry, Nobel Medical Institute
Stockholm, Sweden

For a number of years after the liver alcohol dehydrogenase was crystallized in 1948 (1), it was regarded as being
homogeneous and pure, actually for some fifteen years. Then,
in 1960, Dr. Ungar published a paper (2) saying that there
was some activity on steroid alcohol derivatives present in
the commercial preparations of LADH available; unfortunately
that was published in the Medical Bulletin of the University
of Minnesota, so very few people read it; I am afraid we
did not either. But in 1963-1964, George Waller, who was then
in Stockholm, but is now back in Oklahoma, made a survey of
the substrate specificity of the LADH (3). We knew it was
very broad, so we thought perhaps also steroid alcohols could
be substrates. Some activity was indeed found, and it was
also established that the only substrates were of the 3-β-
hydroxy type. We later found, with Reynier and Sjövall (4),
that the fusion between ring A and B also has to be of the
β-type. That was interesting, because the reaction might
have some connection with the transformation of cholesterol
to bile acids where the 3-α-OH in the cholesterol has to be
transformed to the 3-β-hydroxy in the bile acids.

When we had published that, I had a letter from Howard
Ringold in Shrewsbury, Massachusetts, saying that they had
compared different commercial preparations and found them to
vary considerably in activity towards the steroids. We
therefore took the collected mother liquors from several years
of LADH preparations, and found considerable amounts of steroid active LADH enzymes; so within a couple of months (1966)
we crystallized one of these enzymes (5). It later turned
out to be a hybrid, containing 50-50 amounts of only ethanol
active, respectively, also steroid active parts. I am coming back to this isoenzyme "ES" later.

Regina Pietruszko, working until 1967 in Ringold's group, since then with us in Stockholm, found that the crude liver extracts contain a number of isoenzymes, as demonstrated by gel electrophoresis with activity staining of the bands. Figure 1 shows some results.

To the right we have a rather crude extract. The most acidic fraction is on the top, the most basic at the bottom. Separation on a DEAE cellulose column gave the bottom fraction ("SS", to the right) and the other fractions. To the right, we see "ES", followed by two faint bands, ES' and ES". Then follows a strong band which is the "classical" LADH, (EE), followed by two more acid isoenzymes, EE" and EE". In other experiments, we have also seen the fractions SS' and SS". S-containing fractions partially removed by recrystallization.

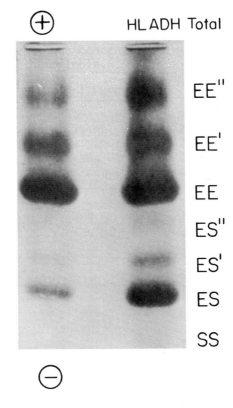

Figure 1.

26

The fractions with primes are spontaneously formed from the fractions without primes at slightly elevated temperature and pH. Von Wartburg (personal communication) has evidence in favor of the reaction being reversible. We shall, in the following, restrict ourselves to the main components EE, ES and SS. The nomenclature may need some explanation. E stands for activity on simple alcohols, like ethanol, without activity on steroids, S for activity on steroids with weaker activity on ethanol. The reason for the two letter names is that the molecules consist of <u>halves</u>, identical for EE and SS, nonidentical for ES.

Vallee's group (6) had some ultracentrifuge results on LADH in 8 M urea and mercaptoethanol indicating dissociation to tetramers of M.W. ∿20,000. In view of later facts, this cannot be maintained, since analysis of the amino acid sequence indicates chains of ∿380 amino acids, corresponding to M.W. ∿40,000. I shall come back to this later.

Another proof came from our dissociation experiments in urea (7). My collaborator Åkeson has prepared practically pure SS, and we could therefore compare EE, ES and SS in the urea-split state and after recombination through dilution and dialysis. Figure 2 shows the results.

EE, on the top, gives one band, slightly more acidic than SS at the bottom. ES (middle), on the contrary, gives two bands, one corresponding to E, the other to S.

The gel electrophoresis picture after recombination is shown in Figure 3.

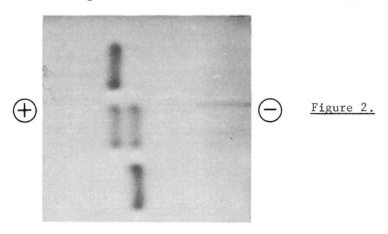

Figure 2.

SS SE EE   SS SE EE
—————————   —————————
Reconst   Controls

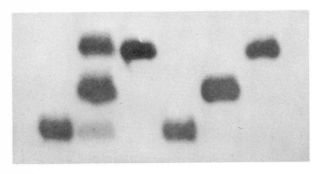

Figure 3.

⊖

EE and SS, when split and recombined, are found in the same positions as the never split controls. ES gives all three, EE, ES and SS, and most of the ES, as statistically expected.

It was, of course, interesting to find out the chemical background of the difference in substrate specificity between the E- and S-type. As soon as we had the ES crystallized three years ago, we compare the amino acid composition of EE and ES. As seen from Table I, they were very closely alike. The only difference slightly outside experimental error was 47 threonine in ES ("LADH$_S$") against 50 in EE. Part of this difference is real, however.

In the last two years, my collaborator Hans Jörnvall, who leared the sequence technique with J.I. Harris in Cambridge, has worked on the primary structure of EE, ES and SS. Since the chains E and S each contain some 380* amino

*Added in proof: November 15, 1970. The total sequences of E and and S are now determined by Jörnvall (Europ. J. Biochem., 16, 41 (1970). The total number of residues is 374 for both E and S. In addition to the mutation Glu → Gln in position 17 there

TABLE I.

Amino Acid Composition of LADH$_S$ and LADH$_E$

Amino Acid	Residues/Molecule[1]	
	LADH$_S$	LADH$_E$
Aspartic acid	54	54
Threonine	47	50
Serine	53	53
Glutamic acid	65	66
Proline	45	46
Glycine	83	83
Alanine	62	61
Valine	84	82
Methionine	19	19
Isoleucine	49	49
Leucine	55	54
Tyrosine	8	8
Phenylalanine	37	38
Tryptophan[2]	4	4
Lysine	63	62
Histidine	15	14
Arginine	24	25

[1] Based on a molecular weight of 83,300.
[2] Calculated from absorption at 280 mμ and tyrosine content.

acids, it has not been easy, but the work is now near its conclusion (8). It is of particular interest to mention that E and S differ only in three positions: number 17 from the acetylated N-terminus has glutamic acid in E, glutamine in S. Near position 130 and 7 residues apart, Jörnvall found threonine in E, isoleucine in S in the first position, and arginine in E, serine in S in the second. All three ex-changes correspond to one-letter differences in the genetic code and are thus most probable mutations.

are five more mutations in the direction E → S: 94, Thr → Ile; 101, Arg → Ser; 110, Phe → Leu; 115, Asp → Ser; and 366, Glu → Lys. These mutations give three extra positive charges to S compared with E, in agreement with the electrophoretic isoen-zyme pattern.

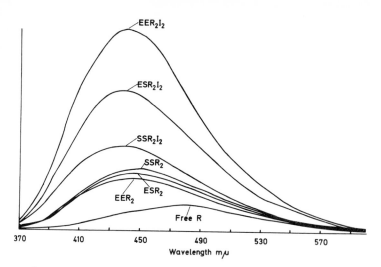

Figure 4. Fluorescence emission spectra of Free NADH (R), binary complexes with EE, ES and SS and ternary complexes with isobutyramide (I). Excitation at 330 mμ.

It is now interesting to find out whether the difference in substrate specificity between E and S is accompanied by differences in the coenzyme binding. We therefore studied, with de Zalenski and B. Kopec (9), the binary complexes of EE, ES and SS with NADH, or the ternary with NADH + isobutyramide. Figure 4 shows a comparison of the fluorescence intensities of these complexes.

It can be seen that the comparatively small fluorescence of free NADH is shifted in wave length and intensified upon coupling to the enzymes, somewhat more for SS than for EE, ES being in between. If now isobutyramide is added, the picture is dramatically changed: the EE-complex gives a very big increase, the SS a rather low increase in fluorescence, ES, as before, in between. This gave us a first hint that the two parts E and S in ES perhaps act independently of one another in spite of the fact that the coupling of E to S is necessary for the activity. I shall bring some further evidence to illustrate this.

Table II compares the fluorescence increase factors at 410 mμ occurring on the coupling of NADH to EE (Q_{ER}), resp. to SS (Q_{SR}) at some different pH values and different buffers. It is seen that they are the same at pH 6 to 8, but differ at higher pH values.

TABLE II.

Q = Fluorescence Intensities of Enzyme-NADH Complexes Compared with NADH at 410 mμ

pH	Q_{ER}		Q_{RS}	
	Phosphate	Tris	Phosphate	Tris
6	10.5	–	11.0	–
7	10.7	10.5	11.0	11.5
8	10.7	10.0	10.0	10.5
9	10.5*	10.6	10.1	15.5
9.35	–	11.0	–	14.0
10	11.0*	21.0	8.8	15.6

*+ glycine.

The dissociation constants between EE and NADH (K_{ER}) and SS and NADH (K_{SR}) determined both by titration, and calculated from the rate constants k_1 and k_2 (E + R $\underset{k_2}{\overset{k_1}{\rightleftarrows}}$ ER, resp. S + R $\underset{k_2}{\overset{k_1}{\rightleftarrows}}$ SR) are seen in Table III.

The dissociation constants were determined by spectro-fluorometric titration of enzyme with NADH at 410 mμ. The "on" velocity rate constants were determined, with de Zalenski (9), by stopped flow. In the case of k_1, enzyme and NADH were mixed and the fluorescence increase measured. For determining the "off" velocities, k_2, ER or SR in one syringe was mixed with a solution of phenanthroline from the other syringe, which prevents the "on" reaction by form-ing a coenzyme-competitive complex with the free enzyme formed in the "off" reaction. Thus the "off" velocity con-stant k_2 can be determined directly from the decrease in fluorescence.

The "on" velocities of R, k_1, are rather similar for EE and SS at all pH values studied. The "off" velocities, k_2, on the contrary, show great variations with pH.

The comparison between the dissociation constants deter-mined either by titration or by calculation from the rate constants, $k_2/k_1 = K$, show good agreement for SS in the whole region investigated, pH 7 to 10.

The titrated values for K_{SR} are lower than those for K_{ER} over the whole pH scale, and the K_{SR} values remain inde-

TABLE III

Rate Constants from Stopped Flow Experiments Compared with Equilibrium Constants

"On" velocity, k_1 $\mu M^{-1} \times sec^{-1}$: EE or SS mixed with NADH (R)

"Off" velocity, k_2, sec^{-1}: EER_2 or SSR_2 mixed with phenanthroline

Equilibrium constants, K_{ER} or K_{ES} from fluorimetric titration

Equilibrium constants, K_{ER} or K_{ES} calculated from $k_2/k_1 = K$

Buffer	pH	k_1		k_2		$k_2/k_1=K$		K Eq. (titr.)	
		E+R	S+R	ER	SR	E	S	E	S
PO$_4$	6	20	26	2	1.4	0.10	–	0.10	0.08
PO$_4$	7	17	16	3	1.4	0.18	0.09	0.10	0.09
PO$_4$	8	14	14	2.9	1.2	0.21	0.09	0.28	0.09
PO$_4$+ glycine	9	10	13	2.3	1.4	0.23	0.11	0.56	0.16
PO$_4$+ glycine	10	10	7.5	4	4	0.40	0.53	5.0	0.52
Tris	9.35	10	10	3	2	0.30	–	0.30	0.08
Tris	10	3?	10	3.5	2	1.2	–	3.3	0.09

pendent of pH from 6 to 10 in Tris buffer at pH 9 to 10, phosphate buffer pH 6 to 8. In glycine buffers at pH 9 to 10, the values are markedly higher than in Tris.

The K_{ER} is much more sensitive to higher pH than K_{SR}. For example, at pH 10 in Tris buffer, K_{ER} = 3.3 μM whereas K_{SR} = 0.09 μM. The differences observed between glycine and Tris buffers at pH 9 to 10 may be caused by ternary complex formation between buffer substances and the binary enzyme-coenzyme complexes.

In summary, the differences between EE and SS both in fluorescence intensities and dissociation constants reflect

TABLE IV

Specific Activities of EE, ES and SS on Different Substrates

Enzyme	Ethanol pH 10	Acetaldehyde pH 7	5β DHT* pH 7
EE	15,500	250,000	< 300
ES	∿10,000	130,000	2,300
SS	∿ 4,000	∿ 7,000	∿5,000

*5β DHT = 5β-Androstane-17β-OH-3-one.

a change in the coenzyme binding sites, in addition to the change in substrate specificity, both caused by the mutations in three amino acid positions out of 370.*

Table IV shows that all three enzymes are active with ethanol as substrate, EE considerably stronger than SS, and ES in the middle. Remarkably enough, SS is practically inactive in the reverse reaction with aldehyde. EE is practically inactive with 5β-DHT. In all cases, the hybrid, ES, is intermediate between EE and SS.

Now, let me show you another experiment we performed in order to test whether there is any interaction between the halves in the hybrid molecule, ES. At pH 10, in Tris buffer, EE and SS have different fluorescence intensity factors, Q_{ER} = 21, Q_{SR} = 15.4, and dissociation constants are 3.3 and 0.09 μM, respectively. We therefore titrated: 1) 0.70 μN ES, and 2) 0.35 μN EE + 0.35 μN SS with increments of 0.38 μM NADH. Identical titration curves were obtained (see Figure 5).

Moreover, the deflections in inches calculated from the Q- and K-values for EE and SS were identical with the observed values (Table V). This would also happen if all three enzymes dissociate into halves at this high dilution. However, in separate experiments, it was found that no dissociation occurs.

*Note added in proof May 10, 1970: The final numbers are six out of 374 (Jörnvall, unpublished results).

33

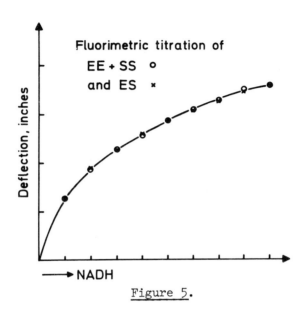

Figure 5.

TABLE V

Fluorimetric Titration

0.35 μN EE + 0.35 μN SS with 9 steps, 0.38 μM R
0.70 μN ES + 0.35 μN SS with 9 steps, 0.38 μM R

Calculated amounts of ER and SR, if K_{ER} = 3.3 μM, Q = 21; K_{SR} = 0.09 μM, Q = 15.4.

Compared with observed deflections in titration 0.70 μN ES.

Step	Total R μM	[ER] calc.	[SR] calc.	[R] free	Inches deflection calc. for EE+SS	Obs. for ES
1	0.38	0.015	0.22	0.15	1.27	1.30
2	0.76	0.040	0.29	0.43	1.91	1.93
3	3.42	0.16	0.34	2.82	3.79	3.72

These results indicate that there is <u>no interaction</u> between the united halves in the molecules EE, ES or SS interfering with the function of the coenzyme binding sites. In other words, since the separated 40,000 units are inactive, the time course of the inactivation in urea has been followed by Pietruszko and myself (7), reactivation occurs through dimerization, and, thereby, it does not matter at all for an E-half, whether it is reactivated by coupling to another E or to an S, or vice versa.

The inactivity of the monomers and the activation by dimerization indicate that the active sites are formed by a conformational change following the dimerization. The three amino acid residues which are different in E and S apparently do not play any role in this process. The X-ray crystallographic work on the structure of EE carried out by Brändén <u>et al</u>. in Uppsala is now nearly finished, so we may hope to know within a short time where the mutations are situated in the molecule.

References

1. Bonnichsen, R., and A. Wassén. Arch. Biochem. Biophys., <u>18</u>, 361 (1948).

2. Ungar, F. Univ. Minn. Med. Bull., <u>31</u>, 226 (1960).

3. Waller, G., H. Theorell, and J. Sjövall, J. Arch. Biochem. Biophys., <u>111</u>, 671 (1965).

4. Reynier, M., H. Theorell, and J. Sjövall. Acta Chem. Scand., <u>23</u>, 1130 (1969).

5. Theorell, H., S. Taniguchi, Å. Åkeson, and L. Skurský. Biochem. Biophys. Res. Comm., <u>24</u>, 603 (1966).

6. Drum, D.E., J.H. Harrison, T.-K. Li, J.L. Bethune, and B.L. Vallee. Proc. U.S. Nat. Acad. Sci., <u>57</u>, 1434 (1967).

7. Pietruszko, R., and H. Theorell. Arch. Biochem. Biophys., <u>131</u>, 288 (1969).

8. Jörnvall, H. Biochem. Biophys. Res. Comm., <u>35</u>, 542 (1969).

9. Theorell, H., Å. Åkeson, B. Liszka-Kopec and C. de Zalenski. Arch. Biochem. Biophys., in press.

THE INTERACTION OF FLUORESCENCE PROBES WITH NATIVE AND DENATURED HORSE LIVER ALCOHOL DEHYDROGENASE*

James R. Gohlke, Ludwig Brand, James R. Heitz
and Richard H. Conrad

McCollum-Pratt Institutes
Johns Hopkins University, Baltimore, Maryland

The basic theme of this symposium has been the application of physical probes to the study of structure and function of macromolecules. Since it is usually impossible to obtain direct physical measurements, such as pH or dielectric constant, at defined loci on a macromolecular structure, it is clearly valulable to use chromophoric molecules as indirect "reporters" of such physical parameters (1). In this report we present evidence that several fluorescent dyes bind non-covalently at sites of interest on horse liver alcohol dehydrogenase.[1] It is suggested that the spectroscopic response of these dyes to changes in their environment have potential in probing the physicochemical nature of the binding site on the protein surface. The effect of solvent environment on the fluorescence of these dyes has been evaluated in some detail, and was presented earlier in this symposium (2). Further, changes in the binding characteristics of these ligands may be useful in monitoring conformational changes in protein structure.

*Supported by NIH GM 11632. Manuscript submitted August 4, 1970.

[1]The following abbreviations are used: 1,8 ANS, 1-anilino-naphthalene-8-sulfonate; 2,6 p-TNA, 2-p-toluidinonaphthalene-6-sulfonate; 2,6-MANS, 2-anilino-N-methylnaphthalene-6-sulfonate; DNS, dimethylaminonaphthalenesulfonate; 2,6 cyclo C_6, 2-cyclohexylaminonaphthalene-6-sulfonate; 2,1 NH_2-SO_4, 2-aminonaphthalene-1-sulfonate; HL-ADH, horse liver alcohol dehydrogenase.

Since the discovery by Weber and Laurence (3) that 1,8 ANS binds to bovine serum albumin with a concomitant fluorescence enhancement, dyes of this class have been shown to bind to several proteins (4). Evidence has already been presented (5) that three positional isomers of ANS as well as rose bengal (a halogenated fluorescein derivative) bind near the active site of HL-ADH.

The evidence for such a specific interaction is based on the stoichiometry (two moles ligand per mole HL-ADH); the competitive binding relationship between the fluorochrome and the coenzymes; and the inhibition of enzymatic activity. The binding of these dyes to HL-ADH is consistent with a model of two equal and independent sites, similar to the binding characteristics of the coenzymes (6). The association constants for the equilibrium binding of a number of fluorochromes to HL-ADH are summarized in Table I.

In addition to the positional isomers of ANS, a number of other compounds related in structure have been shown to bind to HL-ADH. As indicated in Table II, all of the N-arylamino-naphthalenesulfonates exhibit an enhancement of fluorescence and a blue shift of the emission maximum in the presence of HL-ADH. Figure 1 shows typical emission spectra (uncorrected for monochromator and phototube response) for the binding of 2,6 m-TNS to HL-ADH (curve 1) and in phosphate buffer alone (curve 3). In the presence of NAD^+ and pyrazole (curve 4), the fluorescence is decreased relative to that with enzyme

TABLE I.

Binding Constants Between Horse Liver Alcohol
Dehydrogenase and Several Dyes

Rose Bengal	250×10^3
Auramine O	100×10^3
2,6 p-TNS	25×10^3
1,7 ANS	19×10^3
1,8 ANS	8×10^3
1,5 ANS	2.7×10^3

All measurements in 0.1 M sodium phosphate at pH 7.4, room temperature.

TABLE II.

Effect of L-ADH on the Fluorescence of
Several Dyes

Compound	λfree	λbound	Δλ	Enhancement
Auramine O	---	520	--	+
2,6 ANS	472	436	36	+
2,6 MANS	520	450	70	+
2,6 o-TNS	480	435	45	+
2,6 m-TNS	490	442	48	+
2,6 p-TNS	470	450	]0	+
2,6 DNS	444	444	0	-
1,5 DNS	513	510	3	-
2,6 cyclo C_6	434	434	0	-
2,1 NH_2-SO_3	405	405	0	-

+ indicates that enhancement is observed.

alone and the emission is shifted to the red indicating that
the dye is released from the protein. Spectra 2 through 5 (5
is the phosphate buffer base line) have been amplified ten-
fold over spectra 1. The peak at 410 mμ is due to Raman
scattering. The changes in both of these spectral parameters
are consistent with a binding site of low "Z" value (2).
Compounds such as DNS, N-cyclohexylaminonaphthalenesulfonate
and aminonaphthalenesulfonate which have a high quantum
yield in polar solvents are not highly sensitive to solvent environment do not show the marked fluorescence enhancement in
the presence of HL-ADH.

Table II indicates that auramine O, a diphenylmethane
derivative, also interacts with this enzyme to form a fluorescent complex with the emission centered at 520 mμ (7). Auramine O is of particular interest since it does not form a
fluorescent complex with a large number of other proteins including yeast alcohol dehydrogenase. Solvent studies with
this dye indicate that unlike the N-arylaminonaphthalenesul-
fonates, auramine O fluorescence is not sensitive to the solvent polarity but rather the solvent viscosity (11).

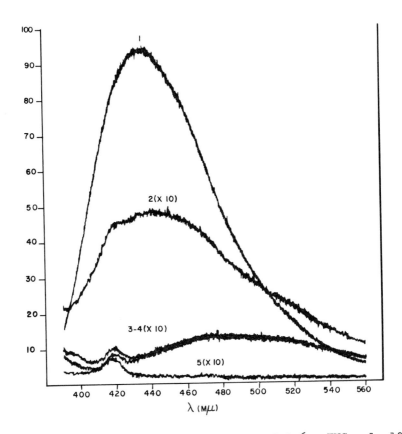

Figure 1. Technical emission spectra of 2,6 m–TNS. 1. 10.0 μM 2,6 m–TNS plus 3.1 μM HL–ADH; 2, with addition of 63.1 μM NAD$^+$ and 49.8 μM pyrazole; 3, 9.6 μM 2,6 m–TNS; 4, with addition of 52.5 μM NAD$^+$ and 41.5 μM pyrazole; 5, 0.1 M phosphate at pH 7.4. Spectra 2-5 are amplified ten fold relative to spectrum 1.

The absence of an increase in quantum yield for some of the dyes listed in Table II does not imply that they do not interact with HL–ADH. Evidence that binding does in fact occur is presented in Figure 2. Part A shows the emission of N-cyclohexylaminonaphthalenesulfonate in phosphate buffer, in the presence of HL–ADH, and in the presence of HL–ADH plus NAD$^+$-pyrazole. The small decrease in fluorescence is due only to dilution. Part B shows an increase in fluorescence

40

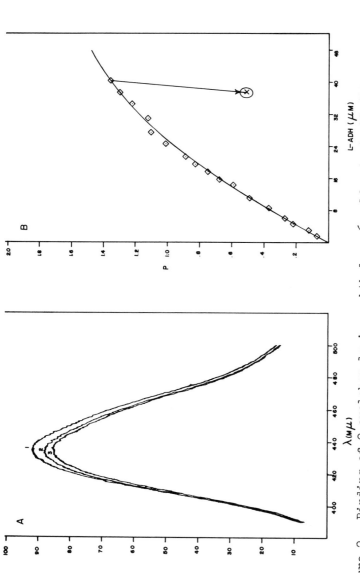

<u>Figure 2.</u> Binding of 2-cyclohexylaminonaphthalene-6-sulfonate to HL-ADH. A: Technical emission spectra of: 1, 25.5 μM dye; 2, with addition of 3.1 μM HL-ADH; 3, with addition of 52.9 μM NAD+ and 41.8 μM pyrazole. B: Polarization of dye fluorescence as a function of added HL-ADH. Dye concentration varied from 5.1 μM to 2.9 ℧M. X represents the polarization after the addition of 97.4 μM NAD+ and 77 μM pyrazole. The vertical axis represents percent polarization times ten.

41

polarization when enzyme is added to the dye. The arrow demonstrates that this increase in polarization can be reversed by the addition of NAD^+ and pyrazole. The coenzyme and pyrazole are known to form a tight ternary complex at the active site of HL-ADH (8). This experiment thus provides evidence that the fluorochrome binds at or near the active site region.

The coenzyme analogs N-benzylnicotinamide chloride (NBZ-NAmCl) and adenosine diphosphate ribose (ADPR) also bind at the active sites of HL-ADH. The effect of these molecules and NAD^+-pyrazole on the fluorescence of enzyme-dye mixtures is shown in Table III. It is of interest to note that although the fluorescence intensity of the auramine-O-HL-ADH complex is decreased by the coenzyme (NAD^+), the association constant is actually doubled. Further, kinetic measurements (Figure 3) indicate that auramine O is a non-competitive inhibitor of

TABLE III

Decrease in Fluorescence in the Presence of
Site Specific Reagents for L-ADH

Compound	NAD-PYR	NBzNAmCl	ADPR
Auramine O	+	+	+
2,6 ANS	+	+	+
2,6 MANS	+	+	+
2,6 o-TNS	+	+	+
2,6 m-TNS	+	+	+
2,6 p-TNS	+	+	+
2,6 p-TNS	+	+	+
2,6 DNS	−	−[a]	−
1,5 DNS	−	−[a]	−
2,6 Cyclo C_6	−	−[a]	−
2,1 NH_2-SO_3	−	−[a]	−

+ indicates that cofactor reverses fluorescence enhancement.
[a]Fluorescence decreases, but it is due to complexing of NBzNAmCl with free dye in solution.

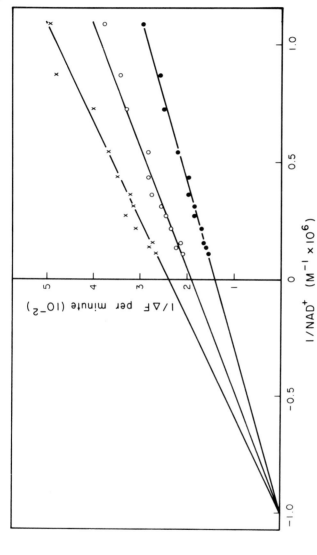

Figure 3. Noncompetitive inhibition of HL–ADH by auramine O. Cuvettes contained 3.0 ml reaction mixtures of 0.1 M phosphate at pH 7.4, and 5.77 mM ethanol. NAD$^+$ concentrations were varied from 92 µM to 920 µM. Curve 1 (●), reaction mixtures contained no inhibitor; Curve 2 (o), reaction mixtures contained 1.78 µM auramine O; Curve 3 (x), reaction mixtures contained 3.57 µM auramine O. Reactions were initiated by the addition of 8.8 x 10^{-12} moles HL–ADH.

HL-ADH activity with respect to both the substrate (ethanol) and the coenzyme (NAD$^+$). Equilibrium dialysis experiments indicate, however, that auramine 0 is released from the protein surface upon the formation of the enzyme-NAD$^+$-pyrazole ternary complex. In addition, auramine 0 is enzymatically competitive with respect to the somewhat bulkier substrate, cyclohexanol. The implication of this type of data is that the auramine 0 binding site is adjacent to, but does not overlap either the coenzyme or catalytic site.

In addition to using these fluorochromes to probe the nature of the enzyme surface, they have proven useful as probes for conformational changes. HL-ADH is denatured at acid pH (9) and, as is shown in Figure 4, there is a marked enhancement and blue shift in the emission of 2,6 p-TNS in the presence of denatured enzyme (curve 2) relative to native enzyme (curve 1). By analogy to experiments in pure solvent systems, the blue shift implies that there is also an increase in the quantum yield of the dye bound to denatured enzyme.

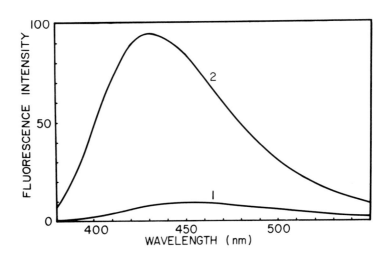

Figure 4. Emission spectra of 2,6 p-TNS in the presence of native and denatured HL-ADH. 1, 22.4 μM 2,6 p-TNS and 4.3 μM HL-ADH in 3.0 ml 0.1 M succinate pH 7.1; 2, 22.4 μM 2,6 o-TNS and 4.3 μM HL-ADH in 3.0 ml succinate pH 4.2.

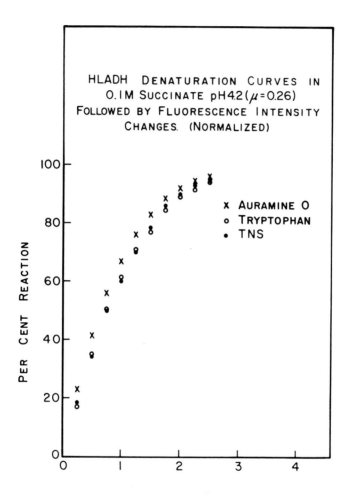

Figure 5. Kinetics of HL-ADH denaturation in acid at high ionic strength ($\mu = 0.26$). Each cuvette contained 0.3 µM HL-ADH in 0.1 M succinate, pH 4.2, at 29°. (o), changes in intrinsic protein fluorescence, λ_{ex} = 290 mµ, λ_{em} = 340 mµ; (x), changes in fluorescence of 4.4 µM auramine 0, λ_{ex} = 390 mµ, λ_{em} = 530 mµ; (●), changes in fluorescence of 6.7 µM p-TNS, λ_{ex} = 370 mµ, λ_{em} = 440 mµ. Ionic strength was increased with sodium chloride.

Since, however, the quantum yield of 2,6 p-TNS bound to native HL-ADH is approximately 0.48, the extent of enhancement is greater than can be accounted for by increases in yield and therefore reflects concomitant changes in the stoichiometry and/or affinity of the dye for the modified protein. The fluorescence observed under these conditions cannot be reduced by NAD^+ even in the presence of pyrazole. In contrast to the N-acrylaminonaphthalenesulfonates, auramine O does not form a fluorescent complex with denatured HL-ADH. Thus one observes a decrease in auramine O emission under denaturing conditions.

In addition to these extrinsic probes, the fluorescence of tryptophan, an intrinsic chromophore, decreases as the enzyme is denatured (10). Such a battery of assays provides an opportunity to study the kinetics of the denaturing process. Figure 5 is a plot of percent denaturation as a function of time, using each of the assays discussed. Under these particular denaturing conditions, all three probes show identical rates of denaturation.

It is hoped that these observations will enable one to obtain information about the topography, surface chemistry and conformation of this enzyme.

References

1. Horten, H.R., and D.E. Koshland, Jr. In Methods in Enzymology, Vol. 11, Academic Press, New York, 1967, p. 856.

2. Brand, L., C. Seliskar and D.C. Turner. Volume I, p.17

3. Weber, G. and D.J.R. Laurence. Biochem. J., 56, 31 (1954).

4. Edelman, G.M. and W.O. McClure. Accounts Chem. Res., 1, 65 (1968).

5. Brand, L., J.R. Gohlke and D.S. Rao. Biochemistry, 6, 3510 (1967).

6. Sund, H. and H. Theorell. In The Enzymes (P.D. Boyer, H. Lardy and K. Myrbäck, eds.), Vol. 7, 2nd Edition, Academic Press, New York, 1963, p. 25.

7. Conrad, R.H., J.R. Heitz and L. Brand. Biochemistry, 9, 1540 (1970).

8. Theorell, H. and T. Yonetani. Biochem. Z., 338, 537 (1963).

9. Blomquist, C.H. Arch. Biochem. Biophys., 122, 24 (1967).

10. Brand, L., Johannes Everse and Nathan O. Kaplan. Biochemistry, 1, 423 (1962).

11. Oster, G. and Y. Nishijima. Fortshor. Hochpolymer-Forsch., 3, 313 (1964).

THE USE OF A SPIN-LABELED ANALOG OF NAD TO STUDY THE MECHANISM OF ACTION OF ALCOHOL DEHYDROGENASE*

Henry Weiner and Albert S. Mildvan

Biochemistry Department, Purdue University
Lafayette, Indiana

and

Institute for Cancer Research
Philadelphia, Pennsylvania

In order to introduce a paramagnetic probe into the active site region of horse liver alcohol dehydrogenase (EC 1.1.1.1), the spin-labeling technique developed by McConnell was employed. Rather than covalently attaching the spin-label to the enzyme, an alternative approach was used. Since it had been shown that ADP-ribose is a potent competitive inhibitor with respect to the coenzyme (1), a spin-labeled analog of NAD was prepared (2). The compound, abbreviated ADP-R·, has its spin located in a region which corresponds to the position occupied by the ribotide bond to the nitrogen atom of the nicotinamide ring in NAD, as can be seen in Figure 1.

ADP-R· was shown to competitively inhibit alcohol dehydrogenase with respect to NAD, $K_I = 5 \pm 2$ μM. The direct interaction of ADP-R· with liver alcohol dehydrogenase was investigated by two methods: EPR line broadening (2) and the enhancement of the relaxation rate of water, $1/T_1$ (3). Both methods revealed that there are two classes of binding

*Supported in part by NSF Grant GB 7263 and USPHS Grants AM-09760 and GM-12446. The work was done during the tenure of an Established Investigatorship (ASM) of the American Heart Association. Journal Paper No. 3672 of the Agricultural Experiment Station, Purdue University, Lafayette, Indiana.

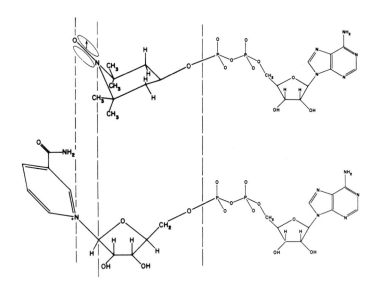

<u>Figure 1.</u>   The structures of ADP-R· and NAD (3).

sites for the radical, each possessing different dissociation
constants.   The dissociation constants are summarized in Table
I.   Since alcohol dehydrogenase has been shown to possess two
active sites, and since ADP-R· can be displaced from the
enzyme by NADH, the two tight binding sites for ADP-R· are
presumably the active sites.

It was observed that ε_b, the enhancement of the bound
ADP-R·, was a function of the amount of analog bound to the
enzyme.   As the occupancy changed from 0 to 2, ε_b decreased
from 81 to 13.   The enhancement remained at 13 when the additional spin labels were bound.   The reduction in ε_b as the
occupancy went from 0 to 2 is suggestive of a site-site interaction not previously detected in the kinetics or thermodynamics
of the reaction.

Additional evidence for site-site interactions came from
a study of the effect of temperature on $1/T_1$.   From this
study, the energy of activation for $1/T_1$ of H_2O can be calculated. E_a (Kcal/mole) changed from 2.9 ± 0.7 to 5.0 ± 0.2
as the occupancy increased from 0.8 to 2 (3).

50

indicates "closing" of H_2O site with increasing occupancy.

TABLE I

Dissociation Constants and Stoichiometry of Binary
Complexes Between Liver Alcohol
Dehydrogenase and ADP-R·

Method	$N_t{}^a$	K_D µM	N_w	K_D µM
EPR	2 ± 0.5	17 ± 8	5.2 ± 0.3	73.3 ± 9.4
PRR[b]	2.3	15 ± 6	5.7 ± 0.7	93 ± 9.6
Kinetics	-	5 ± 2		

[a]N is moles of ADP-R· bound per mole of enzyme in the
tight (N_t) and weak (N_w) sites.
[b]Proton relaxation rate.

No evidence could be found suggesting that one coenzyme
affected the binding of the second coenzyme molecule. However, it appears that an opening of the "water site" occurred
as occupancy of the NAD analog changed from 0 to 2. Such site
openings may be necessary to permit the entry of the substrate.

Substrates such as ethanol and acetaldehyde and an inhibitor for aldehyde, isobutyramide, were found to form ternary
complexes with the enzyme and the NAD analog. The formation
of the ternary complexes were detected by titrating the binary
complex of enzyme-ADP-R· with the above mentioned compounds
and observing the decrease in ε_b. The decrease in ε_b suggests
that the substrate is displacing water from the radical, e.g.

$$E(ADP-R·)(H_2O) + \text{substrate} \rightleftharpoons E(ADP-R·)(\text{substrate}) + H_2O$$

A dissociation constant for ethanol from the ternary complex was calculated to be 1.1 mM, which agrees with a value
determined by Theorell and Yonetani for E(NAD)(ethanol)(4).

To measure directly the binding of substrate to the en-
zyme-ADP-R· complex, the transverse relaxation rate, $1/T_2$, of
the substrate's protons were determined (5). Using the Solomon and Bloembergen equations (6), which relate the transverse
relaxation rate to the distance, r, between the protons and
the spin, it was possible to calculate the distances between
the substrate and the NAD analog.

$$r = 544 \left[T_2 \left(3.5 \, \tau_c + \frac{6.5 \, \tau_c}{1 + 6.15 \, (10^{22}) \, \tau_c^2} \right) \right]^{1/6}$$

τ_c is the correlation time for the dipolar interaction. The constants take into account that the measurements were made at 60 Mc/sec and the spin of the nitroxide radical is 1/2.

The absolute value of τ_c is not known, but estimates for the limiting values of τ_c could be made (5). The values of τ_c used and the distances calculated for the ternary complexes with ethanol are presented in Table II. As can be seen in the Table, the wide range of τ_c used did not greatly affect the calculated distances between the spin and the protons being observed. The mean value of r for the various substrates are presented in Table III.

TABLE II

Distances Calculated Using the Solomon and Bloembergen Equations Between the Unpaired Electron in ADP-R· and the Protons of Ethanol in Ternary Complexes with Liver Alcohol Dehydrogenase

τ_c [a] sec x 10^{12}	Distance in Å	
	$-CH_2-$	$-CH_3$
0.41 [b]	3.1	2.7
8.2 [c]	4.7	4.1
28.0 [d]	5.2	4.6
mean	4.1 ± 1.1	3.6 ± 1.0

[a] τ_c assumed from measurements as indicated below:
[b] From $1/T_1$ measurements with ADP-R· $(H_2O)_4$.
[c] From pyruvate carboxylase–Mn–pyruvate values (7).
[d] From $1/T_1$ measurements with enzyme–ADP-R·–H_2O.

TABLE III

Mean Distances Between the Unpaired Electron in ADP-R· and Protons on Substrates

Proton	r in Å
$\underline{CH_3}-CH_2-OH$	3.6
$CH_3-\underline{CH_2}-OH$	4.1
$CH_3-\underline{CHO}$	3.1
$(\underline{CH_3})_2CH-\overset{\overset{O}{\|}}{C}-NH_2$	≤ 3.9

The distances calculated for the ternary complexes, and their effects on $1/T_1$ of H_2O, are consistent with the structures shown in Figure 2. These structures permit the direct

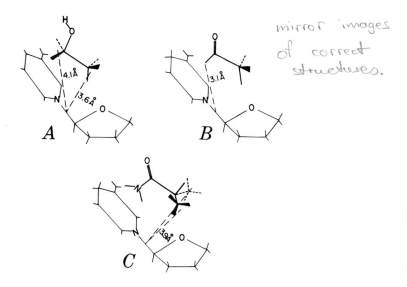

mirror images of correct structures.

Figure 2. Relative positions on liver alcohol dehydrogenase of the appropriate coenzymes and (A) ethanol, (B) acetaldehyde, and (C) isobutyramide, as determined by the distances measured to E-ADP-R· (5).

transfer of hydrogen from substrate to coenzyme. Recently, however, Schellenberg proposed that a tryptophan residue is involved as an intermediate in the hydrogen transfer (8). From our distance calculations, it can be shown that it is not possible to stack the indole ring of tryptophan between the nicotinamide ring and the substrate (5). However, the two rings can be edge-to-edge as shown in Figure 3. This edge-to-edge structure is still able to permit tryptophan to

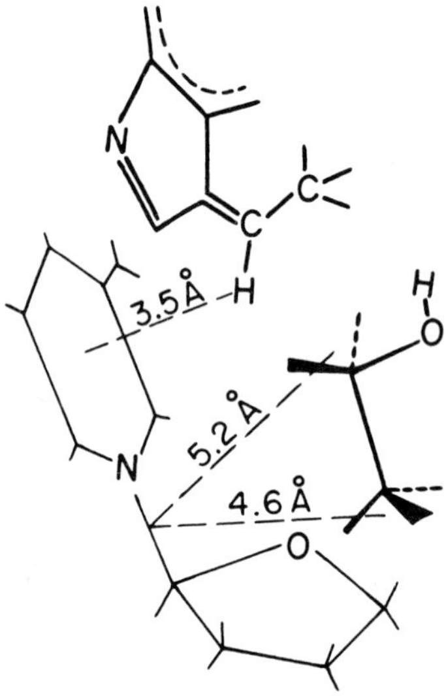

Figure 3. A ternary complex with ethanol which can accommodate a tryptophan residue (5).

be involved in the active site of the dehydrogenase. If, however, the distance between the nicotinamide and the methyl of the substrate were 2.7 Å rather than 4.6 Å (Table II), it would be unlikely that the tryptophan could be involved. Either an exact value of τ_c or an additional spin-labeled compound is needed. The latter would better fix the distance between the substrate and the nicotinamide ring.

From the exchange contribution to $1/T_2$ it would be possible to calculate the exchange rate of substrate into the ternary complex (5). At 25° the exchange rate of ethanol is 610 sec^{-1} which is greater than the turnover number of the enzyme [3.1 sec^{-1} (9)].

From the exchange rate, and the affinity of the ternary complex for ethanol, the rate constant for the combination of ethanol with E–ADP–R· $\leq 5.5 \times 10^5$ $M^{-1}sec^{-1}$ (5). Shore and

Gilleland, using an independent technique (stopped flow spectrophotometry) calculated a second order rate constant of 4.3×10^5 $M^{-1}sec^{-1}$ for the combination of acetaldehyde with E-NADH (10). These rate constants are four orders of magnitude slower than the rate constant expected for a diffusion controlled reaction. The slow binding of substrates to the enzyme-coenzyme and enzyme-analog complexes may be limited by the departure of a water ligand from Zn or by a conformational change (5).

The use of spin-labeled analog with alcohol dehydrogenase permits one to offer a structural explanation for the Theorell-Chance preferred order mechanism in which the coenzyme binds prior to the substrate (9). The enhancement data suggests that the substrate binds on the "water side" of the nicotinamide ring. The site-site effects observed may be necessary to facilitate the entry of substrate into the active site region.

References

1. Dalziel, K. J. Biol. Chem., 238, 1538 (1963).

2. Weiner, H. Biochemistry, 8, 526 (1969).

3. Mildvan, A.S., and H. Weiner. Biochemistry, 8, 552 (1969).

4. Theorell, H., and T. Yonetani. Arch. Biochem. Biophsy., Suppl. 1, 209 (1962).

5. Mildvan, A.S., and H. Weiner. J. Biol. Chem., 244, 2465 (1969).

6. Solomon, I., and N. Bloembergen. J. Chem. Phys., 25, 261 (1956).

7. Mildvan, A.S., and M.C. Scrutton. Biochemistry, 6, 2978 (1967).

8. Schellenberg, K.A. J. Biol. Chem., 240, 1165 (1965); J. Biol. Chem., 241, 2446 (1966).

9. Theorell, H., and B. Chance. Acta Chem. Scand., 5, 1127 (1951).

10. Shore, J.D., and M.J. Gilleland. Fed. Proc., 28, 345 (1969).

DISCUSSION

Theorell: I wonder, Dr. Kierkegaard, whether you care to talk a moment about the X-ray structure work in Uppsala?

Kierkegaard: Yes, I met Dr. Branden only two days ago, and he told me that he is now in the final stage of the data collection work. He has obtained, by means of precession film cameras kept at constant temperature of $+4°C$, data from the native enzyme and four heavy-atom derivatives. The diffraction intensities of the photographs are measured by our automatic film scanner controlled by an IBM 1800 computer. Dr. Branden said that he very likely will be able to produce this summer a structure down to 5.5 Å resolution.

Fabry: I would like to ask Dr. Weiner about the last Figure (see ref. 1). He has four points with large error bars, and from this he calculated the slope of a portion of the curve. How do you do this calculation?

Weiner: True, we have points at only four temperatures, but they definitely cannot be fit to a single straight line. We can fit the data with the Swift-Connick equations. The regions of the curve governed by $1/T_{2m}$ and $1/\tau_m$ can be determined (1). This is the simplest way to fit our data. Since there are no covalent bonds between the spin-label and the protons of ethanol, which are being relaxed, we assume that the hyperfine coupling is zero (1).

Chance: Is this really a question of the validity of the calculation and is there a difference of opinion?

Weiner: Alternative, more elaborate calculations are not justified by the limited data.

Bernhard: Dr. Brand found that several dyes which have very different chemical structures were able to, apparently, occupy the binding site of alcohol dehydrogenase, and I think that this may be due to the fact that the method he is using allows us to measure relatively small binding constants or his dissociation constants, if you want to call them that, so that you can have protein to a concentration of 10^{-4} M, and dye on the order of 10^{-3} M. Especially, I think, if we could measure weak bindings of ligands, we probably will find that binding sites can be occupied by all sorts of things. So, I think, this offers a possibility. Ch?!

Weber: Dr. Weiner has shown that there was a change in the characteristics of the relaxation when the first site was occupied. I think you chose to say that something happened between zero sites and two binding sites occupied, but your Figure was rather in favor of the fact that everything happens before you reach one site occupied. I noticed similar phenomena in several cases; things seem to happen in the process between zero binding and one binding site occupied. We have followed the case of heme-heme interaction where one needs to have a pair of sites occupied. It would be a good thing to start looking into those cases in which you get something happening only between zero and one site occupied to try to find out what that means.

Bernhard: I am going to speak precisely to the point. In choosing chromophores and fluorochromes for enzyme reactions, one should try to choose them so that one can rationally interpret what is going on. In terms of the chemical changes that are occurring in the course of LADH, catalyzed reduction of aldehydes, one can do this very nicely because virtually any aldehyde works as a substrate. Hence, one can choose highly chromophoric or fluorophoric aldehydes. These aldehydes have the added advantage (besides being catalytically reduced as rapidly as acetaldehyde) of being very tightly bound by the enzyme. As a result, one can do an experiment which one could not do with acetaldehyde. If the binding is tight, one can make two extrapolations. The extrapolation at the X-axis indicates the number of equivalents of substrate which are bound to the enzyme. The equivalence at the Y-axis indicates how many equivalents of product are released in the fast burst (as does the slope, dy/dx). The slope of this is 0.5, and the y extrapolation at 1.0 indicates that one half of enzyme sites (or an equivalent weight of 84,000 rather than 42,000) are involved in the rapid burst reaction. The experiment using the excess aldehyde and limiting NADH gives the same result as that illustrated for excess NADH and limiting aldehyde substrate.

Reference

1. Mildvan, A.S. and H.J. Weiner. J. Biol. Chem., 244, 2465 (1969).

FLUORESCENT PROBES FOR GLUTAMATE DEHYDROGENASE

J.R. Brocklehurst and G.K. Radda

Department of Biochemistry, Oxford University

Ox liver glutamate dehydrogenase (GDH) is inhibited by GTP (1), which in the presence of NADH causes a structural transition of the enzyme detectable by optical rotatory dispersion (2). This transition also leads to the dissociation of the enzyme aggregate, present at high enzyme concentrations, to the oligomer. The oligomer is now believed to have a molecular weight of 310,000 (3) to 280,000 (4) and probably consists of six subunits (3), or possibly four (5).

We have used several approaches to elucidate the nature of this transition, and here we present a method for following the conformation equilibria involved.

The Interaction of 1-Anilino-8-Naphthalenesulfonate (ANS) with GDH

ANS interacts with GDH and this results in a large enhancement of its fluorescence (100 ± 20-fold) and a blue shift of 40 nm in its emission maximum (6). A further enhancement is observed in the presence of 0.1 mM NADH and 0.1 mM GTP (Figure 1). This binding is reversible and involves 9 ± 1 binding sites per oligomer (310,000) with an average K_D of 0.22 mM. At low dye concentrations (1-200 μM), ANS had no effect on the sedimentation, optical rotatory dispersion, or catalytic properties of the enzyme.

ANS as a Conformational Probe

When a solution containing 20 μM ANS, 1 mg/ml GDH and NADH is titrated with GTP, the fluorescence enhancement of ANS can be followed (Figure 2). In all ANS experiments, the ANS fluorescence was excited at 410 nm and observed at 550 nm so that contributions from NADH fluorescence were eliminated.

59

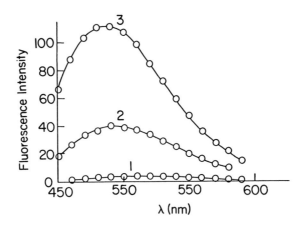

Figure 1. Fluorescence emission spectrum of ANS in the presence and absence of GDH. Curve 1: ANS (20 μM) in 0.1 M phosphate buffer, pH 7.7. Curve 2: ANS (20 μM) + 1 mg/ml GDH. Curve 3: ANS (20 μM) + 1 mg/ml GDH + 0.1 mM NADH + 0.1 mM GTP. The curve is corrected for NADH fluorescence by leaving ANS out of the mixture. Excitation at 410 mμ.

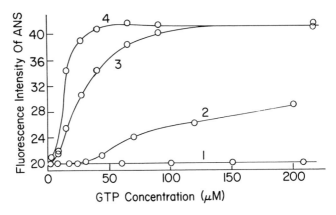

Figure 2. Fluorescence increase of ANS on titration GDH with GTP in the presence of varying amounts of NADH. ANS 2 μM; GDH, 1 mg/ml. Curve 1: no NADH present. Curve 2: 10 μM NADH; Curve 3: 50 μM NADH. Curve 4: 500 μM NADH. Excitation at 410 mμ, emission at 550 mμ.

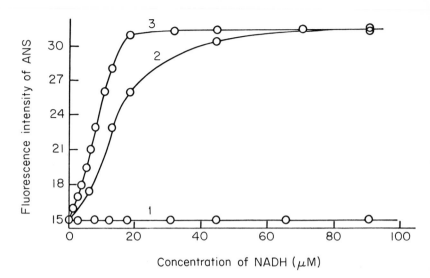

Figure 3. Fluorescence increase of ANS on titrating GDH with NADH in the presence of varying amounts of GTP. ANS, 20 µM; GDH, 1 mg/ml. Wavelengths as in Figure 4. Curve 1: no GTP. Curve 2: 50 uM GTP. Curve 3: 500 µM GTP.

Similar transitions were observed when the solutions were titrated with NADH in the presence of GTP (Figure 3). The transitions detected by the ANS method occur in the same ligand concentration range as was observed by optical rotatory dispersion (2). We have previously presented our arguments to show that the ANS fluorescence is a measure of the conformational equilibria involved (6).

Other Probes*

A variety of other fluorescent molecules may be used to detect the same change (Table I). Of the DNS-amino acids, DNS-glutamate and the DNS derivative of the competitive inhibitor 5-amino-isophthalic acid show a decrease of fluorescence while most other amino acids show an increased fluorescence in the GTP titrations (Figure 4). Mansate (2-(N-methylanilino)naphthalene-6-sulfonate) also gives a positive response while none of its uncharged amides we have studied exhibit fluorescence changes in the GTP titrations.

*DNS = dimethylaminonaphthalenesulfonate; MANS = 2-(N-methylanilino)-naphthalene-6-sulfonate.

TABLE I

Probes for GDH

Probe	Enhancement	
	No GTP	300 μM GTP
DNS–glutamate	5.8	3.2
DNS–5–amino–isophthalate	1.8	1.1
DNS–proline	1.6	1.4
DNS–γ–amino–butyrate	1.0	1.0
DNS–leucine	1.4	1.5
DNS–histidine	1.5	2.3
DNS–norvaline	1.9	2.2
DNS–α–amino–butyrate	1.9	3.4
DNS–serine	2.8	8.1
DNS–alanine	2.6	5.3
DNS–asparagine	3.8	6.0
DNS–methionine	2.0	4.2
DNS–tryptophan	2.5	13.1
MANSate	c2.2	6.8
MANSylamine	8.2	8.2
MANSylethylamide	2.0	2.0
MANSyldecylamide	1.4	1.4

Concentrations: GDH, 1 mg/ml; probe, 50 μM; NADH, 50 μM. DNS derivatives were excited at 400 nm and observed at 550 nm. MANSyl derivatives were excited at 375 nm and observed at 450 nm. When necessary, the contribution from NADH fluorescence was corrected for.

The uncharged amides, however, are still bound by GDH but only ∿2 binding sites are involved. It is thus clear that the sites at which conformational change is detected by the probe require a single negative charge on the probe. If this negative charge is too far away from the chromophore (e.g., DNS–γ–amino–butyric acid), no response is obtained, while if there are two negative charges with the spatial separation of a good substrate or inhibitor (e.g., glutamate and 5–amino–isophthalate), a negative response is observed. In contrast, the DNS derivative of the very poor substrate aspartate behaves like the other negatively charged probes.

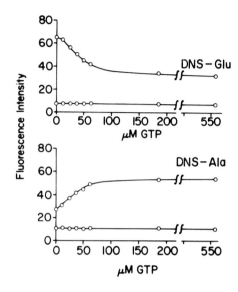

Figure 4. Effect of NADH (10^{-4} M) and GTP on the enhancement of dansyl amino acid fluorescence by GDH (1 mg/ml).

Variation of ANS Response with Substrate Concentrations

The changes in ANS fluorescence brought about by the NADH-GTP heterotropic interaction do not depend on the state of further aggregation of GDH into units larger than the 310,000 molecular weight oligomer. This is supported by the following observations: (1) The transition is clearly observed at protein concentrations varying between 0.1 mg/ml and 2.0 mg/ml, and the values of enhancement of ANS fluorescence remain constant in this range. The ligand concentration required to bring about half the total change ($S_{0.5}$) varies with the protein concentration in a monotonic way (Figure 5). (2) ANS titration curves are normal even with the "fully active, non-associating" acetylated enzyme, prepared by the method of Colman and Frieden (7). (3) When dissociation is brought about by high concentrations of NADH (2) or by 5% dioxane, no enhancement of ANS fluorescence is observed.

On increasing the NADH concentration, the threshold at which GTP effects the change is lowered (Figure 6), but no significant change in the slopes of the titration curves can be observed. At very low NADH concentrations (curves 4 and 5 of Figure 6), the total change in ANS fluorescence is less, indicating that only some of the enzyme molecules have taken

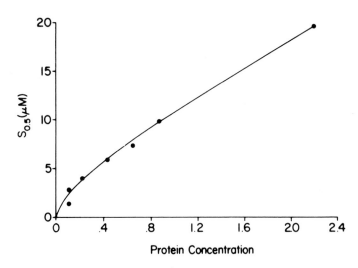

Figure 5. Variation of $S_{0.5}$ (half transition point for NADH) with protein concentration.

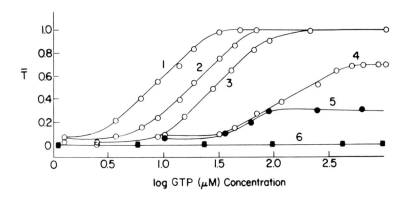

Figure 6. $\bar{T}$ function derived from ANS fluorescence in titration of GDH with GTP. ANS, 20 μM; GDH, 1 mg/ml. Curve 1: in presence of 1 mM NADH. Curve 2: 0.1 mM NADH. Curve 3: 50 μM NADH. Curve 4: 10 μM NADH. Curve 5: 5 μM NADH. Curve 6: no NADH. ($\bar{T}$ = 1-$\bar{R}$, where $\bar{R}$ is the fraction of protein in active form.)

up the new conformation. α-Ketoglutarate decreases further
the half-transition points.

The effect of pH on the allosteric transition must
clearly depend on the pH profiles for NADH and GTP binding
and of the protein conformational equilibrium. It is interesting that the system responds to the ligands most sharply
(i.e., the transition is steepest) at pH 7.8, which is close
to the pH optimum of the enzyme for glutamate oxidation (Figure 7).

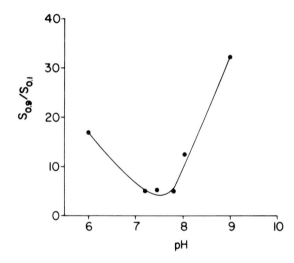

Figure 7. Effect
of pH on the "steep-
ness" of the GTP
titration curves.
GDH, 1 mg/ml; ANS,
40 μM; NADH, 0.2
mM.

The Rate of the Transition

Since ANS fluorescence follows the conformational transition, the rate of this change is directly observable by
stopped flow fluorescence. The rate could be followed either
when GTP (25-400 μM) was rapidly mixed with NADH (25-200 μM),
GDH (0.5-2.0 mg/ml) and ANS (20 μM) or when NADH was mixed
with GTP, GDH and ANS. Under these conditions two distinct
rate processes are observed, which are clearly shown when
the data are plotted on a semilogarithmic scale (Figure 8).
One possible explanation for the two rates is that the first
phase represents a conformational change in the subunits
brought about by the ligand interaction, which is then followed by a slower rearrangement of the subunits into a different geometry in the oligomer.

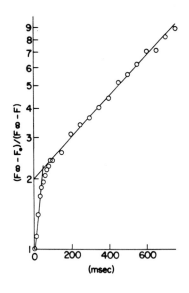

Figure 8. Rate of ANS response. The final concentrations in reaction mixture were: GDH, 1 mg/ml; ANS, 200 μM; NADH, 0.1 mM; GTP, 75 μM. A solution containing GDH, ANS and NADH was mixed rapidly with an equal volume of a solution containing GTP.

Chemical Modification of the Allosteric Site

Modification of GDH in one tyrosine residue per subunit leads to desensitization to the allosteric inhibitor, GTP, without any marked changes in the enzymic activity. Modification may be brought about by either N-acetylimidazole or tetranitromethane (8). The fluorescence titrations using ANS as the probe show that the decreased kinetic response of the modified enzyme to GTP is associated with an increased value of $S_{0.5}$ for GTP, i.e., that higher concentrations of the allosteric ligand are required to bring about the conformational change (Figure 9). From protection and binding experiments, it appears that the modified tyrosine residue forms part of the GTP binding site so that decreased allosteric response is a result of decreased inhibitor binding.

I may conclude by saying that fluorescent probes may be attached to the active site of GDH (9) which, though it leads to inactivation of the enzyme, has enabled us to draw some conclusions about the importance of subunit interactions in this enzyme.

Acknowledgment

We wish to thank the Science Research Council for financial support towards this work.

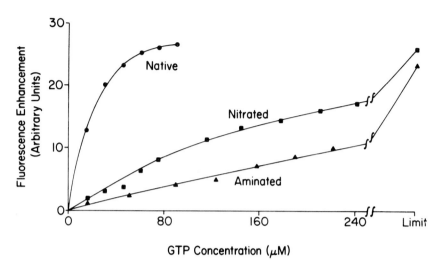

Figure 9. Desensitization of GDH by tyrosine modification as followed by the ANS response. Enzyme, 0.5 mg/ml. The extent of modification was 1±0.2 tyrosines/subunit.

References

1. Frieden, C. J. Biol. Chem., 238, 3286 (1963).

2. Bayley, P.M., and G.K. Radda. Biochem. J., 98, 105 (1966).

3. Eisenberg, H., and G.M. Tomkins. J. Mol. Biol., 31, 37 (1968).

4. Sund, H., and W. Burchard. Europ. J. Biochem., 6, 202 (1968).

5. Minssen, M., and H. Sund. Abs. FEBS Meeting, Madrid, 1969.

6. Dodd, G.H., and G.K. Radda. Biochem. Biophys. Res. Comm., 27, 500 (1967).

7. Colman, R.F. and C. Frieden. J. Biol. Chem., 241, 3652, 3661 (1966).

8. Price, N.C., and G.K. Radda. Biochem. J., 114, 419 (1969).

9. Malcolm, A.D.B., and G.K. Radda. Nature, London, 219, 947 (1968).

DISCUSSION

Chance: It was mentioned by Dr. Anderson that the probe alters the enzyme structure. Do you have any evidence of that happening at this point?

Radda: At the kind of concentration that we used for the probe, neither the activity nor the optical rotatory dispersion properties, nor the sedimentation constants of the enzyme were altered at all. If you use very high probe concentrations, ANS begins to act as a detergent and the enzyme will become less stable.

Mildvan: Have you detected energy transfer between the DPNH and the fluorescent probe?

Radda: We have looked very hard but could not find it.

Hochstrasser: How do you distinguish between looseness or conformation changes of the probe compared with conformational changes of protein?

Radda: The answer is, I think, that in a sense I don't care. The reason why there is a conformational change in the probe must be because the site at which it is bound must have changed its characteristics to such an extent that it will distort the probe. In this sense, it is still a probe no matter what the fluorescence change is due to. In support of what I think is a conformational change in the protein, I can quote a great deal of evidence from optical rotatory dispersion. The changes in the ORD parameters occur at exactly the same range of ligand concentration as is observable by ANS fluorescence.

Hochstrasser: It seems as if you were saying that whenever the polarization changes on addition of substrate, some kind of conformational change has occurred. I, at least, cannot see any necessary relationship between these two things, except perhaps for very simple cases. For a large molecule, with a chromophore on it, it doesn't seem at all obvious that such a change of polarization should immediately imply a conformational change. There are many things that could change the polarization under the conditions that you are suggesting.

Radda: Well, I myself haven't made the statement that because there is a polarization change, there is a conformational change. The polarization changes I observed were in the binding of the coenzyme, and after that I've been using fluorescence intensities of the dye to follow the conformational changes. The fact that the polarization of the dye has changed can be due to a number of reasons, and I don't think I'll elaborate because I don't have the lifetime for the excited states, and until I know this I can't say what the polarization changes are.

Weber: We do know that the facts which intervene are 1) the electronic transitions in excitation and emission, and 2) the lifetime of the excited state. If you know these two things, or if you have some means of checking that they do not change, then you have only the conclusion that the change in the polarization must reflect changes in the protein.

Hochstrasser: I think the approach is certainly justifiable for a free small molecule that has a well-defined transition moment; I don't think the assumptions are justified (before the fact) for a relatively small transition moment attached to a large molecule because the protein-chromophore interactions can change the direction of the transition moment in a number of different ways, both statistically (involving no shift of nuclei) and dynamically (involving changes in nuclear positions or changes after excitation).

Weber: The polarization depends on the angle of the transition moment in absorption and the transition moment in emission. You can determine whether this condition depends on the protein or on the dye by extrapolating to very high viscosities. Your limiting polarization depends upon the angle between the two oscillators and, if this has changed, it will reveal this change. I think there is very good experience that these things change extremely little, if at all; in other words, virtually everytime you absorb a dye onto a protein, the limiting polarization is the same as you would expect from the excitation and emission on ordinary media of very high viscosity. So, for empirical reasons, we can say that this does not enter in our calculations.

Radda: We have measured the polarization of fluorescence spectra, the excitation polarization, of these dyes in proteins at high viscosities, and we always find the same spectrum, which suggests that the angle between the emission and absorption oscillators remains constant.

Weber: That is correct, but this is because the experimentation is very accurate and the theory is still very weak.

Radda: We have much higher polarizations to start with.

Hochstrasser: I realize this. My point is that each case has to be carefully studied with regard to all these mentioned variables. Because a polarization change is caused by a change in conformation in one system does not imply that similar spectral changes will have the same interpretation in another.

Theorell: I want to propose we put this to a committee on biochemical nomenclature to define this smallest change that should be acknowledged as being conformational so we know what we are talking about.

X-RAY DIFFRACTION STUDIES ON THE MALATE DEHYDROGENASES*

L.J. Banaszak, D. Tsernoglou and M. Wade

Department of Physiology and Biophysics, Washington University
School of Medicine, St. Louis, Missouri

This brief discussion describes the preliminary results of an X-ray structural study on the malate dehydrogenases. At this time, the results center almost exclusively on the cytoplasmic malate dehydrogenase (s-MDH) from pig heart. The work on the preparation of suitable crystalline specimens of the mitochondrial malate dehydrogenase (m-MDH) has so far not been successful. However, only a very limited range of crystallizing conditions have been applied. The two different enzymes, present together in many cells, catalyze the same chemical reaction. For each enzyme, the kinetic parameters of the catalytic reaction are quite different (1). In addition, there are notable differences in the amino acid composition and other physical-chemical properties. This study was started with the goal of relating the structural aspects of each molecule which are important to the catalytic activity. Similarities and differences in the structures of s-MDH and m-MDH should be useful in looking for molecular features which are critical to the catalytic mechanism.

The isolation of s-MDH, the early conditions for crystallization, and a limited species survey have already been described (2). The first crystals were obtained from ammonium sulfate buffered with phosphate in the pH range from 5.2 to 6.0. These were labelled Type $\underline{A}$. They are orthorhombic and have the space group $P2_12_12$. The unit cell dimensions were: $\underline{a}$ = 139 Å, $\underline{b}$ = 87 Å, $\underline{c}$ = 59 Å. There are four molecules of s-MDH in the unit cell. This form grew very slowly, usually over several months and could also be obtained from phosphate

*Supported by grants from the U.S.P.H.S. (GM 13925) and the N.S.F. (GB 5888). M.W. is grateful for support received from the PHS Training Grant (GM-714).

alone. In the pH range from 6.1 to about 8.0, a needle-like habit resulted. Although these crystals were often up to several millimeters in length, they were thin and could not be mounted for X-ray examination. At pH 5.0 to 5.1, another form of s-MDH crystals was obtained. These crystals were labelled Type C and have the same space group as Type A crystals. The Type C crystals have a rhomboid shape as compared to the elongated slablike habit of Type A crystals. They were grown in the cold from ammonium sulfate buffered with acetate and containing a ten-fold molar excess of NAD$^+$. An hexagonal habit was also found at a pH slightly above 5.1 with acetate buffer. These crystals are the same as Type C crystals. In addition, it is necessary to point out that some microheterogeneity exists in the enzyme preparations. This can be shown by the application of a shallow gradient during chromatography on DEAE sephadex. It may be that a second peak, which varies considerably from one preparation to another, will result in only Type A crystals. Both forms of the s-MDH produced diffraction data corresponding to Bragg spacings of 2.8 Å. Because of small but definite differences in the unit cell dimensions and noticeable changes in the intensities, Type A and Type C crystals are considered to be two different crystalline forms of the enzyme. However, in general, the scattered X-ray intensities are very similar, indicating that probably both the lattice and the molecular structure are about the same in both forms. All the work which will be described in the remaining sections refers to Type C crystals. This form was chosen for further study because of the comparative ease with which the crystals could be grown and because the crystals are more equi-dimensional.

The density of the Type C crystals and the density of the mother liquor were determined by the usual density gradient columns. A value of 1.227 g/cc was obtained for the crystals and 1.167 g/cc was determined for the mother liquor. If one assumes a solvent content of 43%, the molecular weight of s-MDH was calculated to be 77,000. A solvent content of 50% results in a molecular weight of 68,500. These values are in good agreement with a recently reported value of 73,800 determined by hydrodynamic methods (3).

The structural analysis of s-MDH now centers around the preparation of isomorphous heavy atom derivatives. Blake (4) has recently reviewed the chemistry and application of heavy atom derivatives for solving the X-ray phase problem. The problem of preparing these derivatives is relatively simple to describe. One has to attach a heavy atom (heavier than, say, iodine) preferably to a single site on the enzyme in the crystalline state. This heavy metal must neither change the lattice parameters, and that includes the orientation of the molecule in the lattice, nor change the conformation of the enzyme. There are two chemical approaches to preparing heavy atom derivatives. The first one consists in attaching heavy metals to reactive side chains. The second approach is to use information on some known binding sites of the enzyme molecule. The former method typically involves attaching mercurials to -SH side chains. Its usefulness depends solely on the fact that there must be only a limited number of these reactive side chains. The second method depends on binding sites known to exist on the molecule. An example of such a binding site would be the malate or NAD⁺ binding site on s-MDH. The problem then becomes one of preparing heavy atom tagged analogs of the small compounds or substrates. The third and most used approach is simply one of trial and error. Crystals are soaked or grown in the presence of heavy atom compounds until one is found that produces isomorphous differences. Again, these differences are useful only if they are the result of a limited number of binding sites.

In the case of s-MDH, chemical evidence indicates a limited number of reactive -SH groups on the molecule. For example, beef heart s-MDH has been shown to have about three reactive sulfhydryl groups (5). The reactivity of the sulfhydryl groups of pig heart s-MDH was measured by reaction with 5,5'-dithiobis-2-nitrobenzoic acid (DTNB) according to the method of Ellman (6). The results are shown in Figure 1. The activity of s-MDH is not significantly affected by reacting the first 2 sulfhydryl groups. When the reaction was allowed to continue beyond about 12 hours, it appeared that an appreciable number of additional sulfhydryl groups were reacting but the reaction mixture became turbid preventing further spectrophotometric measurements. When the enzyme was incubated in 4 M urea prior to the addition of DTNB, four to five moles of -SH groups react during the first hour. Preliminary experiments with C^{14} labelled parachloromercuribenzoate indicated that one to two sulfhydryl groups reacted overnight. These results were difficult to interpret, again because of aggregation of the enzyme. X-ray experiments with crystals soaked in 0.05 mM parahydroxymercuribenzoate (PHMB) and parahydroxymercuriphenylsulfonate (PHMS) resulted in definite difference in the diffracted intensities. These results will be discussed in more detail in the last section.

73

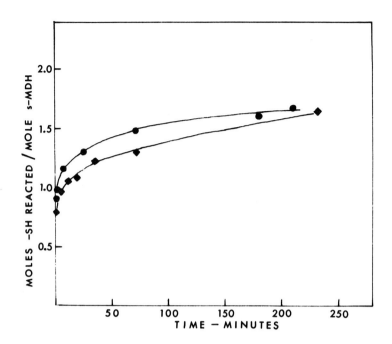

<u>Figure 1.</u> The reaction of DTNB with pig heart s-MDH. The reactions were carried out in 0.03 M KPO_4 buffer, pH 8.0, 25°C. ●, s-MDH concentration was 3.80 x 10^{-6} M, DTNB concentration was 9.09 x 10^{-4} M. ◆, s-MDH concentration was 2.23 x 10^{-6} M, concentration of DTNB was 1 x 10^{-4} M. A molecular weight of 73,900 was used for the s-MDH.

In order to see if the substrate binding site could be used to prepare a heavy atom derivative, a mercuriated form of malic acid was synthesized. This was done using the fact that mercuric salts form addition compounds with olefins (7). The synthesis is described below.

A similar compound with acetate in place of the hydroxyl on the mercury can be obtained by precipitating it with ethanol in the presence of a large excess of acetic acid. The resulting mercuriated form of malic acid is a D,L mixture. Preliminary kinetic studies with pig heart s-MDH indicated that mercuriated malate was an inhibitor. However, because of the mercury group, the compound also will react with -SH groups. In fact, crystals soaked in mercuriated malic acid produce X-ray differences very similar to those of PHMB and PHMS. This occurs at relatively low concentrations, 0.05 mM. If the concentration is raised to 1 mM, the crystals disintegrate. Therefore, in the crystalline state, the mercuriated malic acid will only be useful as an -SH reagent. It may be that making the methyl mercaptide of the malate mercurial will permit its use as an active site label but this has not yet been done.

Our "trial and error" approach to the preparation of heavy atom derivatives was as follows. A single crystal was soaked in a heavy atom containing solution for a 1-3 week period. Then an 8° precession photograph of the hk0 reciprocal lattice plane was taken and compared with that from the native protein. The heavy atom solutions were made up in 70% saturated ammonium sulfate containing 0.05 M acetate, pH 5.0 to 5.1, or, alternatively, the soak medium was 3.2 M phosphate, pH 5.0 to 5.1. Two heavy atom soak concentrations were used at the start, 1 mM and 0.1 mM. The crystals soaked in phosphate solutions seemed to slowly lose their crystallinity, but this took place over several months. If a phosphate medium is to be used in the future, the crystalline stability will have to be investigated in more detail.

In all, 22 compounds were tried in both phosphate and ammonium sulfate. The X-ray results can be categorized as follows: (i) No differences (negligible binding). (ii) Transient loss of crystallinity (binding with either large changes in the lattice or conformation of the protein). (iii) Large non-isomorphous changes (binding with significant changes in the lattice or conformation of the protein). (iv) Isomorphous changes (binding with no detectable changes in the lattice or conformation of the protein). As a typical example of the results described in (ii), CH_3HgCl soaks at 1 mM and 0.1 mM, resulted in crystals showing no X-ray reflections beyond Bragg angles corresponding to 15 Å spacings. At 0.01 mM or less, little or no changes occurred. On the other hand,

$Na_2U_2O_7$ soaks resulted in large but non-isomorphous changes. In addition to the mercurials already described, those heavy atom compounds that were found to result in isomorphous changes, were: $AuCl_4^-$, $PtCl_4^{2-}$, $Pt(NH_3)_4PtCl_4$ and iodophenol.

More recently, flow cell techniques (8) have been used in the search for heavy atom derivatives. Uranyl acetate, $UO_2(C_2H_3O_2)_2$, has now been shown to produce large isomorphous changes with s-MDH. These changes begin to appear at a concentration of $UO_2(C_2H_3O_2)_2$ of about 1 mM and for unknown reasons were missed by the typical soak experiment. It is interesting to point out that using the flow cell and diffractometry, no changes in the X-ray intensities were noted with compounds such as $PtCl_4^{2-}$. This and similar results with other compounds would indicate that in some instances heavy metal binding to s-MDH may be a relatively slow process. Similar slow binding effects have been noted by a number of investigators. This is unlikely to be a diffusion problem. For some coordination complexes, however, ligand exchange reactions may be relatively slow; some may have one-half times of several days. Such exchange reactions may precede binding to the protein, or, alternatively, be involved in the actual binding to the protein. The point of the matter is that time may be an additional parameter in the preparation of some heavy metal derivatives of proteins. It is likely that with s-MDH, there will be "slow" and "fast" heavy atom derivatives. The reaction of -SH groups with mercurials and the binding of uranyl acetate are examples of "fast" reactions. On the other hand, the binding of $AuCl_4^-$ and $PtCl_4^{2-}$ may be very slow; weeks rather than hours or days are required.

The next step in the X-ray structural analysis is locating the heavy atom(s) in the unit cell and optimizing the conditions for heavy atom occupancy and crystal lifetime. This work is in progress, but a brief summary is given here. Type C crystals of s-MDH have three centric projections. Data from only one of these, the hk0, will be described here. This means the heavy atom sites to be described are only tentative. The locations of these sites must be checked with the other two centric projections and eventually with three dimensional data before the structure analysis can proceed.

Figure 2 is a two dimensional $(\Delta F)^2$ Patterson map of the PHMB derivative. The average percent change in intensity used to calculate the map was 23%. The data were obtained by photographic methods. Also indicated on Figure 2 are the

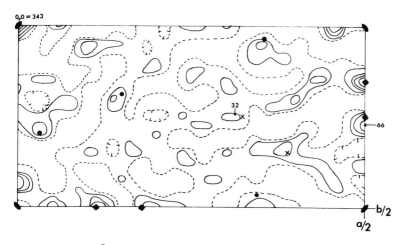

Figure 2. $(\Delta F)^2$ Patterson map of PHMB and s-MDH, hk0 projection at 5.5 Å resolution. Contours are at equal, arbitrary levels with the zero value indicated by the dotted line. One interpretation (2 sites per asymmetric unit) is indicated as follows: x, single weight vector, ◆, double weight vector, ●, cross vectors between sites. Numerical values for three points on the map are given to illustrate the relative heights on the map.

vectors produced by a solution which has two sites per asymmetric unit (or per molecule s-MDH). A comparison of the structure factors calculated from these two sites with the observed amplitude differences gave an R factor (the usual crystallographic residue) of 0.47. Centric signs were obtained and used first to calculate a self Fourier map. No attempt was made to eliminate reflections where a sign crossover might have occurred. The difference Fourier map was very clean. Besides the two peaks at the sites used to calculate the signs, only two other peaks were present. They had a height of 30% of the sites used in the sign calculation. Their inclusion in the refinement resulted in worse agreement between the observed and calculated structure factors.

Figure 3 is a difference Fourier projection using the PHMB signs and the differences from the mercuriated malate derivative. Since the measured differences in the X-ray data were very similar to those of PHMB, this is not a good test of the reliability of the PHMB locations. The two mercuriated

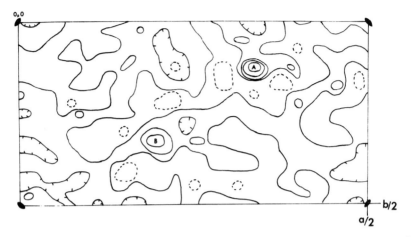

Figure 3. ΔF Fourier map of the mercuriated malate derivative
of s-MDH at 5.5 Å resolution. Contours are at equal, arbitrary levels with negative values marked with dotted lines.
Phases were calculated using PHMB sites indicated by A and B.

malate sites which stand out are the same as those of PHMB
(labelled A and B). No additional sites are apparent.
Clearly mercuriated malate cannot be used as an active site
tag at low concentrations since it is reacting only with an
-SH group. Increasing the concentration with the hope of
labelling the active site was unsuccessful. The crystals disintegrate at concentrations of the mercurial much above 0.05
mM. Although similar calculations have not yet been done for
PHMS, the difference Patterson projection is similar to that
of PHMB. Because of its increased solubility in high concentrations of ammonium sulfate, PHMS is currently being used.

The difference Fourier projection of the $AuCl_4^-$ derivative is shown on Figure 4. On the map, there are three
prominent peaks. Two are very close to the PHMB sites (labelled A and B) that were used in the sign calculation. These
three $AuCl_4^-$ sites result in an R factor of 0.56.

To summarize the results using the PHMB signs and the
derivatives mentioned previously, the cross Fouriers produced
the following results: (i) $PtCl_4^{2-}$ and $PtCl_6^{2-}$ resulted in
four sites in the same positions for both compounds. On the
$PtCl_4^{2-}$ difference Fourier projection, two additional sites at
special positions appeared on the map. None of the sites are
close to the PHMB or $AuCl_4^-$ positions. (ii) The difference

<u>Figure 4</u>. ΔF Fourier map of the $AuCl_4^-$ derivative of s-MDH at 5.5 Å resolution. Contours are at equal, arbitrary levels with negative values marked with dotted lines. Phases were calculated using PHMB sites indicated by <u>A</u> and <u>B</u>.

Fourier map for uranyl acetate had two major peaks and two additional sites weighted at about 60% of the major sites. One of the latter is at the same position as one of the platinum sites. Two high peaks appear at special positions. The final test of this preliminary work is to calculate a difference Fourier projection of PHMB using the signs from one of the forementioned derivatives. This has yet to be done.

The difference Patterson projections have been tentatively solved for the mercurials. Using this information, an attempt has been made to interpret the results obtained with $AuCl_4^-$, $PtCl_4^{2-}$ and uranyl acetate. Current work is directed towards optimizing the heavy atom binding conditions and improving the statistical accuracy of the measurements.

Acknowledgments

The technical assistance of Mrs. Emilia Campos and Miss Barbara Stadtmiller is gratefully acknowledged.

Addendum*

Three dimensional X-ray data to 5.5 Å resolution have been collected for the native and six heavy atom derivatives. Three dimensional Patterson and Fourier analysis of these data has essentially confirmed the results presented here. Further studies on the $PtCl_4^{2-}$ and $PtCl_6^{2-}$ seemed to indicate that only the $PtCl_6^{2-}$ is slow in forming a complex with s-MDH. In both platinum derivatives, the four major binding sites are the same.

References

1. Englard, S., and H.H. Brieger. Biochim. Biophys. Acta, 56, 571 (1962).

2. Banaszak, L.J. J. Mol. Biol., 22, 389 (1966).

3. Gerding, R.K., and R.G. Wolfe. J. Biol. Chem., 244, 1164 (1969).

4. Blake, C.C.F. Adv. in Prot. Chem., 23, 59 (1968).

5. Siegel, L., and S. Englard. Biochim. Biophys. Acta, 64, 101 (1962).

6. Ellman, G.L. Arch. Biochem. Biophys., 82, 70 (1959).

7. Chatt, J. Chem. Rev., 48, 7 (1951).

8. Wyckoff, H.W., M. Doscher, D. Tsernoglou, T. Inagami, L.N. Johnson, K.D. Hardman, N.M. Allewell, D.M. Kelly, and F.M. Richards. J. Mol. Biol., 27, 563 (1967).

*Note added March 14, 1970.

DISCUSSION

Chance: Are the background and contour levels in your figure approximately the same as that which Dr. Schoenborn showed earlier?

Banaszak: I don't really know. I suspect that background level is much lower on this map, but that really tells you nothing because probably the results are more accurate on Dr. Schoenborn's map than mine. In the case of the map I showed, the results are on a purely arbitrary scale. There is the difference in substitution at the two sites, which I cannot account for at all. The molecules in both the S-MDH and M-MDH are thought to consist of identical subunits, and there is no way at all one can explain the apparent differences in substitution. These differences also appear in the PHMB refinement, but phases used here were calculated with equal weight PHMB sites.

Gutfreund: Do you find two -SH groups per 70,000 molecular weight?

Banaszak: Yes. In the solution of the PHMB difference Patterson, we seem to find adequate agreement with only 2 sites. This is in agreement with the results we obtained with Ellman's reagent.

FOUR REACTIVE AMINO ACID RESIDUES
IN 3-PHOSPHOGLYCERALDEHYDE DEHYDROGENASE[*]

Jane Harting Park[+]

Department of Physiology, Vanderbilt University
Medical School, Nashville, Tennessee 37203

Four critical amino acid residues have been located on
the enzyme, 3-phosphoglyceraldehyde dehydrogenase. The
functional significance of these sites is related to the
various reactions catalyzed by the enzyme and the maintenance of the active three dimensional structure of the protein. The purpose of this paper is to discuss the inter-
dependence of the four sites in the catalytic reactions and
the binding of the coenzyme, DPN.

By way of orientation, a brief discussion of the mechanism of the various reactions catalyzed by 3-phosphogly-
ceraldehyde dehydrogenase might be helpful. The early work
of Otto Warburg elegantly demonstrated that 3-phosphogly-
ceraldehyde dehydrogenase oxidized and phosphorylated 3-
phosphoglyceraldehyde to 1,3-diphosphoglyceric acid (1).
In our laboratory we found that a very convenient model
substrate for studying the mechanism of this reaction was
acetaldehyde, which is oxidized in a similar manner to
acetyl phosphate (2). The reaction proceeds via a two step
mechanism in which the aldehyde is first oxidized to an S-
acetyl enzyme in the presence of the enzyme (ESH) and DPN
(Fig. 1). The acetyl enzyme intermediate is then phosphory-
lized to yield acetyl phosphate.

[*]This work was supported by the National Science Foundation,
the U.S. Public Health Service, and the Muscular Dystrophy
Association of America.

[+]Career Development Awardee of the U.S. Public Health Service

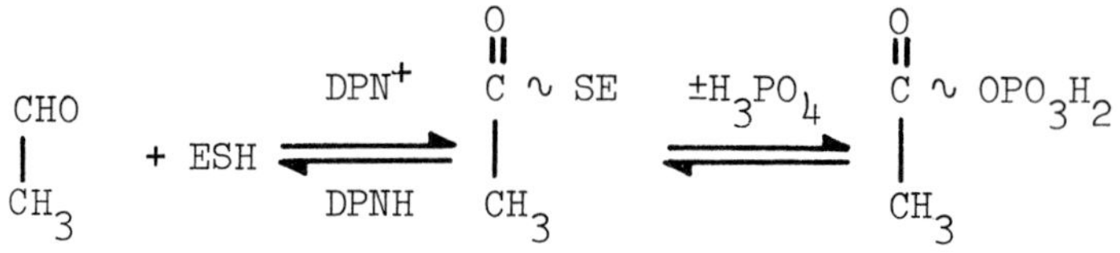

$$\begin{array}{ccc} \underset{|}{CHO} & \xrightarrow[DPNH]{DPN^+} & \overset{O}{\underset{|}{C}} \sim SE & \xrightarrow[\,]{\pm H_3PO_4} & \overset{O}{\underset{|}{C}} \sim OPO_3H_2 \\ CH_3 & + ESH \rightleftharpoons & CH_3 & \rightleftharpoons & CH_3 \end{array}$$

Figure 1. Dehydrogenase Reaction. The two step mechanism for the oxidation and phsophorylation reactions catalyzed by 3-phosphoglyceraldehyde dehydrogenase.

On the basis of this work, we were then able to demonstrate two other reactions catalyzed by the dehydrogenase that also proceed through the same common acetyl-enzyme intermediate (Fig. 2 and Fig. 3): first, the transacetylase reaction in which acetyl group of acetyl phosphate is transferred to coenzyme A to give S-acetyl coenzyme A and, secondly, the esterase reaction in which p-nitrophenyl acetate is hydrolyzed to acetic acid and p-nitrophenol (3, 4).

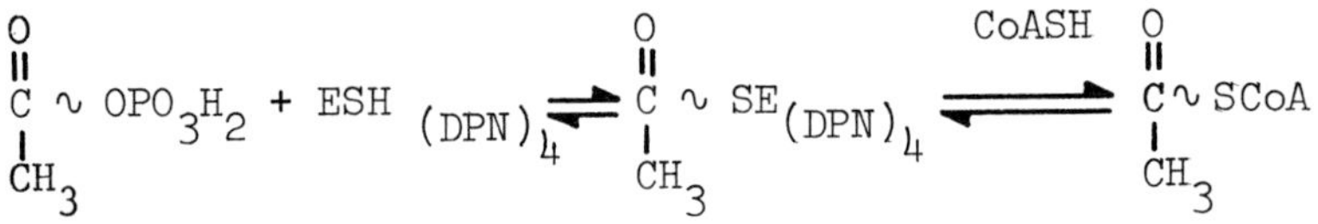

$$\begin{array}{ccc} \overset{O}{\underset{|}{C}} \sim OPO_3H_2 + ESH & \underset{(DPN)_4}{\rightleftharpoons} & \overset{O}{\underset{|}{C}} \sim SE & \overset{CoASH}{\underset{(DPN)_4}{\rightleftharpoons}} & \overset{O}{\underset{|}{C}} \sim SCoA \\ CH_3 & & CH_3 & & CH_3 \end{array}$$

Figure 2. Transacetylase Reaction. The transacetylase reaction catalyzed by a two step mechanism involving an S-acetyl enzyme intermediate. The enzyme is designated as $ESH(DPN)_4$ to indicate the requirements of SH groups and DPN for this reaction.

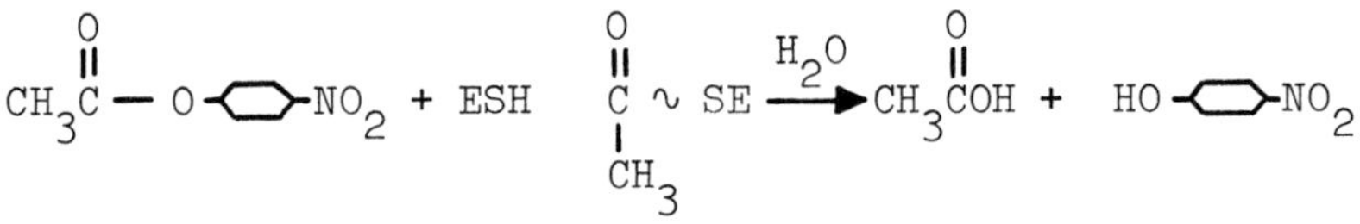

$$CH_3\overset{O}{\underset{|}{C}} - 0 \bigcirc NO_2 + ESH \quad \overset{O}{\underset{|}{C}} \sim SE \xrightarrow{H_2O} CH_3\overset{O}{\underset{|}{C}}OH + HO \bigcirc NO_2$$
$$CH_3$$

Figure 3. Esterase Reaction. The esterase reaction, which is not a DPN dependent reaction, also proceeds through the S-acetyl enzyme complex.

The common S-acetyl enzyme intermediate was prepared with ^{14}C-acetyl phosphate or ^{14}C-p-nitrophenyl acetate as substrate and then in collaboration with Dr. Ieuan Harris of Cambridge England the sequence of the labeled octadecapeptide was established (5). There are four identical S-acetyl sites on the tetrameric enzyme with a molecular weight of 140,000 (5,6).

Sequence of the Active Site

$$^{14}COCH_3$$

Lys·Ileu·Val·Ser·Asn·Ala·Ser·Cys·Thr·Thr·

Asn·Cys·Leu·Ala·Pro·Leu·Ala·Lys

Figure 4. The amino acid sequence of the octadecapeptide surrounding the reactive cysteine residue at the active site of 3-phosphoglyceraldehyde dehydrogenase. A second unreactive cysteine residue, separated by only three amino acids, is not acetylated or carboxymethylated under these conditions at pH 7.0 and 0°.

The S-acetyl enzyme intermediate is formed by a direct nucleophilic attack of the specific reactive cysteine on the carbonyl carbon of the acetate moiety of p-nitrophenyl acetate or acetyl phosphate as formulated by Olson and Park (7). The question now arises as to how the acetyl group is released from the enzyme. On the basis of kinetic data (7), histidine appeared to be a very good candidate for a participant in the deacylation of the S-acetyl enzyme intermediate. In order to modify the histidine residues, the enzyme was photooxidized in the presence of Rose Bengal (8). Under the proper conditions, the acetylation of the enzyme with p-nitrophenylacetate was not affected, but the deacylation was 50 to 60% inhibited (Figure 5).

Peptide mapping shows that this selective inhibition of deacylation was due to the alteration of a specific histidine residue. The identification and sequence of the peptide containing the photosensitive histidine was determined in our laboratory with the assistance of Dr. Richard Perham (9).

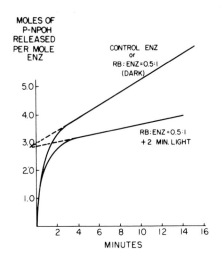

<u>Figure 5</u>. Spectrophotometric measurements of the enzymatic hydrolysis of p-nitrophenyl acetate illustrating the formation and breakdown of the S-acetyl enzyme complex. The reaction mixture for the assay contained 0.02 μmole of DPN-free 3-phosphoglyceraldehyde dehydrogenase in veronal buffer, pH 8.0 at 4°C. The reaction was started by the addition of 1.2 μmole of p-nitrophenyl acetate. The hydrolysis of p-nitrophenyl acetate is followed spectrophotometrically by measuring the increase in absorption at 400 mμ due to the liberated phenol. The top curve for the control enzyme shows an initial rapid release of p-nitrophenol due to the formation of the S-acetyl enzyme complex. This initial "burst" of activity is followed by a linear and slower p-nitrophenol liberation caused by the deacetylation and reformation of the acetyl enzyme complex. Extrapolation of the linear portion of the curve back to zero time gives an intercept which is a measure of the moles of acetyl groups bound per mole of enzyme. In the example shown this corresponds to 2.6 moles of acetyl groups per mole of enzyme under these particular conditions.

The lower curve shows the effect of a 2 min photooxidation in the presence of Rose Bengal (RB) on the acetylation and deacetylation of the enzyme. With the photooxidizing conditions optimal for selective inhibition of deacetylation (molar ratio of RB:Enzyme = 0.5:1) there was a 58% reduction in the linear rate of hydrolysis of the S-acetyl enzyme compou and no effect on the initial rate or the extent of acetylatior of the dehydrogenase. In the absence of light, Rose Bengal had no effect on either acetylation or the deacetylation of the enzyme (top curve).

The Peptide Containing the Photosensitive Histidine Residue

Val·Asp·Val·Val·Ala·Ile·Asn·Asp·Pro·

Phe·Ile·Asp·Leu·His·Tyr

Figure 6. The amino acid sequence around the photosensitive histidine residue involved in the deacylation of the common S-acetyl enzyme intermediate.

Since Perham and Harris have worked out the entire sequence of the 322 amino acids of the dehydrogenase isolated from pig muscle and found it essentially identical to the rabbit muscle enzyme (10), it is now possible to identify the residues involved in the acetylation and deacetylation steps as Cysteine 149 and Histidine 38. The mechanism by which the histidine residue facilitates deacetylation has not been determined, but it could function as a nucleophile for the thioester, a general base catalyst, or as a conformational factor (7,11,12). The general importance of this histidine residue is supported by the finding that photooxidation causes a 50% inhibition of four reactions catalyzed by the enzyme: namely, dehydrogenase, transferase, esterase and phosphatase (9,13).

The third important residue was discovered in the course of characterizing the pH curve for the acetylation of the reactive cysteine residue. It was noted that as the pH was shifted from pH 7.0 to 8.5, a lysine residue was acetylated (14). The acetylation of an ε-amino group occurred by an S-N migration reaction in which the acetyl group is transferred from the reactive cysteine residue to a specific lysine residue (15). The properties of this remarkable migration are

The Tridecapaptide Containing the N-Acetylated Lysine Residue

$$HN^{14}COCH_3$$
Ala·Thr·Gln·Lys·Thr·Val·Asp·Gly·Pro·

Ser·Gly·Lys·Leu

Figure 7. The amino acid sequence surrounding the lysine residue which is specifically labeled by acetyl phosphate or p-nitrophenyl acetate. Harris and Polgar have shown that the same lysine residue in the pig enzyme is acetylated with ^{14}C-p-nitrophenyl acetate (17).

similar to those of the intramolecular S-N transfer reaction of S-acetyl-β-mercaptoethylamine or S-acetyl-γ-mercaptopropylamine (16). The reactive lysine is one out of a total of 26 lysine residues per monomer or 4 out of a total of 94 residues on a tetrameric basis.

The N-acetyl-lysine moiety is catalytically inactive and cannot be removed by any type of enzymatic deacylation; however, the lysine residue may be involved in DPN binding. The S-N transfer reaction is inhibited by the coenzyme, DPN, which blocks this dead end street. Conversely, N-acetylation interferes with DPN binding (15,18). This type of competition for the lysine residue indicates that the ε-amino group participates in DPN binding. In addition, DPN is bound more tightly to this dehydrogenase than to any other dehydrogenase. Consequently, the coenzyme can promote optimal enzymatic activity by channeling the substrate down the proper pathway and thereby preventing inhibitory side reactions.

When the acetylating procedures are further modified, still another residue can be acetylated. At pH 7.0 and 37°, in the absence of DPN, there is multiple acetylation of -SH groups and as many as 9 or 10 out of a total of 16 cysteine residues can be acetylated with ^{14}C-p-nitrophenyl acetate (19). Analysis of the radioactive peptides obtained from the labeled enzyme shows a new sequence in which the acetyl group was in thioester linkage.

Peptide Containing Another Acetylated Cysteine Residue

$$\overset{\overset{\displaystyle ^{14}COCH_3}{|}}{Glu \cdot Asp \cdot Gln \cdot Val \cdot Val \cdot Ser \cdot Cys \cdot Asp}$$

Figure 8. The amino acid sequence surrounding the cysteine residue in the fourth critical peptide.

By analogy with the S-N transfer reaction, it appears that this cysteine moiety is probably acetylated by an S-S transfer reaction, and the acetyl group migrates from the cysteine in the active center to another cysteine some 132 residues away. The second cysteine is enzymatically inactive as the acetyl group cannot be removed by the usual techniques of catalytic deacylation. However, this cysteine has a function with regard to structural maintainance. With multiple acetylation of 9 or 10 cysteine residues, the protein rapidly

precipitates in about one minute. The precipitation can be
prevented by addition of DPN or iodoacetic acid which inhibits
the acetylation of the active site cysteine with p-nitrophenyl
acetate and the subsequent S-S transfer reaction. Moreover,
yeast dehydrogenase which contains only the two cysteine
residues of the active center (Cysteine 149 and 153) does not
precipitate when acetylated at pH 7.0 and 37°. Thus the
"outside" cysteine number 281 appears to be essential for
structural maintainance of the muscle enzyme (20).

In conclusion, these findings may be summarized by a
diagrammatic representation of the enzyme topography.

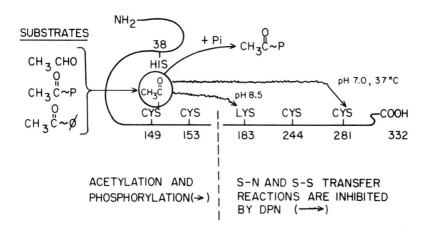

Figure 9.

The active site occurs at residue 149 which is about the
middle of the 332 residues in one monomer of the tetrameric
enzyme. We have shown this as the -SH group which is nearest
to the N-terminal end of the molecule and is acetylated by a
variety of substrates at pH 7.0 and 0°. The deacylation of
the active site is facilitated by the histidine residue at
position 38, the presence of the coenzyme and an acceptor such
as inorganic phosphate. Upon raising the pH to 8.5, an acetyl
group on the cysteine residue can migrate from the active site
to an ϵ-NH_2 group of a specific lysine, residue 183, some 33
amino acid distance as indicated by the wavy line. Now, if
one starts back again at the active site, which is labeled at
pH 7.0 and 0°, and raises the temperature to 37°, the acetyl

group migrates to another cysteine residue at <u>position 281</u>. The coenzyme prevents these S–N and S–S transfer reactions and thereby functions in a newly discovered role to protect the enzyme from inactivation.

References

1. Warburg, O. and W. Christian. Biochem. Z., <u>303</u>, 40 (1939).

2. Harting, J. and S.F. Velick. Fed. Proc., <u>10</u>, 195 (1951); J. Biol. Chem., <u>207</u>, 867 (1954).

3. Harting, J. and S.F. Velick. Fed. Proc., <u>11</u>, 226 (1952); J. Biol. Chem., <u>207</u>, 867 (1954).

4. Park, H.J., B.P. Meriwether, P. Clodfelder and L.W. Cunningham. J. Biol. Chem., <u>236</u>, 136 (1961).

5. Harris, J.I., B.P. Meriwether and J.H. Park. Nature, <u>197</u>, 154 (1963).

6. Dandliker, W.B. and J.B. Fox. J. Biol. Chem., <u>214</u>, 275 (1955).

7. Olson, E.J., and J.H. Park. J. Biol. Chem., <u>239</u>, 2316 (1964).

8. Bond, J.S. and J.H. Park. Fed. Proc., <u>26</u>, 1707 (1967).

9. Halcomb, S., J.S. Bond, R. Kloepper and J.H. Park. Fed. Proc., <u>27</u>, 292 (1968).

10. Harris, J.I. and R.N. Perham. Nature, <u>219</u>, 1025 (1968).

11. Behme, M.T.A. and E.H. Cordes. J. Biol. Chem., <u>242</u>, 5500 (1967).

12. Lindquist, R.N. and E.H. Cordes. J. Biol. Chem., <u>243</u>, 5837 (1968).

13. Bond, J.S., S.H. Francis and J.H. Park. J. Biol. Chem., <u>245</u>, 1041 (1970).

14. Mathew, E., C.F. Agnello and J.H. Park. J. Biol. Chem., <u>240</u>, PC 3233 (1965).

15. Park, J.H., C.F. Agnello and E. Mathew. J. Biol. Chem., <u>241</u>, 769 (1966).

16. Wieland, T. and H. Hornig. Ann. Chem., <u>600</u>, 12 (1956).

17. Harris, J.I. and L. Polgar. J. Mol. Biol., <u>14</u>, 630 (1965)

18. Mathew, E., B.P. Meriwether and J.H. Park. J. Biol. Chem. <u>242</u>, 5024 (1967).

19. Park, J.H. In Current Aspects of Biochemical Energetics N.O. Kaplan and E. Kennedy, eds.), Academic Press, 1966, p. 299.

20. Park, J.H., B.P. Meriwether and S. Halcomb. Seventh International Congress of Biochemistry, Tokyo, Vol. IV, 819 (1967).

THE FUNCTIONALITY OF E·DPN COMPOUND IN
3-PHOSPHOGLYCERALDEHYDE DEHYDROGENASE

B. Chance and J. Park

Johnson Research Foundation, School of Medicine
University of Pennsylvania, Philadelphia, Pennsylvania 19104

In order to determine the catalytic role of the E·DPN compound for the 3-phosphoglyceraldehyde dehydrogenase crystallized from rabbit muscle, we have compared the kinetics of the DPNH formation with the decomposition of the E·DPN compound (Figure 1).

14 µM enzyme is titrated with three additions of 7.9 µM DPN to give a total of 24 µM E·DPN corresponding to an occupancy of approximately half the total sites on the enzyme. Complete occupancy was undesirable in this case as the experiment was designed so that the results refer exclusively to the bound and not the free DPN. The regular increases of absorbancy recorded differentially at 412–495 nm verify the formation of the E·DPN compound; further titration indicated that additional DPN would be tightly bound. At this point, 58 µM 3-phosphoglyceraldehyde were added in order to activate the turnover of the enzyme. The initiation of rapid DPN reduction by the addition of substrate is indicated by the abrupt jump in the fluorescence trace which occurs in the mixing time. It is apparent that a biphasic reaction occurs and the remainder of the NAD reduction proceeds in a slower phase extending over an interval of 30 sec.

The calibrations indicate that 24 µM DPN is reduced. Thus the jump in the fluorescence trace corresponds to 7.6 µM DPNH or equal to a change of E·DPN concentration corresponding to each one of the steps of the titration. It is apparent from the E·DPN concentration kinetics that no jump of corresponding size has occurred; in fact, it seems that the decrease of E·DPN concentration is synchronized only with a slow phase of the reduction. Since the DPN is tightly bound to the enzyme under these conditions (enzyme sites being in excess over DPN), these data clearly suggest that very little DPN could have been bound to non-absorbing sites responsible

93

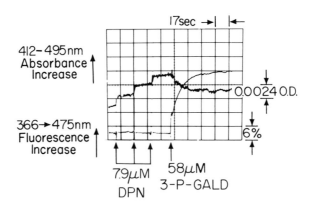

<u>Figure 1</u>.   Reaction mixture contained 14 µM Norite treated,
DPN-free enzyme, 1 mM versene, 0.05 M NaAsO$_4$, pH 7.4.

for the enzyme activity. Thus, this record appears to confirm the viewpoint put forward by us in 1954 (1) and elabora-
ted in more detail recently (2) that evidence for the func-
tionality of the E·DPN compound in the enzymatic activity
of the yeast and mammalian 3-phosphoglyceraldehyde dehydrogenases is presently lacking.

References

1. Harting, J. and B. Chance. McCollum-Pratt Symposium,
    September 1953.

2. Chance, B. and J.H. Park. J. Biol. Chem., <u>242</u>, 5093
    (1967).

THE RELATIONSHIP BETWEEN STRUCTURE AND ACTIVITY OF GLYCERALDEHYDE-3-PHOSPHATE DEHYDROGENASES

Richard N. Perham

Department of Biochemistry
Cambridge University, Cambridge

Glyceraldehyde-3-phosphate dehydrogenase (GPDH)(EC 1.2.1.12) has been the subject of much interest for many years, and many aspects have been covered in recent reviews (1,2). However, with the recent determination of the primary structure of the enzyme from lobster (3) and pig (4) muscle, the time is now ripe for a further evaluation of the connection between structure and activity. Some of this ground has been explored before (4), but further results are now available and can also be included.

The complete amino acid sequences of the lobster and pig enzymes are shown in Figure 1. These results prove conclusively that these enzymes comprise four identical polypeptide chains, each of molecular weight approximately 36,000, yielding a tetrameric molecule of molecular weight 144,000. (The occurrence of electrophoretically distinct forms of the enzyme in some organisms has been discussed elsewhere (5,6) and will not be considered further here.) The sequence homology between the lobster and pig enzymes is most impressive, 72% of the residues being in identical positions and with many of the changes also highly conservative. This argues very strongly for a close similarity in the three-dimensional structures. Of the 90 recorded changes, 54 are one-base, 35 are two-base, and only one (Met-Cys at position 130) is a three-base change in the amino acid codon dictionary established for E. coli (7), which is clear evidence of descent from a common genetic ancestor by an accumulation of single base changes. Many of the correlations between structure and activity that can be made by analysis of primary structure are of a negative kind, assuming the enzymes from different organisms utilize a common reaction mechanism (which appears very reasonable in view of the high conservation of sequence). Replacement of a residue

```
                      10                      20
Ac. Val. Lys. Val. Gly. Val. Asp. Gly. Phe. Gly. Arg. Ile. Gly. Arg. Leu. Val. Thr. Arg. Ala. Ala. Phe. Asn. Ser. Gly.
    Ser.                                          Ile.       Ile.            Leu.                 Leu. Ile. Ser.
                                                                                                       Ile.

                      30                      40
Lys. Val. Asp. Ile. Val. Asn. Asp. Pro. Phe. Ile. Asp. Leu. His. Tyr. Met. Val. Tyr. Met. Phe. Gln. Tyr. Asp.
     Ala. Gln. Val.                         Ala.                                              Gln.       Lys.
          Val.

 50                                            60                                 70
Ser. Thr. His. Gly. Lys. Phe. His. Val. Lys. Thr. Lys. Ala. Glu. Asp. Gly. Lys. Lys. Asp. Gly. Lys. Ala. Ile.
     His. Lys.                              Met.                 Ser.                Val.            Ala. Ile.
                                                                                                         Lys.
                                                                      80                      90
Thr. Ile. Phe. Gln. Glu. Ala. Asn. Ile. Lys. Trp. Gly. Asp. Ala. His. Leu. Lys. Tyr. Val. Val. Glu. Ser.
     Val.      Asn.      Met. Lys.            Ser. Pro.           Ala. Glu.       Tyr.           Ile.
                                                                 100
Thr. Gly. Val. Phe. Thr. Met. Glu. Gly. Ala. His. Lys. Gly. Gly. Ala. Lys. Arg. Val. Ile. Ile. Ser.
                    Ile.      Ser.                      Phe.                 Lys.            Val.
                                                    110                      Lys.
                                                                                  140
120                                           130
Ala. Pro. Ser. Ala. Asp. Ala. Pro. Met. Phe. Val. Met. Gly. Val. Asn. His. Glu. Lys. Tyr. Asp. Asn. Ser. Leu. Lys. Ile.
                                        Met.           Cys.              Leu.           Ser. Lys. Asp. Met. Thr. Val.
                                                                                                         160
Val. Ser. Asn. Ala. Ser. Cys. Thr. Thr. Asn. Cys. Leu. Ala. Pro. Leu. Ala. Lys. Val. Ile. His. Asp. His. Phe. Gly. Ile.
                                                              Val.                     Leu.       Glu. Asn.       Glu.
 170                              180                                                                       190
Val. Glu. Gly. Leu. Met. Thr. Val. His. Ala. Ile. Thr. Ala. Thr. Gly. Lys. Thr. Val. Asp. Gly. Pro. Ser. Gly. Lys.
                                              Val.                                    Val.                      Ala.
                                  200                                     210
Leu. Trp. Arg. Asp. Gly. Arg. Gly. Ala. Ala. Gln. Asn. Ile. Ile. Pro. Ala. Ser. Thr. Gly. Ala. Ala. Lys. Ala. Val. Gly.
Asp.           Gly.                                                            Ser.
          220                                                  230
Lys. Val. Ile. Pro. Glu. Leu. Asp. Gly. Lys. Leu. Thr. Gly. Met. Ala. Phe. Arg. Val. Pro. Thr. Pro. Asn. Val. Ser. Val.
                                                                                                            Asp.
 240                              270                      260
Val. Asp. Leu. Thr. Cys. Arg. Leu. Glu. Lys. Pro. Ala. Lys. Tyr. Asp. Asp. Ile. Lys. Lys. Val. Val. Gln. Ala. Ser.
                    Val.                Gly. Glu. Cys. Ser.                                  Ala. Ala. Met.      Thr.
                              270                                     280
Glu. Gly. Pro. Leu. Lys. Gly. Ile. Leu. Gly. Tyr. Thr. Glu. Asp. Gln. Val. Val. Ser. Cys. Asp. Phe. Asn. Ser. Thr. His.
                         Gln. Phe.                          Asp.                   Ser. Ser.      Ile. Gly. Asp. Asn. Arg.
                                                                               310
290                           300
Ser. Thr. Phe. Asp. Ala. Gly. Ala. Gly. Ile. Ala. Leu. Asn. Asp. His. Phe. Val. Lys. Leu. Ile. Ser. Trp. Tyr.
Ile.                Lys.                        Gln.         Ser. Lys. Thr.          Val. Val.
                                                          330
                              320
Asp. Asn. Glu. Phe. Gly. Tyr. Ser. Asn. Arg. Val. Val. Asp. Leu. Met. Val. His. Met. Ala. Ser. Lys. Glu.
                                        Gln.                       Ile.                Gln. Lys. Val. Asp. Ser. Ala.
                                                                                                Leu. Lys.
```

Figure 1. Amino acid sequence of GPDH from pig and lobster muscle. The upper sequence is that of the pig enzyme; that of the lobster is identical except where indicated below the line. Residue 24 is represented as a deletion in the lobster sequence so as to maximize the homology between the two chains.

suggests that it cannot have a critical role in the catalysis effected by the enzyme. For example, between pig and lobster only two out of five cysteine residues are conserved (positions 149 and 153) and five out of 11 histidines (positions 50, 108, 162, 176, 327). Conversely, the three tryptophan residues in each enzyme occur in corresponding positions, but that at position 310 is in the most highly conserved environment. Since a tryptophan residue has been implicated in a charge-transfer complex with coenzyme, NAD (2), it may well be that it is this particular tryptophan residue which is concerned. Studies of this type are being extended to the enzymes isolated from other organisms. For example, the amino acid sequence of the yeast enzyme (8) is nearing completion and is clearly going to be closely homologous with the pig and lobster proteins. Further, the GPDHs from mammalian sources are almost identical (6) and one is forced to conclude that there is here a very close connection between primary structure, three-dimensional structure, and enzyme activity.

The cysteine residue at position 149 is known to be the site of substrate attachment in thiol-ester linkage (9). Some chemical clues about the three-dimensional structure of the molecule are coming from analysis of intramolecular trans-acylation reaction involving this residue. Thus, it has been found that acetyl groups can migrate from cysteine-149 to the ϵ-amino group of lysine-183 (10,11) and to the thiol group of cysteine-281 (10), suggesting that these residues are in relatively close proximity at the catalytic site. The N-acetylation of lysine-183 interferes with binding of NAD, indicating that this residue may play an important part in coenzyme binding (10). On the other hand, cysteine-281 is absent in both yeast (12) and monkey (6) GPDH, which would seem to deny it a critical role in the enzyme mechanism. Unfortunately, these experiments do not allow an answer to the question of whether these residues are all contributed by one polypeptide chain or whether the catalytic site contains elements of more than one polypeptide chain in the tetramer (13). It is hoped that chemical cross-linking experiments will help solve this problem. Obviously, a definite answer will become available with the completion of an X-ray crystallographic analysis (14).

Another interesting problem yet to be solved is how the tetrameric molecule composed of identical subunits produces non-equivalence of NAD binding sites (15) and asymmetry of reaction with acyl phosphate substrates (16). It seems likely at this time that X-ray crystallographic analysis will be required to help resolve this situation.

We should also recognize that the remarkable conservation of amino acid sequence in the GPDHs requires explanation. One of the clear rules to emerge strongly from correlations of primary and three-dimensional structures of proteins (17) is the severe restriction on the nature of residues best described as "internal"; that is such residues are non-polar, few alterations are allowed and those alterations that do occur are generally highly conservative, e.g. isoleucine for valine. On the other hand, three dimensional structure can survive many changes in "surface" residues. Clearly, as we move to larger and larger proteins, particularly those that are oligomeric with a need, therefore, to maintain subunit contacts, the ratio of volume to surface must increase and the proportion of internal (conserved) residues must also rise. None the less, it seems unlikely that this can entirely account for the very high conservation of sequence in GPDHs, particularly those from mammals. We are forced, therefore, to take refuge in the fact that GPDH is a highly specific enzyme, forming part of a critical metabolic pathway, and that the specificity and possib. interactions of the enzyme necessitate a highly conserved surface structure.

In conclusion, it is worth remarking that it is impossible to decide at this stage what selection pressures have caused the stabilization of amino acid sequence changes in the GPDHs from different organisms (6). The turnover numbers of many of the enzymes are essentially the same in vitro (18) and, hence, other reasons must be sought. Perhaps there is necessity to interact with other molecules in which complementary changes have occurred. No explanation yet advanced is wholly convincing.

References

1. Velick, S.F., and C. Furfine. In The Enzymes (P.D. Boyer, H. Lardy and K. Myrback, eds.), Vol. 7, Academic Press, New York, 1963, p. 243.

2. Colowick, S.P., J. van Eys, and J.H. Park. In Comparative Biochemistry (M. Florkin, and E.H. Stotz, eds.), Vol. 14, Elsevier, Amsterdam, 1966, p. 1.

3. Davidson, B.E., M. Sajgo, H.F. Noller, and J.I. Harris. Nature, 216, 1181 (1967).

4. Harris, J.I., and R.N. Perham. Nature, 219, 1025 (1968).

5. Lebherz, H.G., and W.J. Rutter. Science, 157, 1198 (1967).

6. Perham, R.N. Biochem. J., 111, 17 (1969).

7. Crick, F.H.C. Cold Spring Harbor Symp. Quant. Biol., 31, 1 (1966).

8. Jones, G.M.T., and J.I. Harris. Abstr. 5th Meet. Europ. Biochem. Soc., Prague, p. 185 (1968).

9. Harris, I., B.P. Meriwether, and J.H. Park. Nature, 197, 154 (1963).

10. Mathew, E., B.P. Meriwether, and J.H. Park. J. Biol. Chem., 242, 5024 (1967).

11. Polgar, L. Biochim. Biophys. Acta, 118, 276 (1966).

12. Harris, J.I., and R.N. Perham. J. Mol. Biol., 13, 876 (1965).

13. Perham, R.N. Biochem. J., 99, 14C (1966).

14. Watson, H.C., and L.J. Banaszak. Nature, 204, 918 (1964).

15. Chance, B., and J.H. Park. J. Biol. Chem., 242, 5093 (1967).

16. Malhotra, O.P., and S.A. Bernhard. J. Biol. Chem., 243, 1243 (1968).

17. Perutz, M.F., J.C. Kendrew, and H.C. Watson. J. Mol. Biol., 13, 669 (1965).

18. Allison, W.S., and N.O. Kaplan. J. Biol. Chem., 239, 2140 (1964).

THE BINDING OF NAD$^+$ TO GLYCERALDEHYDEPHOSPHATE
DEHYDROGENASE ISOLATED FROM RABBIT AND LOBSTER MUSCLE

J.J.M. De Vijlder, W. Boers, B.J.M. Marmsen and E.C. Slater

Laboratory of Biochemistry, B.C.P. Jansen Institute
University of Amsterdam, Amsterdam, The Netherlands

and

Laboratory of Physical Chemistry, University of Nymogen
Nymogen, The Netherlands

Since the primary structure of lobster muscle glyceraldehydephosphate dehydrogenase has been determined by the
Cambridge group (1) and a start has been made on determining
its tertiary structure by X-ray crystallographic analysis
(2), we have recently extended our studies on the rabbit
muscle enzyme (3-6) to the lobster enzyme (7).

As was to be expected from the tetrameric structure of
the enzyme, it binds four molecules of NAD$^+$ (Table I). However, like the rabbit enzyme (4,8,9), the binding constants
differ greatly. The first two molecules are bound very
strongly, the third less strongly and the fourth quite weakly.

As is also the case with the rabbit enzyme, only three
of the four molecules contribute to the charge transfer band
at 360 mμ (Figure 1).* In this respect, the two muscle
enzymes appear to differ from the yeast (10,11).

Since the four protomers in the enzyme have identical
primary structure, the anti cooperative effects on the binding of NAD$^+$ and the lack of a charge transfer band with the
fourth molecule are presumably due to conformation changes in
the oligomer induced by binding of NAD$^+$ to a protomer. However,

*The slight increase in absorbance at 360 mμ after three
molecules of NAD$^+$ are bound continues after four molecules
are bound. It is probably due to contact charge transfer
absorption with free NAD$^+$.

101

TABLE I

Dissociation Constants of NAD^+ Bound to Glyceraldehydephosphate Dehydrogenase

Dissociation constants*	Lobster (Equilibrium dialysis) (7)	Rabbit Ultra- centrifugation (4)	Equilibrium dialysis (8,9)
K_1 (M)	$<5 \times 10^{-9}$	$<5 \times 10^{-8}$	$<10^{-11}$
K_2 (M)	$<5 \times 10^{-9}$	$<5 \times 10^{-8}$	$<10^{-9}$
K_3 (M)	$6 \times 10^{-7} \pm 2 \times 10^{-7**}$	4×10^{-6}	3×10^{-7}
K_4 (M)	$13 \times 10^{-6} \pm 3 \times 10^{-7**}$	35×10^{-6}	26×10^{-6}

*As defined in (4).
**Standard error of the mean.

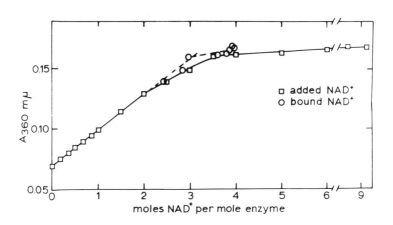

<u>Figure 1</u>. Titration at 360 mμ with NAD^+ of charcoal-treated glyceraldehydephosphate dehydrogenase (28.3 μM) isolated from lobster muscle. The enzyme was dissolved in 100 mM Tris-HCl buffer (pH 8.2) containing 5 mM EDTA. Temperature, 23°. □, added NAD^+; o, bound NAD^+ calculated from the dissociation constants in Table I.

no evidence could be obtained by circular dichoism for a change in the α-helix content of the rabbit enzyme on binding NAD$^+$ (5). In agreement of the ORD data of Listowsky et al. (12) and in disagreement with Bolotina et al. (13), we found no difference in the CD spectra of NAD$^+$-free and NAD$^+$-containing enzyme in the region of 200–250 mμ. The 360 mμ band in the holoenzyme is optically active, giving a very broad positive CD band centered at 350 mμ, with a molecular ellipticity of 18,000 per mole enzyme (Figure 2). The only other changes observed on adding NAD$^+$ to apo-enzyme are also in the region where NAD$^+$ has a positive CD band. No change is seen in the positive band at 299 mμ, also lying in the aromatic amino acid wavelength region. Thus, all the effects of NAD$^+$ on the CD spectrum can be explained by an extrinsic effect of bound NAD$^+$. Thus, changes in the Moffitt-Yang parameter b_0 will reflect these extrinsic Cotton effects rather than an overall change in protein conformation. The dependence of $-b_0$ on the amount of added NAD$^+$ (Figure 3) is very similar to

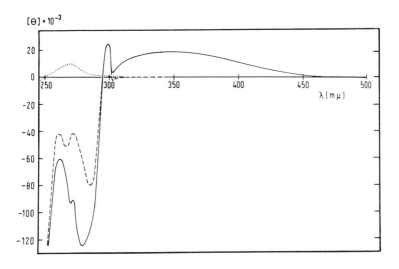

Figure 2. Circular dichroism of glyceraldehydephosphate dehydrogenase. ---, spectrum of the apo-enzyme; ———, spectrum of the holo-enzyme (7.5 moles added NAD$^+$ per mole of enzyme), corrected for free NAD$^+$ (3.5 moles free NAD$^+$ per mole enzyme, as calculated from dissociation constants; ···, spectrum of NAD$^+$, 152 μM enzyme in 1 M Tris-HCl buffer (pH 8.2) containing 2 mM EDTA.

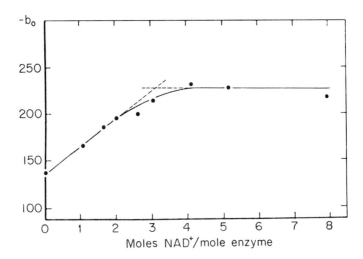

Figure 3. Change of the Moffit-Yang parameter b_o of glyceraldehydephosphate dehydrogenase (158 μM) upon adding NAD$^+$. The enzyme was dissolved in 0.1 M Tris-HCl buffer (pH 8.2) containing 2 mM EDTA.

that on the absorption spectrum. Although the $-b_o$ values measured here for apo- and holoenzyme are similar to those of Listowsky et al (12), we do not support their finding that 70% of the maximal increase was already reached with only one mole of NAD$^+$ per mole enzyme.

There is a remarkable difference between the rabbit and the lobster enzymes with respect to the kinetics of the absorbancy change at 360 mμ on mixing NAD$^+$ and enzyme in a stopped-flow apparatus. In the case of the rabbit enzyme, the first molecule is completely incorporated within 3-5 msec, corresponding to a second order rate constant greater than 10^8 M^{-1}sec^{-1}, but on the addition of more than one molecule, rapid and slow phases were observed (Figure 4). With the lobster enzyme, however, only the rapid phase was observed even with excess NAD$^+$ (Figure 5) and at 10°, corresponding to a second order rate constant greater than 10^{10} M^{-1}sec^{-1}.

Whatever the mechanism of the conformation change underlying the anti cooperative effects observed on binding NAD$^+$ to the muscle enzymes, it seems to make sense from the point of view of the mechanism of enzyme action. The close association between the protein -SH group and NAD$^+$ promotes

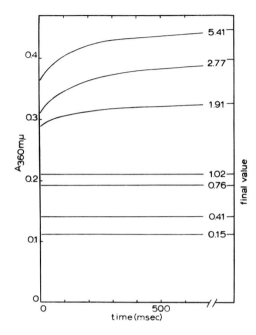

Figure 4. Reaction of NAD⁺ with rabbit muscle glyceraldehydephosphate dehydrogenase as measured in the Durrum stopped-flow apparatus. One of the two syringes contained 99 μM enzyme in 0.1 M Tris-HCl buffer-5 mM EDTA (pH 8.2), and the other various amounts of NAD⁺ in the same buffer with 0.5% serum albumin. Temperature, 10°; light path, 2 cm.

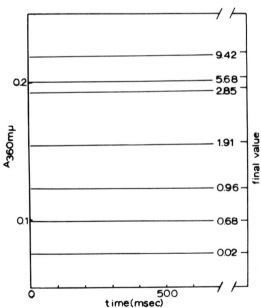

Figure 5. Reaction of NAD⁺ with lobster muscle glyceraldehydephosphate dehydrogenase as measured in the Durrum stopped-flow apparatus. The experiment was carried out in the same way as described in Figure 4 with 46 μM enzyme in 0.1 M Tris-HCl buffer-5 mM EDTA (pH 8.2) and various NAD⁺ concentrations. Temperature, 10°

hydride ion transfer to the NAD^+ and thiol ester formation. NAD^+ is required for all reactions of the enzyme with acyl phosphate, including its reduction by NADH (3). At least one place on the tetramer, however, needs to be kept available for binding of NADH in glucogenesis. In agreement with this picture, maximum activity in the reduction of acyl phosphate is obtained with 3 moles NAD^+ per mole enzyme, and excess NAD^+ inhibits competitively with respect to NADH. The K_i is 45 µM, similar to the binding constant of the fourth molecule (6).

References

1. Harris, J.I., and R.N. Perham. J. Mol. Biol., 13, 876 (1965).

2. Watson, H.C., and L.J. Banaszak. Nature, 204, 918 (1964).

3. Hilvers, A.G., K. Van Dam, and E.C. Slater. Biochim. Biophys. Acta, 85, 206 (1964).

4. De Vijlder, J.J.M., and E.C. Slater. Biochim. Biophys. Acta, 167, 23 (1968).

5. De Vijlder, J.J.M., and B.J.M. Harmsen. To be submitted.

6. De Vijlder, J.J.M., and Hilvers, A.G. To be submitted.

7. De Vijlder, J.J.M., W. Boers, and E.C. Slater. To be submitted.

8. Koshland, D.E., Jr., A. Conway, and M.E. Kirtley. In Regulation of Enzyme Activity and Allosteric Interactions (E. Kvamme and A. Pihl, eds.), Universitetsforlaget and Academic Press, Oslo and London, 1968, p. 131.

9. Conway, A., and D.E. Koshland, Jr. Biochemistry, 7, 4011 (1968).

10. Chance, B., and J. Harting Park. J. Biol. Chem., 242, 5093 (1967).

11. Kirschner, K., M. Eigen, R. BIttman, and B. Voigt. Proc. Natl. Acad. Sci. U.S.A., 56, 1661 (1966).

12. Listowsky, I., C.S. Furfine, J.J. Betheil and S. England. J. Biol. Chem., 240, 4253 (1965).

13. Bolotina, I.A., D.S. Markovich, M.V. Volkenstein, and P. Zavodsky. Biochim. Biophys. Acta, 132, 260 (1967).

THE CATALYTIC ROLE OF NAD$^+$ IN THE REACTION CATALYZED BY D-GLYCERALDEHYDE-3-PHOSPHATE DEHYDROGENASE

D.R. Trentham

Molecular Enzymology Laboratory,
Department of Biochemistry, The Medical School
University of Bristol, England

It has been proposed that NAD$^+$ has a catalytic role in the reaction catatalyzed by GPDH* (1). The reversible reaction scheme indicates that NAD$^+$ is bound to the enzymes at any time during the turnover of GPDH when there is formation or cleavage of the covalent bond between the active centre essential thiol and the C_1 carbon atom of the substrate.

$$E\overset{S.CO.R}{\underset{NADH}{}} + H^+ \rightleftharpoons E\overset{S.CH(OH).R}{\underset{NAD^+}{}} \rightleftharpoons E_{NAD^+} + R.CHO$$

$$E_{NAD^+} + R.CO.OPO_3^= \rightleftharpoons E\overset{S.CO.R}{\underset{NAD^+}{}} + HPO_4^=$$

$$E\overset{S.CO.R}{\underset{NAD^+}{}} \rightleftharpoons E\overset{S.CO.R}{} + NAD^+$$

$$E\overset{S.CO.R}{} + NADH \rightleftharpoons E\overset{S.CO.R}{\underset{NADH}{}}$$

Since NAD$^+$ binds to the free enzyme more tightly than NADH (2), it is a priori likely that the NAD$^+$-NADH exchange which must occur during each turnover of the enzyme will take place when the enzyme is acylated. The experiments

*Abbreviation: GPDH, D-glyceraldehyde-3-phosphate dehydrogenase (EC 1.2.1.12). RCHO is D-glyceraldehyde-3-phosphate.

described here are designed to answer whether this is indeed the case, and, if so, does NAD$^+$ catalyze sulphur carbon bond formation and cleavage?

Transient kinetic studies of the reduction of 1,3-diphosphoglycerate were performed using sturgeon muscle GPDH. This enzyme is readily crystallized and is free of bound NAD$^+$ (3). The stopped flow traces (Figure 1) show that when GPDH, preincubated with NADH, is mixed with 1,3-diphosphoglycerate, then the rate of reduction of 1,3-diphosphoglycerate initially is much slower than the steady state rate, while if the enzyme is preincubated also with NAD$^+$, the reaction proceeds immediately at the steady state rate. Either NAD$^+$ catalyzes the

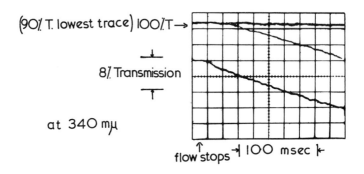

Figure 1. The effect of NAD$^+$ on the reaction of NADH and 1,3-diphosphoglycerate catalysed by sturgeon muscle GPDH at 20°. Concentrations refer to those in the reaction chamber. In the upper base line trace, syringe A contained ATP (1.7 mM), MgCl$_2$ (0.7 mM), 3-phosphoglycerate (1.7 mM) and syringe B contained GPDH (1 x 10^{-5} M sites), NADH (8 x 10^{-5} M), 2-mercaptoethanol (10^{-4} M). Both syringes contained EDTA (2 mM) and triethanolamine hydrochloride (0.16 M) treated with NH$_3$ to yield a final reaction pH value of 7.3. In the centre trace, ATP, MgCl$_2$ and 3-phosphoglycerate were preincubated with 1,3-diphosphoglycerate kinase at pH 6.5 to generate 1,3-diphosphoglycerate prior to adding EDTA and buffer. In the lowest trace (in which the transmission scale is displaced by 10%), reactants are those for the centre trace except that NAD$^+$ (6 x 10^{-5} M) was added to syringe B.

formation of the acyl enzyme or there is a rate limiting conformation change, induced by NAD⁺, analogous to the T to R
transformation of the yeast enzyme which must occur before
the enzyme is active as a catalyst (4). There are three arguments for supposing that a conformation change is not a rate
limiting process of the first turnover. First, if sturgeon
GPDH and arsenate are mixed in a stopped flow apparatus with
NAD⁺ and D-glyceraldehyde-3-phosphate at pH 8, there is a
burst of NADH production associated with the first turnover
of the enzyme followed by the steady state rate. This is in
contrast to the slow first turnover of yeast GPDH (4). Secondly, the optical rotary dispersion change observed in going
from the rabbit muscle holoenzyme (NAD⁺ bound) to the apoenzyme is reversed by NADH addition to the apoenzyme (5).
Thirdly, at least one mole of NAD⁺ per mole of enzyme sites
is necessary to reverse the lag phase (Figure 1). The major
optical rotary dispersion change observed in the rabbit enzyme in going from the apoenzyme to the holoenzyme is observed when the first mole of NAD⁺ is added per mole of
tetramer (5).

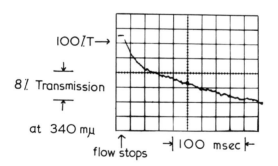

Figure 2. The reaction of 1,3-diphosphoglycerate (preincubated with GPDH) and NADH catalysed by lobster muscle GPDH at
20°. Concentrations refer to those in the reaction chamber.
Syringe A contained 1,3-diphosphoglycerate generated under
the same conditions as in Figure 1, GPDH (1.08 x 10⁻⁵ M sites),
2-mercaptoethanol (1 mM), and Syringe B contained NADH (7 x
10⁻⁵ M). Both syringes contained EDTA (1 mM) and sodium citrate (0.16 M) treated with HCl to yield a final reaction pH
value of 6.05.

The role of NAD^+ as a catalyst may be studied by an independent method. The stopped flow trace (Figure 2) illustrates the reaction between 1,3-diphosphoglycerate and NADH catalysed by NAD^+ bound lobster muscle GPDH. In this experiment, 1,3-diphosphoglycerate is initially preincubated with GPDH so that the acyl enzyme is formed (6). *The biphasic reaction shows that probably aldehyde release or acyl enzyme formation is the rate limiting process at pH 6. If the experiment is carried out at pH 8, zero order kinetics are observed with no transient phase showing that there has been a change of rate determining step. A steady state analysis at 25° of the reduction of 1,3-diphosphoglycerate by NADH (1.5×10^{-4} M) shows that at pH 8 the rate of production of NAD^+ during the first 0.2 extinction change at 340 mμ is constant and is unaffected by the initial addition of NAD^+ (2.0×10^{-5} M). However, at pH 5.5 when aldehyde release or the acylation reaction is rate limiting the rate of production of NAD^+ increases during the first 0.2 extinction change showing that the reaction is autocatalytic. Moreover, if the initial reaction mixture contains NAD^+ (2.0×10^{-5} M), then the initial rate of NAD^+ production is the same as the rate when 2.0×10^{-5} M NAD^+ has been generated in the previous experiment. Addition of NAD^+ (1.0×10^{-4} M) increases the initial rate of reaction 5-fold. It should be noted that ATP (3.4 mM) is present in the assay medium so that NAD^+ may be competing for enzyme sites not only with NADH but with ATP. However, similar results are obtained when a solution of 1,3-diphosphoglycerate is used that contains no ATP.

*These results show that NAD^+ production is responsible for the autocatalysis. The acceleration by NAD^+ indicates that the binding of NAD^+ is involved in the rate limiting step at low pH. It is proposed that, if NAD^+ dissociates from the aldehyde-NAD^+-enzyme ternary complex formed during the redox step prior to aldehyde release, then a dead-end, aldehyde-enzyme complex is formed.

References

1. Trentham, D.R. Biochem. J., 109, 603 (1968).

2. Velick, S.F. and C. Furfine. In The Enzymes (P.D. Boyer, H. Lardy, and K. Myrbäck, eds.), 2nd ed., vol. 7, Academic Press, New York, 1963, p. 243.

*Manuscript revised November 5, 1970.

3. Allison, W.S., and N.O. Kaplan. J. Biol. Chem., 239, 2140 (1964).

4. Kirschner, K. In Regulation of Enzyme Activity and Allosteric Interactions, Fed. Europ. Biochem. Soc. Colloq. (E. Kvamme, and A. Pihl, eds.), Oslo University Press, Oslo, 1967, p. 39.

5. Listowsky, I., C.S. Furfine, J.J. Betheil, and S. Englard. J. Biol. Chem., 240, 4253 (1965).

6. Krimsky, I., and E. Racker. Science, 122, 319 (1955).

DISCUSSION

Eigen: Let me report briefly on the state of affairs of what somebody in this audience has called "allohysteria." You all know that essentially two models have been proposed which are shown in Figure 1. Actually, this picture is self-explanatory

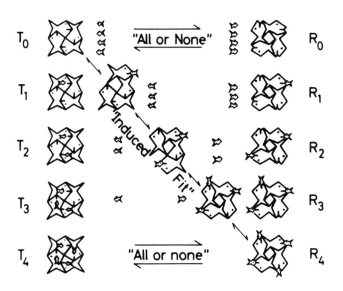

Figure 1. Limiting cases of allosteric interaction.

and—since you all know the model of Monod, Changeux and Wyman as well as that of Koshland, Nemethy and Filmer—we may immediately proceed to Figure 2. This picture shows that both models depicted in Figure 1 are the two meaningful limiting cases of the same general mechanism.

112

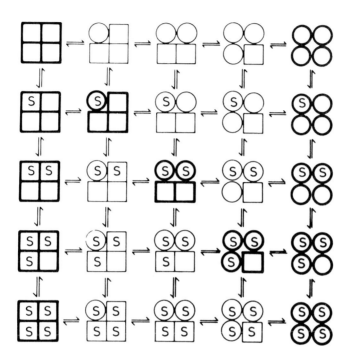

Figure 2. General reaction scheme of allosteric binding to a four subunit protein (from ref. 2).

If there are states of the enzyme, and you can identify them, it turns out that the higher the resolution of the method, the more states you will see. Then there is a limiting number of possibilities for the mechanism. If you have four of these subunits (Figure 1), you can make two conformations: a tight one, the T conformation in which the fish are not very hungry and the affinity to the substrate will be low. Second, you might have a relaxed or R conformation.

As you see, in the R conformation, the subunits can eat small fish quite well, can easily form gefilte fish, or something like that. Now, if you write down all the possible states here, and allow all types of transitions, you could write down 44 different states of these conformations if you chart all the intricacies. There are two limiting cases; if all these 44 states were populated, and you could do a relaxation experiment, you would see possibly 43 relaxation times, possibly much less. If you see very few relaxation times, you may be sure that the intermediates are not populated and you see only some extreme of these mechanisms. There are two possible extremes when you have only two conformations: the Monod and the Koshland mechanisms (Figure 1). That is, either you saturate along these two roads, and make all-or-none transitions between the different states, then you get in each row five states which you can saturate, and possibly four relaxation times; here again are four relaxation times for the other states of saturation, and then you have possibly one relaxation time for the whole conformation change, if these are fast compared to the conformation change.

The other limit would be that you get this induced-fit mechanism so that you go from the T state in a sequential mechanism over to the saturated state, which is in the other conformation, and there you can see as a maximum four relaxation times, possibly three. Now if you can be sure that the Monod mechanism is valid, and you can see three relaxation times, then it is most probable that the different binding sites are degenerate. In other words, that they behave identically with respect to binding. The question then is which real systems come close to any of these limits.

Glyceraldehydephosphate dehydrogenase (GAPDH) from yeast is still the best representative of Monod's model we know. I suppose the work of Kirschner (1) and coworkers is known, so I shall mention only some facts.

1. The relaxation spectrum of GAPDH with NAD contains only three measurable time constants which are related to the mechanism in Figure 1 or Figure 2. A fourth relaxation time in the microsecond region has been found recently, which is probably related to the temperature dependence of the chromophore spectrum.

2. The concentration dependencies of these three relaxation times are only compatible with the "all or none" mechanism of Figure 1, they do not agree with any sequential mechanism. Since only three time constants are present, none of the more complicated schemes following from Figure 2 can

apply (i.e. hybrid states are not populated to a measurable extent.).

3. The analogy to Figure 1 reaches very far. The less affine state is the slowly reacting one. Here rather high enthalpy and entropy increments are found indicating a "diffuse" T-state. The conformation change is quite slow (slower than any of the NAD-uptake reactions). Again, high enthalpy and entropy increments indicate a phase-transition-like process involving many groups of the enzyme molecule.

4. The enzyme can be split at high citrate concentration into two dimers (2), but not into monomers. Thus, symmetry is somehow limited.

5. The signal to noise ratio of the relaxation experiments is now so good (about 10^4:1) that all dependences can be determined very precisely. They follow correctly the predictions of theory for the "all or none" model (3). Also, recent determinations of physical parameters related to structure show that there is no 1:1 relationship between NAD binding and conformation change as to be expected for any sequential mechanism, but rather the relation suggested by the "all or none" model. However, this case so far is the only one which shows such a clear picture. Hemoglobin will be discussed in more detail tomorrow. It certainly does not belong to this type of model and might be closer to the other limit. Phosphorylase a and b exhibit more complex behavior, and Fred Kayne will show you tomorrow that pyruvate kinase-- even in absence of any other substance--shows a multiplicity of states, indicating a more complex behavior than any of the models depicted in Figures 1 and 2. Thus, for other systems, we still have to rely essentially on experimental facts, before we can make any prediction of mechanism.

The models which have been introduced are very useful to tell us what the limiting cases are, but reality seems mostly to prefer intermediate states, and which one is present and which one is not present is solely to be determined by the free energy which you can write down for each of these states (Figures 3 and 4).

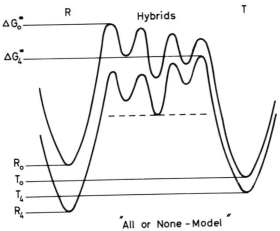

Figure 3. Free energy diagram for the Monod Changeux Wyman model.

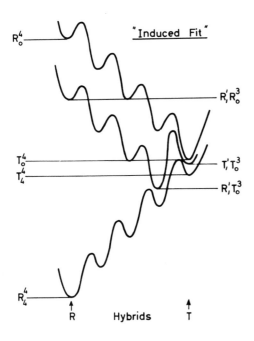

Figure 4. Free energy diagram for the Koshland model.

Chance: At one time, Kirschner had some evidence for a spec-
troscopically distinct form of the compound.

Eigen: The fast isomerization process I mentioned is related
to this form. We cannot really say what kind of change this
is.

Velick: We find that we cannot uniquely determine binding
parameters by fluorescence because we don't understand what
happens when we add coenzyme to the four-site protein in
which Dr. Chance suggests that one side does all the work.
Because of the strong temperature dependence of binding, we
may plot the dissociation constant against reciprocal temperature. It is calculated that there should be a large negative
enthalpy change when DPN is bound to the protein and DPNH
behaves similarly. I won't discuss it in any detail. (It
turns out that Dr. Slater and I brought the same type of data.)
This is circular dichroism as a function of DPN added, and
again you notice the equivalence point is very close to three,
and the only difference with his data is that we did it at the
two temperatures at 5° and 25°. Our circular dicroism, unlike
his, went all the way down to 205 mμ and there is absolutely
no change below 250 mμ as a function of DPN concentration and
all these effects are what he described as extrinsic effects.
Now, as a result of the very large apparent ΔH's that we calculated for coenzyme binding, we decided to undertake some
direct calorimetric measurements and these are still in progress. They're being done with Dr. Julian Sturtevant at Yale
by microflow calorimetry.

The best data we have now indicates a limit very close
to three sites and there seems to be very little enthalpy
change with DPN binding at the fourth site, and, in this particular series of experiments, the total enthalpy change is
very near the value of the sum of the values for the three
sites that we calculated from the Van't Hoff data. I needn't
now discuss all of the assumptions involved in applying Van't
Hoff plots to data of this type. What I might emphasize at
this point is that the properties of the enzymes are very
sensitive to temperature and that the fact that the lobster
enzymes have quite different kinetics of DPN addition to its
site than the rabbit muscle enzyme. This probably reflects
the fact that it lives and functions at 10°C instead of 36°C.
I think the properties of this enzyme are like those of the
elephant studied by the blind men. It is very important in
any physical work that you specify the ionic strength, the

ionic medium, the pH, temperature and so forth, because I think a different story can arise with each set of conditions.

Eigen: There are some other cases of binding here; such as the problem of the enzyme-NAD binding to substrate. The equilibrium constant for acylation is cooperatively dependent upon the concentration of NAD, this makes it even more complicated. That may be the fault of the elephant.

Gutfreund: Are you talking about the yeast enzyme or the muscle enzyme?

Eigen: The cooperativity for NAD binding showed up very clearly in the yeast enzyme. The muscle enzyme might be quite different.

References

1. Kirschner, K., M. Eigen, R. Bittman and B. Boigt. Proc. Nat. Acad. Sci., 56, 1661 (1966).

2. Schuster, . Unpublished data.

3. Eigen, M. Nobel Symposium, No. 5, p.

LACTATE DEHYDROGENASE AS A MODEL FOR THE STUDY OF TRANSIENTS IN NAD-LINKED DEHYDROGENASE REACTIONS*

H. Gutfreund

Molecular Enzymology Laboratory, Department of Biochemistry
University of Bristol, Bristol, England

Two methods of analysis have been used to interpret the measurements of transients and relaxations of the reactions of lactate dehydrogenase with its substrates. These methods are considered here for their general application to NAD-linked dehydrogenases also, and they have, of course, a much wider application. The first method involves the analysis of the time course of changes in reactant (substrates and products) concentrations, during the transient period between mixing enzyme and substrate and the establishment of steady-state product formation. The second method involves the analysis of the time course of the formation and decomposition of different enzyme intermediates during approach to steady-state. If specific signals can be monitored during the interconversion of enzyme-reactant complexes, these can be interpreted in terms of chemical changes as well as rates.

The rate of formation of products during the transient period has three phases. How many of these phases can be detected and how many rate constants can be calculated from each phase depends on the ratios of the rates of transformation of the intermediates. The three phases correspond to:

(1) The rate of approach to the steady-state concentration of all enzyme intermediates prior to transformation of substrate to product results in an acceleration of the rate of formation total product (enzyme bound and free).

(2) The interconversion of enzyme bound substrate to enzyme bound product can result in a rapid product formation prior to its steady state release.

*Supported by Science Research Council and U.S.P.H.S.

(3) Steady state rate determined by regeneration of free enzyme.

The simplest model compatible with the kinetic data so far obtained for the pig heart lactate dehydrogenase system involves the following intermediates:

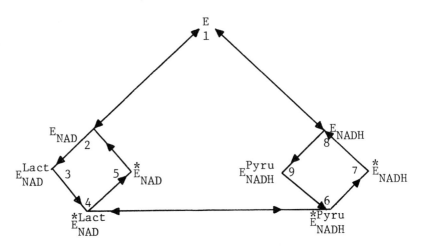

The directions of the arrows indicate the directions of the flux during transients and steady state. The kinetic information summarized below and discussed in greater detail by Heck, McMurray and Gutfreund (1,2) is more sensibly interpreted in terms of relative rates of different pathways than in terms of random, compulsory or alternative order pathways.

For the lactate-NAD to pyruvate-NADH reaction above pH 5.5, the formation of the three postulated enzyme-NADH complexes is more rapid than their decomposition. This results in an initial rapid formation of four moles of NADH (enzyme bound) per mole of tetrameric enzyme (70 sec^{-1} at pH 5.5 with pig heart LDH). This rate increases rapidly with increasing pH. Above pH 6, the resolution of our stopped-flow technique (1 msec) does not permit the study of the time course of this burst. Accurate evaluation of the number of sites (four per tetramer) is, however, achieved. The initial transient in the reverse direction, when enzyme is mixed with pyruvate and NADH, has an inverse pH rate profile.

Studies of the rapid formation of NADH during the pre-steady state period have provided similar preliminary information with glyceraldehyde-3-phosphate dehydrogenase and malic dehydrogenase in our laboratory. Glutamate dehydrogenase was investigated by Iwatsubo and Pantaloni (3). The interpretation of the data becomes more detailed in connection with information obtained from the use of the other approach: the observation of spectral and fluorescence changes during the association of enzymes with the nucleotides in the absence of the other substrate partner. Heck et al (1,4) found that these binary complexes, enzyme–NAD and enzyme–NADH, are both formed in a single second order step (10^8 $M^{-1}sec^{-1}$). The dissociation of NADH occurs at a rate of 40 sec^{-1} and that of NAD of approximately 2500 sec^{-1}. The fact that an additional step is involved in the dissociation of the nucleotides from the enzyme after a complete catalytic turnover can be deduced from the following evidence. At pH values when the steady state rate of enzyme catalyzed overall reactions are slower than the rates of dissociation of the respective nucleotide-enzyme complex (in the absence of other substrates) equivalent bursts of NADH or NAD are still observed during the transient. The delay in the dissociation of the nucleotide cannot be due to slow dissociation of the other products (pyruvate or lactate), which have to come off first. Steps 6 to 7 and 4 to 5 must be fast. Since bursts in both directions can be observed, the oxidation/reduction step must be made essentially irreversible by the rapid dissociation of pyruvate or lactate, respectively.

The kinetic analysis of the approach to the steady state and of the transient signals due to the formation and decomposition of enzyme intermediates presents us with a framework which has to fit any proposed chemical mechanism. In this symposium on probes, it is pertinent to draw attention to the fact that substrates can act as probes and provide chemical information about the events during the interconversions of enzyme intermediates. Three types of reversible processes are found during enzyme-substrate reactions: rapid absorption of substrate, rearrangement to form the reactive complex and catalysis. We have found kinetic evidence for, and determined the rates of, the rearrangement step in such diverse enzymes as lactate dehydrogenase and alkaline phosphatase (5). This is the most distinctly enzymatic process. In nonenzymic reaction, one also finds absorption and catalysis, but only enzymes and related biological devices respond to the binding of a specific ligand. The use of probes to obtain information about events during the rearrangement of the enzyme substrate complex thus presents a challenging topic. Up to now, the interpretation of environmental changes from spectral and fluorescence changes during transients has been very tentative. It is to be hoped that the study of probes bound to known structural segments of crystalline enzymes will provide a sound basis for the interpretation of signals from unknown environments.

References

1. Heck, H. d'A., C.H. McMurray, and H. Gutfreund. Biochem. J., 108, 793 (1968).

2. Criddle, R.S., C.H. McMurray, and H. Gutfreund. Nature, 220, 1091 (1968).

3. Iwatsubo, M., and D. Pantaloni. Bull. Soc. Chim. Biol. Paris, 49, 1563 (1967).

4. Heck, H. d'A. J. Biol. Chem., 244, 4375 (1969).

5. Halford, S.E., N.G. Bennett, D.R. Trentham, and H. Gutfreund. Biochem. J. 114, 243 (1969).

STUDIES OF CONFORMATIONAL TRANSITIONS OF HOMOSERINE DEHYDROGENASE I USING THE FLUORESCENCE OF BOUND PYRIDINE NUCLEOTIDE*

Harold J. Bright[+]

Department of Biochemistry, School of Medicine
University of Pennsylvania, Philadelphia, Pennsylvania 19104

I should like to describe some of our studies of the allosteric homoserine dehydrogenase I from E. coli, in which we have used the fluorescence of the enzyme-bound TPNH as an indicator of conformational states of the enzyme. It has been shown that the enzyme requires K^+ for catalytic activity and is subject to allosteric inhibition by threonine (1). Figure 1 shows that the fluorescence enhancement (290 mμ excitation) of the bound TPNH is much greater in the presence of K^+ (R state, curve A) than in the presence of threonine (T state, curve B) although the titration end-points indicate that there are three TPNH sites per mole enzyme in each case. Similar experiments were recently published by Truffa-Bachi et al. (1) and we show Figure 1 to emphasize that in the absence of both K^+ and threonine, the fluorescence enhancement (curve B) is indistinguishable from that observed in the presence of threonine. Figure 1 also demonstrates that the K^+- and threonine-induced changes in fluorescence of the bound TPNH are fully reversible. Similar patterns are obtained by 340 mμ excitation, although the enhancement values are smaller than those observed with 290 mμ excitation. Fluorescence titrations at lower enzyme concentrations show that the dissociation constants for TPNH are 0.34 μM and 0.52 μM for the R and T states, respectively, at 25°. Similar experiments in which TPN is also present show that the dissociation constants for TPN are 20 μM and 5.5 μM for the R and T states, respectively.

*Supported by NIH GM-11040 from the USPHS.

[+]Career Development Awardee (5-K3-GM-34960) from the USPHS.

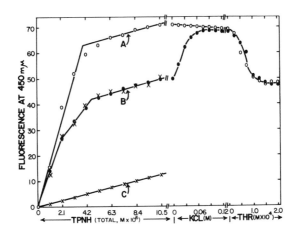

Figure 1. Fluorimetric titration of enzyme with TPNH in the presence and absence of K^+ at 25°. Curve A: 1.4 µM enzyme plus 0.15 M KCl. Curve B (-o-): 1.4 µM enzyme, no KCl. Curve B (-x-): 1.4 µM enzyme, no KCl, 0.5 mM threonine. Curve C: no enzyme, no KCl, no threonine. The enzyme from experiment B was then titrated with K^+ while that from experiment A was titrated with an equal volume of water. Subsequently, both solutions were titrated with threonine. The buffer was 0.05 M Tris-HCl, pH 7.0, excitation was at 290 mµ and the fluorescence is in arbitrary units.

The increase in fluorescence enhancement of the bound TPNH caused by K^+ is shown at different levels of threonine in Figure 2. The interaction of K^+ with the enzyme is highly cooperative and threonine acts as an antagonist of K^+. The initial velocity patterns for the overall reaction are similar to the patterns of Figure 2.

When the R → T transition is studied by mixing the K^+-activated enzyme with threonine and substrates (aspartic semialdehyde and TPNH) in the stopped flow apparatus, there is a burst of TPNH disappearance (Figure 3, trace D) corresponding to the R → T transition induced by threonine. As shown previously (3), the rate ($k_{1,obs}$) of the R → T transition is a hyperbolic function of the threonine concentration whereas the extent of the burst can be shown to be $K[E]/k_{1,obs}$ (see Figure 3), where [E] denotes the concentration of K^+-induced R state immediately after mixing and K is a constant composed of steady state kinetic coefficients. This result confirms the

124

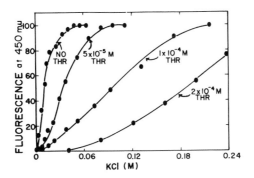

Figure 2. Fluorimetric titration of ENZYME-(TPNH)$_3$ with K$^+$ at different threonine concentrations. Each titration was carried out with 1.55 μM enzyme, 17 μM TPNH and the indicated levels of threonine. Other conditions were as indicated in Figure 1.

conclusion drawn from the titration data of Figure 1, namely that the E-TPNH complex exists predominantly, if not exclusively, in the catalytically inactive T state in the absence of threonine and K$^+$.

The R → T transition can also be monitored by stopped flow measurements of the decrease in flourescence of the bound TPNH caused by the binding of threonine. An example of such an experiment is trace A of Figure 4. The rate of fluorescence quenching has the same hyperbolic dependence on threonine concentration as the rate of inhibition of the enzyme computed from burst experiments (3) such as curve D of Figure 3.

Similarly, the T → R transition can be resolved in stopped-flow experiments. Curve C of Figure 3 shows the lag in the overall reaction corresponding to this transition when enzyme in the presence of threonine (T state) is mixed with K$^+$ and substrates. The first order rate constant, k_a, describing this process is in excellent agreement with the first order rate constant associated with the K$^+$ induced increase in fluorescence enhancement of the bound TPNH shown in trace B of Figure 4.

The picture of the allosteric regulation of homoserine dehydrogenase I which emerges from these studies is that the

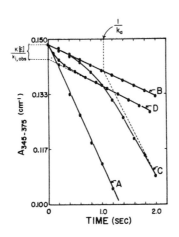

Figure 3. The T → R (trace C) and R → T (trace D) transitions of the enzyme at 25° as measured by the kinetics of TPNH disappearance in the overall reaction. Two kinds of enzyme solution (E_{AD} and E_{BC}) and substrate solution (S_{AC} and S_{BD}) were used. E_{AD} contained 2.8 µM enzyme and 0.2 M KCl, while E_{BC} contained 2.8 µM enzyme and 0.5 mM threonine. S_{AC} contained 0.17 M KCl, 1.43 mM aspartic semialdehyde and 43 µM TPNH, while S_{BD} was the same except for the addition of 0.84 mM threonine. Trace A (E_{AD} + S_{AC}) represents the fully active enzyme (R state). Trace B (E_{BC} + S $_{BD}$) represents inhibited enzyme (T state). Trace C (E_{BC} + S_{AC}) represents the time dependence of the T → R transition. Trace D (E_{AD} + S_{BD}) represents the time dependence of the R → T transition. The buffer was 0.04 M imidazole-HCl, pH 7.0. The experiments were carried out on the regenerative flow apparatus in collaboration with Prof. Britton Chance and involved a 90-fold dilution of the enzyme solution, after mixing, in each case.

enzyme can exist in two major conformations, R and T, with the T state being by far the more stable. The catalytically active R state is induced and stabilized by K⁺, whereas the catalytically inactive T state is stabilized by threonine. The rates of interconversion of the two states under our experimental conditions are much smaller than the maximum turnover number (about 350 sec⁻¹) in the direction of TPNH oxidation.

126

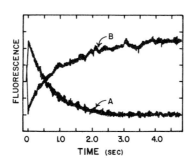

Figure 4. The R → T (trace A) and T → R (trace B) transitions as measured at 25° by stopped flow measurements of the fluorescence changes of enzyme-bound TPNH. For trace A, 8 μM enzyme, containing 30 μM TPNH and 0.12 M KCl, was mixed with an equal volume of 0.4 mM threonine and 30 μM TPNH. For trace B, 8 μM enzyme, containing 30 μM TPNH and 0.1 mM threonine, was mixed with an equal volume of 0.3 M KCl and 30 μM TPNH. The buffer was 0.05 M Tris-HCl, pH 7.0 and excitation was at 340 mμ.

References

1. Truffa-Bachi, P., R. van Rappenbusch, J. Janin, C. Gros, and G.N. Cohen. Europ. J. Biochem., 5, 73 (1968).

2. Janin, R., R. van Rappenbusch, P. Truffa-Bachi, and G.N. Cohen. Europ. J. Biochem., 8, 128 (1969).

3. Barber, E.D., and H.J. Bright. Proc. Nat. Acad. Sci., 60, 1363 (1968).

DISCUSSION

Perham: It occurs to me that another aspect of the problem might be mentioned at this stage; I would like to use the aldolase protein as an example. One of the more interesting questions one would like to answer in this system, as in other oligomeric enzymes, is whether the active site exists between or within protein subunits? In the absence of crystallographic evidence, one possible approach would be to try to find a way of keeping the subunits apart and seeing whether enzyme activity exists in a subunit. The standard method of taking protein subunits apart is a rather strong treatment--extremes of pH, guanidine hydrochloride, etc. A possible way around this is to use a protein blocking group that would force the protein subunits apart, as a result of which it might be possible to induce them to stay apart and to look at the enzyme activity. The reagent we chose is 2-methyl maleic anhydride which reacts reversibly with protein amino groups (1). It is interesting because when it reacts, with a protein amino group, the conditions under which it comes off are rather less stringent than those required for maleic anhydride. Thus, one can readily regenerate the amino groups, and the protein should reform. Trying this with aldolase, it turns out that the amount of chemical modification required to get the subunits apart is really quite high. With modification of up to 50% of the protein amino groups, little dissociation is noted. As the substitution is increased to 70% or so, dissociation occurs and can be made complete. Unfortunately, 70% blocking of the amino groups is sufficient to unfold the polypeptide chains too much for our purposes, but it is worth mentioning because some of you may be interested in taking proteins apart reversibly, and 2-methyl maleic citraconic anhydride could be a very useful reagent for this purpose.

Gutfreund: Dr. Bright, do I understand that these switchings on and off are slow compared with the turnover of the enzyme?

Bright: The inhibition has a maximum rate of 17 sec^{-1}, which is one-tenth the maximum turnover rate.

1. PROBES OF ENZYMES

Bernhard: We have been studying single turnovers in a variety of enzyme reactions by preparing chromophoric and fluorophoric pseudo-substrates in which, hopefully, we can make a positive identification of the enzyme target by the way the enzyme-substrate covalent bond behaves, and by identifying the spectrum or the fluorescence emission spectrum, of the enzyme-substrate covalent bond. Among the enzymes which we have studied in this way is GAPDH, which forms, as an intermediate, an acyl enzyme. Oxidation of the aldehyde gives a thiol-acyl-ester enzyme and by making a chromophoric or particular aromatic acyl enzyme, we can positively identify and quantitate the formation of this aryl-thiol-ester; we know how much we have, and we can identify it as a thol ester.

In every species of GAPDH which we have investigated, which includes beef, lobster, and rabbit muscle, if we take an excess of an aryl phosphate, which is an analog of a reactant or reaction product, with an excess of this and expost it to the enzyme, we get a very rapid reaction which leads to the thol ester which, under favorable conditions gives a very special spectrum which can be identified. In all cases, this leads to a di-acyl enzyme tetramer, which is composed of four identical subunits; nevertheless, the maximum yield, no matter what the conditions, gives a di-aryl enzyme. This is rather surprising, since we know from other work, such as the classical information of Dr. Racker's on acylation, that there are four equivalent especially reactive sulfhydryls, on each of these enzymes. So that, we have examined that alkylation reaction, and we have special conditions which give us less than four alkylation sites as well as acylation sites. Under the same conditions where we get only two acyl groups, we get four alkylation sites, and we presume these four to be the same cysteine residue as those of the enzyme; we have that number of acyl groups.

Czerlinski: I would like to comment briefly on various dehydrogenases which I have investigated previously with chemical relaxation techniques. Table I summarizes the results.

129

TABLE I. Temperature jump experiments on dehydrogenase systems, using fluorescence detection.

Dehydrogenase (=DH)	Other Reactants	Reaction Type		Bimolec. k $M^{-1}sec^{-1}$	Monom. k sec^{-1}	Ref.
Liver Alcohol DH	R+I	EIR	EIR'	—	500	1
Lactate DH	R	ER	ER'	—	300 & 1000	2
Malate DH	R	E+R	ER	6.8×10^9	240	3
Liver Alcohol DH	R+A	cyclic		?	~ 3.2, etc.	4,5

Key: R = DPNH (or NADH), I = Imidazole, A = Acetaldehyde, E = Enzyme site.

1. Czerlinski, G. Biochim. Biophys. Acta, 64, 199 (1962).

2. Czerlinski, G. and G. Sihreck. J. Biol. Chem., 239, 1913 (1964).

3. Czerlinski, G. and G. Shireck. Biochemistry, 3, 89 (1964).

4. Czerlinski, G. In Pyridinenucleotide-dependent Dehydrogenases (Silmert, G.W., ed.), University of Kentucky Press, Lexington, 1969.

5. Czerlinski, G. J. Theoret. Biol., 21, 408 (1968).

STRUCTURE AND ROLE OF THE METAL COMPLEXES OF AN ALLOSTERIC
PYRUVATE KINASE FROM YEAST, AND A NON-ALLOSTERIC
PYRUVATE KINASE FROM MUSCLE*

A.S. Mildvan[+], J.R. Hunsley, and C.H. Suelter[‡]

Johnson Research Foundation, School of Medicine
University of Pennsylvania, Philadelphia, Pennsylvania 19111

and

Department of Biochemistry
Michigan State University, East Lansing, Michigan 48823

Summary

Rapid quenching experiments of muscle pyruvate kinase,
activated by the slowly dissociating Ni^{2+} ion, suggest that
the active enzyme-Ni-ADP bridge complex forms by multiple
pathways as had been previously proposed for the Mn and Mg
complexes. Managanese and substrate binding to the allosteric yeast pyruvate kinase have been studied by EPR and by
the enhancement (ε) of the proton relaxation rate of water.
Unlike the muscle enzyme, which binds 2 Mn with equal affinity, the yeast enzyme binds up to 6 Mn with apparent affinities which vary by three orders of magnitude. The unusual
nature of the binding curve is probably due to interaction
among the Mn binding sites. All substrates, as well as the
activator FDP, decrease ε from 15 to values as low as 5, suggesting the formation of EMS complexes as in the muscle enzyme.
The allosteric activator FDP raises the affinity of the

*Supported in part by U.S.P.H.S. Grants AM-09760, AM-13351,
GM-12246 and by N.S.F. Grants NSF-GB-8579 and NSF-GB-7780.
[+]This work was done during the tenure of an Established Investigatorship from the American Heart Association. Present
address: The Institute for Cancer Research, Fox Chase, Philadelphia, Pennsylvania 19111.
[‡]Research Career Development Awardee from the National Institutes of Health.

131

enzyme-Mn complex for PEP in agreement with kinetic observations only when more than one Mn binding site is occupied. When less than one Mn is available to the enzyme, FDP lowers the affinity for PEP. The activator K^+ raises the affinity of the enzyme-Mn complex for ADP. A comparison of the binding and kinetic data for the yeast enzyme suggests that there may be a preferred order binding of PEP prior to ADP.

It is concluded that the ternary complexes and the reaction mechanisms of the muscle and yeast enzymes are similar, but that the yeast enzyme is complicated by site-site interactions among the Mn binding sites which alters their affinity for phosphoenolpyruvate. It is suggested that the affinity of the binding site for PEP may be controlled by the distance between two subsites: the Mn, which binds the phosphoryl group, and a hydrogen bond donor, which binds the carboxylate group.

Introduction

Allosteric enzymes are here understood to possess, in addition to a catalytic site, a second, non-overlapping site, either catalytic or non-catalytic, which binds substrates or other substances and which interacts with the catalytic site to influence the enzyme activity (1). The essential characteristic of an allosteric enzyme is non-overlap or "action at a distance" between two sites on an enzyme.

An approach to understanding the mechanism of action of allosteric enzymes would be to compare the interaction of substrates with allosteric and non-allosteric enzymes which catalyze the same reaction. This approach may be limited by the difficulty of finding a non-allosteric enzyme. Site-site interaction may occur but may not be detected until the appropriate techniques are used. Recently, interaction has been detected between the two NAD binding sites of an apparently non-allosteric enzyme, liver alcohol dehydrogenase, using the techniques of nuclear magnetic relaxation (2).

A fruitful comparison of properties seems possible in the case of the pyruvate kinase reaction. The enzyme from muscle shows hyperbolic substrate kinetics (3,4) and is not activated by FDP while the enzyme from yeast (5-7) and liver (8) show sigmoidal substrate kinetics (with phosphoenolpyruvate) and are activated by FDP. Yet, it may be incorrect to assume that the muscle enzyme is completely "non-allosteric."

132

The muscle enzyme binds two molecules of phosphoenolpyruvate (3) and two Mn^{2+} ions per mole (9). Potassium, which is required for activity (10), has been shown to alter the structure of the muscle enzyme (11,12), its binary complex with Mn (11,13), and its ternary E-M-S complexes (13). Although activation by FDP has not been demonstrated, the binding of FDP has been detected by a change in the electrophoretic mobility of the muscle enzyme.* At present, it is safe to say that there is no kinetic evidence for allosteric activation of muscle and of leukocyte pyruvate kinase by FDP but that FDP behaves kinetically like an allosteric activator of the yeast and liver enzymes. With these limitations in mind, we have studied the interaction of Mn^{2+} and substrates with the allosteric yeast enzyme in the presence and absence of FDP and have compared the structural, kinetic, and thermodynamic properties of binary and ternay complexes with those previously found for the muscle enzyme.

We have also reinvestigated the kinetic role of the E-M complex, of free ADP, and of the ternary E-M-ADP complex in muscle pyruvate kinase because of a recent suggestion (14) that the combination of M-ADP with free enzyme was the only pathway by which the kinetically active quaternary complex could form.

Materials and Methods

Yeast pyruvate kinase was prepared as previously outlined (7), and assayed by the coupled assay of Bücher and Pfleiderer (15). The purified enzyme migrated as a single component in acrylamide gel electrophoresis, when applied in 50% glycerol. Kinetic studies were also carried out by the coupled assay procedure (15). The binding of Mn to yeast pyruvate kinase was studied at 30° by EPR and proton relaxation rate measurements as previously described for the muscle enzyme (9). The binding of substrates to the E-Mn complex was studied by the PRR method and the data were analyzed by procedures I and III as previously described (16). Hill plots of the binding data give dissociation constants which agreed within experimental error with those obtained by the above procedures. The rapid quenching experiments were done in the instrument previously described (17), using a narrower reaction tube permitting more rapid quenching (< 43 msec) with 1 M HCl. The pyruvate formed was assayed fluorimetrically using lactic dehydrogenase (18).

*W.A. Susor and W.J. Rutter. Personal communication.

Results

Nature of the Complexes Formed by Muscle Pyruvate Kinase.
Table I lists the binary and ternary complexes formed by
rabbit muscle pyruvate kinase, Mn, and substrates as detected
by binding studies using EPR, nuclear relaxation, the kinetic
protection method, and U.V. difference spectroscopy (9,11,16,
19). In all cases studied, the enhancement of the water
proton relaxation rate of ternary E-Mn-S complexes is less
than that of the binary E-Mn complex. Moreover, the "phosphorylated" substrates (phosphoenolpyruvate, FPO_3^{2-}, ATP) reduce ε more than the corresponding non-phosphorylated substrate (pyruvate, ADP, F^-), suggesting that the metal serves
as a bridge between the enzyme and the phosphoryl group which
is transferred. Direct evidence for such a role of Mn was
obtained by the demonstration (17)(by fluorine NMR) of an
$E-Mn-O_3PF$ complex which dissociates fast enough to participate
in the fluorokinase reaction of pyruvate kinase (20):

$$ ATP + F^- \xrightarrow[HCO_3^-]{Mn^{2+} \ K^+} ADP + FPO_3^{2-} $$

Since FPO_3^{2-} is displaced from the $E-Mn-O_3PF$ complex by
phosphoenolpyruvate, an analogous complex may form with the
latter compound (19).

Kinetic Role of the Complexes Formed by Muscle Pyruvate
Kinase. Table I defines and lists the dissociation constants
of the binary and ternary complexes formed by manganese activated rabbit muscle pyruvate kinase as measured by EPR, PRR,
the kinetic protection method, and U.V. difference spectroscopy (9,11,16,19). It also lists, for comparison, the dissociation constants obtained by an analysis of the substrate
kinetics of the Mn activated enzyme, using the general rate
equation for a metal activated enzyme which permits all pathways to operate to form each ternary EMS complex (21):

$$ v = \cfrac{V}{1 + \cfrac{K_A'}{[Mn]_f} + \cfrac{K_A'K_S}{K_D[ADP]_f} + \cfrac{K_SK_A'}{[Mn]_f[ADP]_f}} \tag{1A} $$

$$ = \cfrac{V}{1 + \cfrac{K_A'+K_2}{[Mn]_t} + \cfrac{K_A'K_S}{K_D[ADP]_f} + \cfrac{K_A'[ADP]_f}{K_1[Mn]_t} + \cfrac{K_A'K_S}{[Mn]_t[ADP]_f}} \tag{1B} $$

TABLE I

Binary and Ternary Complexes of Muscle Pyruvate Kinase
with Manganese and Substrates

Complex		Dissociation Constants[a] from		Enhancement[a] of
		Binding Studies (μM)	Kinetics (μM)	PRR
$E(Mn)_2$	K_D	$75,63^b$	71	32.7
E-ADP	K_S	140	148	–
E-PEP	K_S	$33,80^b$	41	–
E-Mn-ADP	K_2	50	31	19.9
	K_3	68	44	
	$K_A{}'$	37	21	
E-Mn-PEP	K_2	0.6	1.0	2.2
	K_3	$10,^b15$	27	
	$K_A{}'$	34	48	
E-Mn-ATP	K_2	160	335	13.1
	K_3	31	66	
E-Mn-Pyruvate	K_2	0.92^c	1.0	>19.9
(+ ADP)	K_3	3.5×10^3	4.0×10^3	
$E\overset{Mn}{\underset{F}{\diagdown}}$	K_3	2.8×10^4	6×10^5	25
$E-Mn-FPO_3$	K_3	2.6×10^3	3.4×10^3	1.5

Definition of Dissociation Constants:

$$K_D = \frac{(E)(M)}{(EM)}; \quad K_S = \frac{(E)(S)}{(ES)}; \quad K_1 = \frac{(M)(S)}{(MS)}; \quad K_2 = \frac{(E)(MS)}{(EMS)};$$

$$K_3 = \frac{(EM)(S)}{(EMS)}; \quad \text{and} \quad K_A{}' = \frac{(ES)(M)}{(EMS)}.$$

[a]From ref. 9, 16, and 19, except as indicated;
[b]From reference 11.
[c]Corrected from ref. 16, which was based on an incorrect K_1 of Mn-pyruvate. The correct value is 284 mM (K. Berman, unpublished).

The dissociation constants, obtained by the binding studies, are seen to be in reasonable agreement with those obtained by the kinetic analysis. However, when tight MS complexes are present, substrate and activator kinetics alone are often inconclusive in defining the catalytic role of EM and EMS complexes (9). Equation 1 predicts inhibition by high $[ADP]_f$ and by high $[Mn]_t$. Inhibition at high $[Mn]_t$ has been observed (13), but inhibition by ADP in excess of Mn has not been observed.

Cleland, in a review (14), has criticized the proposal of the random binding of Mn and ADP (16) on four grounds:

(a) that the most general rate equation for an enzyme, metal and substrate interaction was not used. This criticism, which is based on the reviewer's misunderstanding of the symbol used for free Mn, may be dismissed as incorrect.

(b) that "nucleoside di- and triphosphates and free and combined pyrophosphates always react in the form of their magnesium complexes in vivo and that the role of the metal in these complexes is solely to complex the reactant and facilitate the catalytic step. If the uncomplexed molecules are allowed to combine with the enzyme under non-physiological conditions, they act as inhibitors" (14). This criticism is speculative, and while it may fit the observations on creatine kinase, is inapplicable to pyruvate kinase, where inhibition by ADP in excess of the divalent cation has not been observed.

(c) that a slope replot of the kinetic data at low ADP and Mn fails to detect a $1/[ADP]_f$ term in the rate equation, which is equivalent to the statement that the K_M of $[ADP]_f$ approaches zero as $[Mn]_t$ becomes infinite. This criticism is correct if our double reciprocal plots are truly linear. However, one can rarely rule out curvature in such cases, especially when making distant extrapolations.

(d) that a more restrictive kinetic scheme provides a better fit to the data:

$$E + MnADP \rightleftharpoons E \cdot MnADP \rightleftharpoons etc. \qquad (2)$$

which would yield a rate equation (14):

$$v = \frac{V}{1 + \dfrac{K_2}{[Mn]_t} + \dfrac{K_1 K_2}{[ADP]_f [Mn]_t}} \qquad (3)$$

Equation 3, however, fails to predict the observed inhibition by excess Mn and the observed agreement (whether one uses Mn or Mg) in the K_M of free ADP extrapolated to zero divalent cation.

Equation 3, which defines this limiting K_M as the dissociation constant (K_1) of M-ADP, would predict a five-fold lower value for the Mn activated enzyme, as compared to the Mg activated enzyme. Equation 1, which defines this limiting K_M to be the dissociation constant (K_S) of E-ADP, would predict the observed agreement and gives a K_S(ADP) which agrees with the value obtained in a binding study (Table I)(16).

Hence, the kinetic data with Mn and Mg appear consistent with equation 1 if one extrapolates the K_M of ADP to $[M^{2+}] = 0$. The data on Mn appear consistent with equation 3 if one extrapolates the K_M(ADP) to infinite $[M^{2+}]$. The dissociation constants calculated from the kinetic data by the use of equation 1 are consistent with those found in the binding experiments. While the agreement between the kinetic and binding data suggest multiple pathways, as originally proposed (16), substrate and activator kinetics are inconclusive, and a different kinetic approach is required to determine whether E-M and free ADP can participate in the catalysis. For this reason, rapid quenching experiments were carried out in which the amount of pyruvate formed from ADP and PEP in one enzyme turnover was determined. The slowly dissociating Ni^{2+} ion (Table II) was used as the divalent activator at 3°C. The reaction was begun by mixing Ni-ADP with free enzyme or the Ni-enzyme complex with free ADP. The thermodynamic and kinetic properties of the binary Ni-enzyme and Ni-ADP complexes are given in Table II. From the binding constants, it may be calculated that 67% of the Ni^{2+} was complexed with enzyme or that 78% was complexed with ADP prior to mixing. During the time of mixing (4 msec) and the reaction time (calculated to be < 43 msec from the extent of the reaction) there was little opportunity for the equilibration of Ni^{2+} between the E-Ni and Ni-ADP complexes. The results (Table III) indicate no effect on the nature of the initial binary metal complex (E-Ni or Ni-ADP) on the amount of pyruvate formed. Pre-equilibrating the PEP with the enzyme resulted in a greater amount of pyruvate formed, but this amount was also independent of the nature of the initial Ni complex.

It is concluded that the reaction pathways, 4A and 4B, are equally productive.

137

TABLE II

Thermodynamic and Limiting Kinetic Properties of the
Nickel Complexes of ADP and Pyruvate Kinase

Complex	K_D	k_{on}	k_{off}	% Ni in complex before mixing	$t_{\frac{1}{2}off}$[e]	$t_{\frac{1}{2}on}$[e]
	μM	$\mu M^{-1}sec^{-1}$	sec^{-1}		msec	msec
Ni-ADP	31.6[a]	<4.0	<126[d]	78	> 5.5	> 2.8
Ni-pyruvate kinase	65 ± 19[b]	<0.021[c]	< 1.36	67	>508	>633

[a]Reference 27.
[b]Determined at 4° by competition with Mn^{2+} using EPR and PRR.
[c]Measured for Ni-glycylglycine at 25° (28) and assuming that the rate of formation of Ni complexes are limited by the rate of dissociation of water from Ni as found (29), and confirmed for Ni-pyruvate kinase by F. Kayne (personal communication).
[d]Calculated from k_{off} for Ni-ATP at 25° (29) using the factor $K_1(Ni-ADP)/K_1(Ni-ATP) = 3.3$.
[e]Half-time calculated for the conditions of the rapid quenching experiment.

$$E-Ni + ADP \longrightarrow \longrightarrow products \qquad (4A)$$

$$E + Ni-ADP \longrightarrow \longrightarrow products \qquad (4B)$$

The pathways, 5A and 5B, are also equally productive, but yield more pyruvate than the 4A and 4B pathways.

$$E-Ni-PEP + ADP \longrightarrow \longrightarrow products \qquad (5A)$$

$$E-PEP + Ni-ADP \longrightarrow \longrightarrow products \qquad (5B)$$

Hence, the rate of binding of PEP to the enzyme and to the enzyme-Ni complex may be slower than the rate of combination of the nucleotide with the enzyme. Previous work has indicated that in the combination of PEP with E-Mn, a conformational change takes place (16,22). This process might introduce a slow step in the reaction. Although these results are preliminary and must await confirmation by ATP analysis , the view that equation 2 is the only pathway for nucleotide binding is probably an over-simplification.

TABLE III

Rapid Quenching of the Pyruvate Kinase Reaction

Method of Mixing		Pyruvate formed
Syringe 1	Syringe 2	in 4 msec
		(μM)
PK, Ni	ADP, PEP	4.5 ± 0.6
PK	Ni, ADP, PEP	5.2 ± 0.9
PK, Ni, PEP	ADP	8.9 ± 0.4
PK, PEP	Ni, ADP	8.1 ± 0.7

Final Reaction mixture contained: [ADP] = 63 μM; [PEP] = 58 μM; [MPK] = 52 μM; [Ni] = 5.8 μM; [KCl] = 115 mM; [Tris-HCl] = 46 mM. Quenching was with 1 M HCl. T = 3°C.

Binary Complexes Formed by Yeast Pyruvate Kinase and Mn. Unlike muscle pyruvate kinase, which at 0.1-0.2 M KCl binds 2 Mn independently and identically per mole of protein (Table I), the allosteric yeast enzyme shows a very complicated titration curve with Mn as determined by two independent methods: EPR and PRR studies (Figure 1).

A total of approximately 6 binding sites are detected per molecule of yeast enzyme, the molecular weight of which (166,000) is 70% of that found (23) for the muscle enzyme.

The tightest binding site titrates atypically, giving a dissociation constant, K_D = 3 μM, for the first 1/3 of the site and a K_D = 72 μM for the remaining 2/3. The next two sites are weaker (K_D = 740 μM), and the last three sites are weakest (K_D = 1.8 mM). Alternative methods of fitting the data are currently under investigation, but all schemes devised so far require an atypical mode of binding of Mn.

The non-integral and low value of the stoichiometry of Mn binding might be explained in three ways:
1) Heterogeneity of, or impurities in, the yeast enzyme preparation.
2) Aggregation of the protein by Mn.
3) Strong site-site interaction among the Mn binding sites.

Heterogeneity, or impurities in the preparation, seem unlikely because of the reproducibility of the enzyme preparation, the observation that it shows a symmetrical peak in the

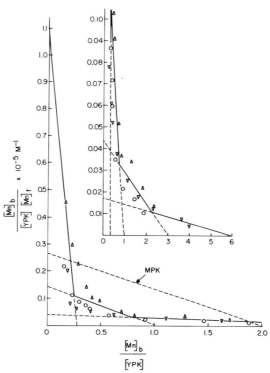

Figure 1. Scatchard plot of the binding of manganese to yeast pyruvate kinase, in 0.2 M KCl, 0.05 M Tris-HCl, pH 7.5, T = 30°. The bound and free Mn were determined by the proton relaxation rate of water using a value of ε_b = 15.1 as obtained by EPR and proton relaxation measurements. The curve is arbitrarily fit by four straight line segments. MPK is the titration curve which was obtained for muscle pyruvate kinase (9).

analytical ultracentriguge,* and a single band in acrylamide gel electrophoresis. However, when the enzyme is inactivated at low temperatures, many bands are observed in acrylamide gel electrophoresis. Although multiple types of subunits may therefore exist, the native enzyme appears to be a unique aggregate. Muscle pyruvate kinase consists of 4 very similar, if not identical, subunits (24). Aggregation of the enzyme by Mn was not detected by its sedimentation in the analytical ultracentrifuge, under the same conditions as the binding studies. One is left, therefore, with the third alternative site-site interaction, as has been detected kinetically (5,6,7). At

*R.T. Kuczenski and C.H. Suelter, unpublished observations.

low concentrations of K^+, a Hill coefficient of 4 has been observed for Mn (Table IV). The site-site interaction detected in the Mn binding studies manifests itself only in the affinity but not in the enhancement of the bound manganese, which remains at 15.1 ± 1.8 as the occupancy of the binding sites increases from 0 to 6. The corresponding value of ε_b for the muscle enzyme is 32.7 ± 3.2 (Table I).

When the monovalent activator, K^+, is replaced by the inert tetramethylammonium (TMA) cation in the Mn^{2+} titration, the muscle enzyme binds 4 Mn per mole instead of two with little change in affinity (11,13) or enhancement (13). By contrast, the replacement of K^+ by TMA with the yeast enzyme has no significant effect on the apparent stoichiometry of binding of Mn, the enhancement factor (17.0 ± 1.2) or the apparent affinities. The presence of FDP lowered the ε value of the yeast enzyme-Mn complex indicating an interaction with Mn or a structural change in the environment of the metal upon binding of the activator.

Ternary E-Mn-S Complexes of Yeast Pyruvate Kinase. The addition of all substrates to the yeast pyruvate kinase-Mn complex decreased the enhanced effect of Mn on the proton relaxation rate of water (Table IV), suggesting the formation of E-M-S bridge complexes as had been found for the muscle enzyme (16,19).

Substrate titrations, measuring the relaxation rate of water protons, were carried out at low [Mn]/[E] (0.5 to 0.9) and at high [Mn]/[E] (3.5). The resulting values of K_3 and ε are given in Table IV as E(Mn)S and $E(Mn)_3S$ complexes, respectively. Table IV also gives the $K_{0.5}$ values and Hill coefficients obtained by kinetic studies of the Mn^{2+} activated enzyme.

No significant effect of FDP on the binding or the kinetic behavior of ADP was detected at low or high Mn. The dissociation constants of ADP measured in the binding studies are significantly greater than those estimated from kinetics. This may be due to the absence of PEP in the binding studies. Haeckel et al. (6) found a halving of the $K_{0.5}$ of ADP at high levels of PEP.

At low [Mn], FDP weakened the binding of PEP contrary to the behavior of the kinetic data. At high [Mn], FDP tightened the binding of PEP in qualitative and quantitative agreement with the kinetic observations. Hence $E(Mn)_3$ shows a similar response to the allosteric activator, FDP, in both the binding

TABLE IV

Binary and Ternary Complexes of Yeast Pyruvate Kinase
with Manganese and Substrates

Complex	Activator[a]	Dissociation Constant Binding K_D or K_3	Dissociation Constant Kinetics $K_{0.5}$[b]	Hill Coefficient (Kinetics)	ε
		(μM)	(μM)		
$E(Mn)_6$	–	Variable (4–7200)	–	–	17.0
	K^+	Variable (3–1800)	<760	3.8	15.1
	FDP	Variable (10–1870)	<120	1.4	10.6
$E(Mn)ADP$	K^+	200	–	–	13.0
	$K^+ + FDP$	260	–	–	≥ 5.5
$E(Mn)_3ADP$	–	1270	–	–	<17.0
	K^+	332	54[c]	1.3	<15.1
	FDP	<1210	48[d]	1.2	<10.6
$E(Mn)PEP$	K^+	21	–	–	12.5
	$K^+ + FDP$	460	–	–	5.2
$E(Mn)_3PEP$	–	376	–	–	<17.0
	K^+	324	160	2.6	<15.1
	$K^+ + FDP$	72	83	1.0	<10.6
$E(Mn)ATP$	K^+	32	–	–	<15.1
$E(Mn)Pyruvate$	K^+	765	–	–	<15.1

[a]The activators were 0.1–0.2 M KCl or 1.1 mM FDP or both. In ab
sence of activator, 0.1 M tetramethyl ammonium chloride was used
Other components present were 0.05 M Tris-HCl buffer, pH 7.5. T
30 ± 1°C.
[b]$K_{0.5}$ is the concentration of the substance required for half ma
imal activity at saturating concentrations of all other componen
[c]Expressed in terms of total ADP in presence of 1 mM Mn and 1 mM
[d]Expressed in terms of total ADP in presence of 1 mM Mn, PEP, an
FDP.

and the kinetic studies while E(Mn) does not. It is concluded that the occupancy of more than one Mn binding site is required to demonstrate an effect of the allosteric activator, FDP, on the binding of the substrate, PEP.

The fact that FDP itself lowers ε when added to E(Mn) or $E(Mn)_3$ ($\varepsilon_T = 10.6 \pm 1.3$) suggests that FDP, like the substrates, alters the environment of enzyme bound manganese.

The agreement of the dissociation constants obtained by reciprocal plots (16) of the enhancement data with the kinetically determined $K_{0.5}$ values for PEP, and the lack of agreement of the corresponding parameters for ADP suggest that the yeast enzyme-metal complex may bind PEP prior to the binding of ADP, in a preferred order. This differs from muscle pyruvate kinase which binds substrates in a random order (3,16) but is similar to PEP carboxykinase (25,26).

The monovalent activator K^+ appears to raise the affinity of the $E(Mn)_3$ complex for ADP as found by kinetics (6) but has no effect on the binding of PEP. This contrasts with the muscle enzyme in which K^+ has little effect on the binding of either substrate (13).

Discussion

The substrate kinetics, binding data, and rapid quenching experiments of muscle pyruvate kinase suggest that the active quaternary $E-M-S_1-S_2$ complex forms by multiple pathways.

The active quaternary complex of muscle pyruvate kinase appears to involve a metal bridge between the enzyme and the phosphoryl group which is transferred (16,19).

The enhancement data suggest that the allosteric yeast pyruvate kinase forms structurally similar ternary E-M-S complexes as does the muscle enzyme.

The yeast enzyme is, however, more complicated by the apparent site-site interaction among the Mn binding sites as detected by the atypical Mn titration curve (Figure 1). This interaction among the Mn binding sites appears to be necessary for the allosteric action of FDP since an effect of FDP on the binding of phosphoenolpyruvate could be demonstrated only when more than one Mn binding site was occupied (Table IV). The qualitative and quantitative agreement of the binding and kinetic data support the view that the $E(Mn)_3PEP$ complexes

143

detected in the binding studies are the kinetically active ones. A detailed analysis of the binding data in terms of the various allosteric models is in progress.

At present, it appears justified to conclude that the ternary complexes and the mechanism of action of the muscle and yeast enzymes are similar, but that the latter is complicated by site-site interactions among the Mn binding sites which affects their affinity for phosphoenolpyruvate.

The binding and enhancement data on the muscle and yeast enzymes suggest a possible mechanism for this cooperative effect. Although the ternary complexes of FPO_3^{2-} and PEP have the same enhancement ($\sim$ 2) in the muscle enzyme, the latter substrate is bound more tightly by 2.8 KCal/mole (16,19). PEP also binds tightly to the Mn complex of the yeast enzyme (Table IV). Hence the PEP binding site may consist of at least two subsites: the Mn, which binds the phosphoryl group (19), and possibly a hydrogen bond donor which interacts with the carboxylate group and adds 2.8 KCal to the binding energy. The affinity of the enzyme-Mn complex for PEP would be expected to depend on the distance between these two subsites. The affinity would diminish if this distance became significantly larger or smaller than the optimum value which may be $\sim$6Å. From the X-ray coordinates of doubly protonated PEP in the crystalline state kindly provided by Mrs. O. Kennard and Dr. D.G. Watson, Dr. Jenny Glusker has calculated the distance between the protonated phosphate oxygen and the protonated carboxyl oxygen to be 4.04 Å. Assuming the same conformation of PEP to be the most stable in solution as in the crystalline state, and replacing the two protons on PEP with Mn and a hydrogen bond donor, respectively, we obtain a value of 6 Å as the optimum distance between the subsites.

We suggest that on the non-allosteric muscle pyruvate kinase, this distance is fixed, while on the allosteric yeast pyruvate kinase this distance is adjustable. The binding of the first molecule of PEP would optomize the distance between subsites for the next molecule of PEP. If the two subsites for a single molecule of PEP were on different subunits of the enzyme, the effect of PEP would manifest itself as changes in subunit interaction. The activator, FDP, might mimic the action of PEP, since the enhancement data on the yeast enzyme suggest that both PEP and FDP interact with enzyme bound Mn.

Addendum*

Recent observations by J. Reuben, M. Cohn, G.L. Cottam, and A.S. Mildvan suggest that muscle pyruvate kinase binds 4 Mn/mole under all conditions.

Acknowledgments

We are grateful to Professors Mildred Cohn, Britton Chance and John Williamson for making their instruments available to us, to Mrs. B. Corkey for performing the pyruvate analysis, and to Miss Marie L. Smoes for capable technical assistance.

References

1. Monod, J., J.P. Changeux, and F. Jacob. J. Mol. Biol., 6, 306 (1963).

2. Mildvan, A.S., and H. Weiner. Biochemistry, 8, 552 (1969).

3. Reynard, A.M., L.F. Hass, D.D. Jacobsen, and P.D. Boyer. J. Biol. Chem., 236, 2277 (1961).

4. Boyer, P.D. In The Enzymes (P.D. Boyer, H. Lardy, and K. Myrback, eds.), Vol. VI, 2nd ed., Academic Press, Inc., New York, 1962, p. 95.

5. Hess, B., R. Haeckel, and K. Brand. Biochem. Biophys. Res. Comm., 24, 824 (1966).

6. Haeckel, R., B. Hess, W. Lauterborn, and K.H. Wüster. Hoppe-Syeler's Z. Physiol. Chem., 349, 699 (1968).

7. Hunsley, J., and C.H. Suelter. Fed. Proc., 26, 559 (1967).

8. Tanaka, T., Y. Harano, F. Sue, and H. Morimura. J. Biochem. (Tokyo), 62, 71 (1967).

9. Mildvan, A.S., and M. Cohn. J. Biol. Chem., 240, 238 (1965).

10. Boyer, P.D., H.A. Lardy, and P.H. Phillips. J. Biol. Chem., 146, 673 (1942).

11. Suelter, C.H., R. Singleton, F.J. Kayne, S. Arrington, and A.S. Mildvan. Biochemistry, 5, 131 (1966).

*Note added in proof: October 1970.

12. Sorger, G.J., R.E. Ford, and H.J. Evans. Proc. Natl. Acad. Sci. U.S.A., 54, 1614 (1965).

13. Mildvan, A.S., and M. Cohn. Abs. VI Int. Cong. Biochem., IV, 111 (1964).

14. Cleland, W.W. Ann. Rev. Biochem., 36, 77 (1967).

15. Bücher, T., and G. Pfleiderer. In Methods in Enzymology (S.P. Colowick and N.O. Kaplan, eds.), Vol. I, Academic Press, New York, 1955, p. 435.

16. Mildvan, A.S., and M. Cohn. J. Biol. Chem., 241, 1178 (1966).

17. Chance, B. In Currents in Biochemical Research (D.E. Green, ed.), Interscience, 1956, p. 308.

18. Williamson, J.R., and B.E. Herczeg. In Methods in Enzymology (J.M. Lowenstein, ed.), Vol. 13, Academic Press, New York, 1969, p. 434.

19. Mildvan, A.S., J.S. Leigh, Jr., and M. Cohn. Biochemistry, 6, 1805 (1967).

20. Tietz, A., and S. Ochoa. Arch. Biochem. Biophys., 78, 477 (1958).

21. Dixon, M., and E.C. Webb. Enzymes, 2nd ed., Academic Press, New York, 1964, p. 438.

22. Kayne, F. and C.H. Suelter. J. Amer. Chem. Soc., 87, 897 (1965).

23. Warner, R.C. Arch. Biochem. Biophys., 78, 494 (1958).

24. Cottam, G.L., P.F. Hollenberg, and M.J. Coon, J. Biol. Chem., 244, 1481 (1969).

25. Miller, R.S., and M.D. Lane. J. Biol. Chem., 243, 6041 (1968).

26. Miller, R.S., A.S. Mildvan, H.-C. Chang, R. Easterday, H. Maruyama, and M.D. Lane. J. Biol. Chem., 243, 6039 (1968).

27. Taqui Khan, M.M., and A.E. Martell. J. Amer. Chem. Soc., 84, 3037 (1962).

28. Hammes, G.G., and Steinfield. J. Amer. Chem. Soc., 84, 4639 (1962).

29. Hammes, G.G., and S.A. Levison. Biochemistry, 3, 1504 (1964).

A KINETIC STUDY OF THE CONFORMATIONAL CHANGES IN PYRUVATE KINASE

F.J. Kayne*

Max-Planck-Institut für physikalische Chemie
Göttingen, Germany

A series of previous papers (1-3) has described equilibrium studies of apparent conformational changes in rabbit muscle pyruvate kinase (E.C. 2.7.1.40). The most sensitive indication for a conformational change was the UV difference spectrum which showed a maximum at 295 nm. Only very small, or in some cases the absence of changes, were demonstrated in the optical rotatory dispersion, sedimentation velocity and chemical reactivity of various groups (3). The conclusion from these studies was that the thermodynamic data obtained from the difference spectrum could be explained by a simple equilibrium between two states of the protein. The ΔH for the process was about 35 kcal/mole, and the equilibrium could be shifted by changing various ligands, solvent properties or temperature. Although the muscle enzyme does not exhibit classical "allosteric" behavior, protein-mediated effects are seen in the interaction of binding monovalent and divalent cations (2). In contrast to the muscle enzyme, yeast pyruvate kinase exhibits a number of significant cooperative effects (4,5).

The purpose of this study was to take a careful look at the rapid kinetics of a "model" for a relatively limited (no large net change), conformational change by direct observation of the protein itself; to correlate these results with those from the equilibrium studies and, if possible, to extend the studies to the more interesting systems where the protein isomerization is recognized to affect the catalytic activity. [This may also be the case with muscle pyruvate kinase (3)].

*Present address: Johnson Research Foundation, University of Pennsylvania, Philadelphia, Pennsylvania 19104.

Methods

The preparation of muscle pyruvate kinase, assay and general methodology have been described previously (3). T-jump studies were carried out on a commercial instrument (Messanlagen, Sg. m.b.H.) modified to double-beam operation by C.R. Rabl, and using the slow T-jump device developed by F. Pohl (6). Stopped flow studies were carried out with the use of the Durrum-Gibson instrument extensively modified by K. Kirschner.

Results

Figure 1 illustrates, in general, the results obtained when the enzyme solution is placed in a 2.0 ml T-jump cell and the absorption changes at 297 nm recorded after jumps at various temperatures in the transition region. A qualitative description is as follows.

One first notes that the apparent amplitudes (with a constant T-jump) first increase to a maximum and then begin to decrease as one goes through the midpoint of the transition. The maximum total amplitude is just at the temperature expected from the equilibrium data. There is apparently a rapid initial increase in O.D. (faster than 5 μsec) of almost constant amplitude which is most probably the intrinsic absorption change of the protein chromophores with temperature, but could also contain contributions from very rapid isomerizations or protonation. Most important of all, however, is that the relaxation effects cannot be described by a single exponential. The portion of the relaxation curves shown in this figure can be separated into two exponentials in a semi-log plot. Treatment of these data as a spectrum of relaxation times by the use of τ*, the mean relaxation time, does not seem to be of value since there appears to be a separation of about 5-10 fold in the values for τ. Mathematical separation is probably not justified due to the relatively low signal to noise ratio.

Figure 1 suggests additional effects may be present. First, the total amplitudes observed do not correspond to the absorption differences found in the equilibrium study, lacking some 25-40% of the expected ΔO.D. Also, at the end of the oscilloscope trace, some 2 seconds after the T-jump, one sees little or no change in O.D., where cooling of the

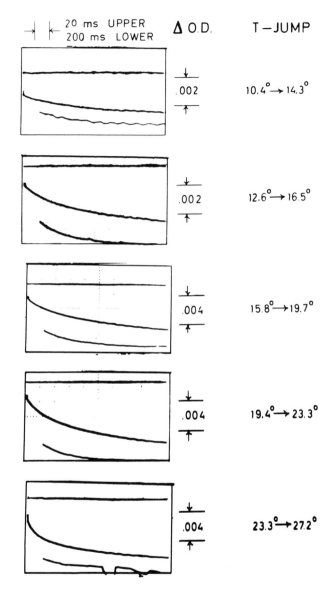

Figure 1. Oscilloscope trace of temperature jumps of pyruvate kinase, 3.9 mg/ml enzyme in solvent described in Figure 2. Wavelength 297 nm. Initial trace represents absorption at starting temperature. O.D. increases towards bottom of figure.

solution in the light path would be expected to show a considerable return (<u>ca</u>. 7%) of the absorption to the initial value.

Experiments using the slow T-jump apparatus showed a single relaxation time in the range of 1-2 min. This would not account for the compensation of the cooling effect just described, therefore the system was studied using the stopped flow apparatus to effect a ligand jump. Mn (II) ($\sim$ 1 mM) was used at temperatures in the transition range, and two additional relaxation times were found in the range of 1-10 sec. This very rich relaxation spectrum, for what was regarded previously as a single step process, is shown in Figure 2. The relaxation times arbitrarily denoted τ_3 and τ_4 were observed only in the presence of Mn (II) since these were in the time range covered only by the stopped flow. τ_3 is apparently the effect responsible for the compensation of the cooling seen in Figure 1. The relative amplitudes are only approximate and may vary somewhat at different points on the transition curve. It is very difficult to make either amplitude or τ determinations away from the midpoint of the transition since the total difference becomes smaller and still must be divided into the **various** components shown here. The small effect shown by the dotted line in Figure 2 may be an additional relaxation time seen only at higher temperatures.

Discussion

The analysis of such a spectrum is made most difficult because of the lack of suitable "handles" in determining what processes give rise to the various times. First of all, from the figure one concludes that, at least in the case of Mn (II), the ligand does not have a great influence on the spectrum of relaxation times. Apparently the τ's are all somewhat decreased in the presence of Mn (II), however, relative amplitudes are not greatly affected (Mn (II) shifts the midpoint of the transition curve some 10° higher).

τ_1, τ_2 and τ_5 show little or no temperature dependence. Indeed, plots of ln $1/\zeta$ vs $1/T$ exhibit slopes <u>corresponding</u> to energies of activation of less than 10 kcal/mole. The temperature dependencies of τ_3 and τ_4 are also probably low, however the data at present is not sufficient to say this with certainty.

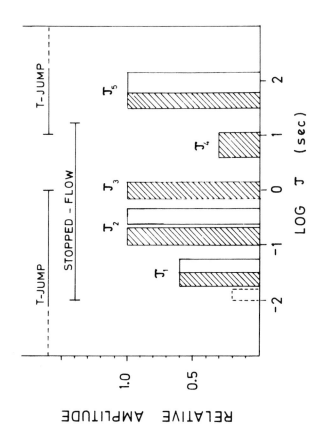

Figure 2. Relaxation spectrum of rabbit muscle pyruvate kinase. Relaxation times resolved between 10^{-6} and 10^3 sec by the methods indicated. Shaded bars represent those effects observed in the presence of 1 mM Mn (II). τ values are approximate for the conditions around 20°, 0.1 M tetramethylammonium cacodylate, pH 7.3–7.5, 0.05 M tetramethylammonium chloride.

The wavelength dependence of the relaxation times should be expected to give additional information, however in this case, the difference spectrum is so relatively large and sharp around 295-297 nm that the individual relaxation times cannot be resolved at other wavelengths in order to make the comparison.

Although the relaxation spectrum as it stands does not allow us to say much until additional handles can be found, some discussion would appear to be in order. First of all, the enzyme used in these experiments is apparently better than 95-98% homogeneous with respect to the usual protein analytical criteria. The method of detecting the relaxation times utilizes the UV difference spectrum characteristic of tryptophan, and since there are only 14 tryptophans/molecule of protein (1) (ca. 2000 amino acid residues), we are observing only a very limited portion of the molecule. The enzyme is known to be a tetramer (237,000 M.W.) (8), and no subunit heterogeneity has yet been detected. However, with apparently two active sites (8), one would expect some assymetry in the molecule. No aggregation or dissociation processes are known to take place in the concentration range studied, and preliminary studies of the concentration dependence of the relaxation times have shown no effects.

From a consideration of the above points, it is most probable that the relaxation spectrum observed for the conformational changes does represent intermediate states in the isomerization process. The amplitudes associated with the relaxation times would seem to indicate that the various states are populated to a similar extent.

Two schemes which could most likely lead to this behavior would be:

I Sequential reactions

$$A \rightleftharpoons X_1 \cdots\cdots\cdots X_n \rightleftharpoons B \qquad \text{or}$$

II Coupled "parallel" reactions

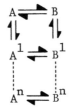

152

or a combination of the two schemes.

Calculation of the relaxation times for case I would be identical to the method proposed by Hammes and Schimmel (9) for sequential reactions, and sequential reactions coupled by rapid equilibria to parallel paths. Eigen (10) has discussed the general solution for cases analogous to case II.

Explicit equations for the various τ's in such cases can only be given with assumptions as to the order in which the various equilibria are coupled. These assumptions are not yet justified on the basis of the data in hand.

Preliminary experiments have been carried out in a similar manner with yeast pyruvate kinase prepared by the method of Hunsley (11). In the presence of approximately 0.4 mM FDP, one observes at least two relaxation times in the ranges of 50–200 msec and 30 sec observing the 292 nm maximum of the difference spectrum. The amplitudes of the effects are so small, however, that useable data is nearly impossible to obtain. A substrate analog 4, –SH–UDP was also used in an attempt to obtain data on both binding and conformational changes using a near-UV absorption band; however, in this case also, the difference spectra obtained were not large enough to utilize.

The results indicate clearly that caution should be used in assuming that limited intramolecular isomerizations in proteins are simple processes and in the interpretation of relaxation spectra involving binding steps coupled to the isomerizations. The need for T-jump instrumentation accurately covering the range of 0.1–10 sec is also apparent. It is hoped that further developments along this line would allow solution of the muscle pyruvate kinase relaxation spectrum by a careful analysis of the temperature dependence of the relaxation times.

Acknowledgments

The author wishes to express his thanks for postdoctoral support during the course of this work from NATO and the Max-Planck Society, and especially, to Dr. Manfred Eigen and his colleagues in Göttingen.

References

1. Kayne, F.J., and C.H. Suelter. J. Am. Chem. Soc., 87, 897 (1965).

2. Suelter, C.H., R. Singleton, Jr., F.J. Kayne, S. Arrington, J. Glass, and A.S. Mildvan. Biochemistry, 5, 131 (1966).

3. Kayne, F.J., and C.H. Suelter. Biochemistry, 7, 1678 (1968).

4. Haeckel, R., B. Hess, W. Lauterborn, and K.-H. Wüster. Hoppe-Seyler's Z. Physiol. Chem., 349, 699 (1968).

5. Hunsley, J.R., and C.H. Suleter. Fed. Proc., 26, 559 (1967).

6. Pohl, F.M. Europ. J. Biochem., 4, 373 (1968).

7. Steinmetz, M.A., and W.C. Deal, Jr. Biochemistry, 5, 1399 (1966).

8. Reynard, A., L. Hass, D. Jacobsen, and P.D. Boyer. J. Biol. Chem., 236, 2277 (1961).

9. Hammes, G.G., and P.R. Schimmel. J. Phys. Chem., 71, 917 (1967).

10. Eigen, M. Nobel Symposium 5 on Fast Reactions and Primary Processes in Chemical Kinetics (S. Claesson, ed.), Interscience, 1967, p. 333.

11. Hunsley, J.R., and C.H. Suelter. J. Biol. Chem., 244, 4815 (1969).

MECHANISM OF ACTION OF TRANSALDOLASE*

K. Brand and B.L. Horecker

Max-Planck-Institut für Ernährungsphysiologie
Dortmund, Germany

and

Department of Molecular Biology
Albert Einstein College of Medicine, New York

Transaldolase (TA) has been shown (1,2) to catalyze the transfer of "active dihydroxyacetone" derived from fructose-6-phosphate (F-6-P) by aldol cleavage to an acceptor aldehyde, erythrose-4-phosphate (E-4-P)(Figure 1). In the absence of this acceptor, a stable enzymatically active intermediate accumulates (3,4) which has been identified as a Schiff base containing dihydroxyacetone (DHA) linked to the ε-amino group of a lysine residue at the active site of the enzyme. The structure of this intermediate was established by reduction of the Schiff base intermediate with sodium borohydride followed by hydrolysis of the protein which yielded the amino acid derivative N^6-β-glycerlylysine (5). Recently, we have obtained further evidence for the structure of the intermediate by demonstrating the addition of HCN to the isolated transaldolase-dihydroxyacetone complex (TA-DHA)(6). Cyanide causes a reversible inactivation by forming an inactive aminonitril derivative.

In the presence of E-4-P, an aldol condensation between this aldehyde acceptor and the transferred dihydroacetone occurs, yielding sedoheptulose-7-phosphate.

It should be pointed out that only one mole of dihydroxyacetone can be incorporated into one mole of complex when transaldolase from Candida utilis is incubated with fructose-6-phosphate (3,7) although this enzyme is composed of two identical subunits (ββ) as has been shown by hybridization (8) and ultracentrifugation (9) studies.

*Supported by NIH GM 11301 and NSF GB 7140. Communication No. 202 from Joan and Lester Avnet Institute of Molecular Biology.

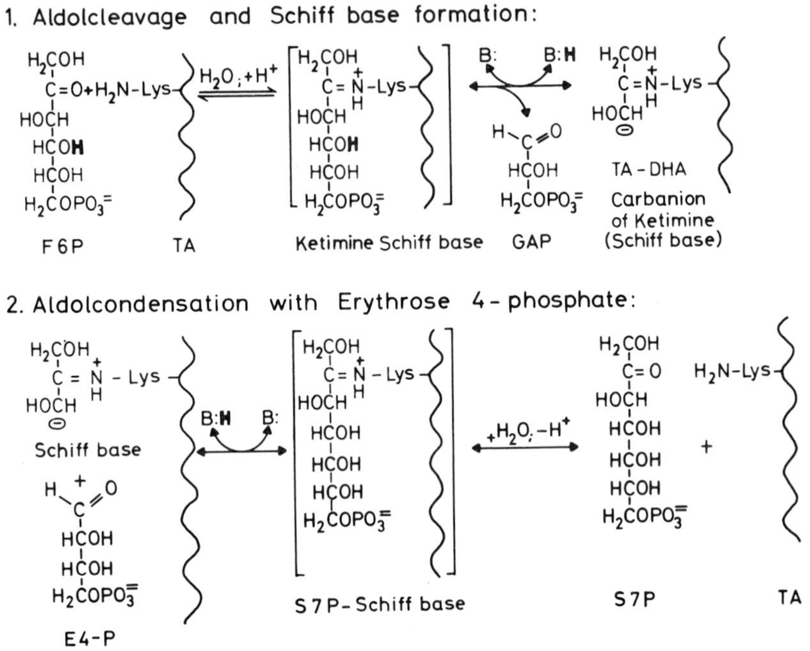

1. Aldolcleavage and Schiff base formation:

H₂COH C=O+H₂N-Lys HOCH HCOH HCOH H₂COPO₃⁼ **F 6 P** —TA— $H_2O; +H^+$ ⇌ [H₂COH C=N-Lys HOCH HCOH HCOH H₂COPO₃⁼] **Ketimine Schiff base** B: B:H GAP H₂COH C=N-Lys HOCH HCOH H₂COPO₃⁼ **TA – DHA Carbanion of Ketimine (Schiff base)**

2. Aldolcondensation with Erythrose 4 - phosphate:

H₂COH C=N-Lys HOCH **Schiff base** HCOH HCOH H₂COPO₃⁼ **E 4-P** B:H B: [H₂COH C=N-Lys HOCH HCOH HCOH HCOH H₂COPO₃⁼] **S 7 P-Schiff base** $+H_2O; -H^+$ H₂COH C=O HOCH HCOH HCOH H₂COPO₃⁼ **S 7 P** H₂N-Lys + **TA**

Figure 1. Reaction of transaldolase.

Since a proton has to be transferred or exchanged from the C-4 hydroxyl group of F-6-P during the aldol cleavage reaction, a histidine residue of the active site could be involved in this proton transfer as has been shown in other enzymes.

In order to find out if histidine is essential for the catalytic activity of transaldolase, photooxidation experiments were carried out. Photooxidation in the presence of rose Bengal has been shown to be a useful and specific method for the identification of functional histidine residues in a number of enzymes (10,11,12,13,14). When transaldolase was exposed to photooxidation at neutral pH in the presence of excess dye, there was a loss of catalytic activity which followed first order kinetics (15). The photoinactivation of transaldolase was contrary to that of aldolase (11), i.e., markedly dependent on pH. At lower pH, when the histidines are protonated, transaldolase was more stable to photooxidation than at more alkaline pH (Figure 2). The susceptibility of transaldolase to photoinactivation was found to follow

closely the curve obtained by Westhead (10) for the photooxidation of imidazole (Figure 2). The results suggest that
photoinactivation is associated with the destruction of one
or more essential histidine residues with pK values corresponding to that of free imidazole.

The loss of histidine by photooxidation was confirmed by
amino acid analysis. Native transaldolase was found to contain two histidine residues as reported earlier (16), and the
loss of activity was complete when one of this was destroyed
(Figure 3). This indicates that there is only one critical
histidine per mole of enzyme, although we have two identical
subunits (ββ) in transaldolase from Candida utilis, each of
them containing one histidine (Figure 7).

The photooxidized transaldolase which has lost one
histidine is not   able to form a Schiff base when incubated with F-6-P, indicating that the loss of a histidine
residue abolishes not only overall catalytic activity, but
also the ability of the enzyme to form the Schiff base.

In order to get more information about the role of this
critical histidine, tritium labelling experiments were carried out. Figure 4 shows schematically the experimental procedures and the results obtained. When transaldolase was
incubated with F-6-P in the presence of tritiated water in
order to label all OH groups of the substrate, tritium was
found to be incorporated into the isolated transaldolase-
dihydroxyacetone complex. To determine the quantity and
location of tritium, the isolated complex was reduced with
sodium borohydride. The reduced complex was found to contain
0.56 moles of tritium per mole of complex, and, after acid
hydrolysis, nearly all of the radioactivity was recovered in
the C-3 position of β-glyceryl-lysine. The fact that the
amount of tritium incorporated was somewhat less than the
amount of Schiff base formed may be due to isotope discrimi-
nation, or to a slow exchange of tritium with protons from
water during the isolation and washing of the complex (15).
The tritium labelling experiments demonstrate clearly that
there is a specific incorporation of tritium into the transaldolase-dihydroxyacetone complex when this is formed from
F-6-P in tritiated water. The tritium in the complex must
either be attached to the carbon atom 3 of dihydroxyacetone
or held by a group which is very close to the presumed carbanion.

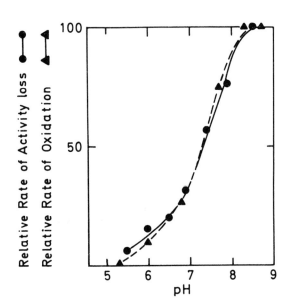

Figure 2. Effect of pH on the rate of photooxidation of
transaldolase. The initial rates of inactivation of transaldolase calculated from first order kinetic constants
(●--●) are compared with the photooxidation of imidazole
(▲--▲). The data for the latter curve were taken from Westhead (10). Conditions: 23°C; triethanolamine buffer, 0.02 M,
containing 0.002 M EDTA; transaldolase, 1.55 mg/ml; molar
ratio of enzyme to dye, 7:1.

In order to show to which group of the complex this tritium is attached, the isolated complex was exposed to photooxidation at pH 7.9 under the same conditions; the photooxidation was carried out with native enzyme. If histidine
in the complex is protonated at neutral or alkaline pH also,
it should be more stable to photooxidation than the native
enzyme. This was indeed the case. A linear relationship
between the amount of complex present in the reaction mixture

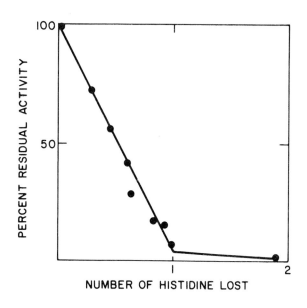

Figure 3. Correspondence between loss of histidine residues in transaldolase and degree of photoinactivation. The rate and extent of photooxidation were controlled by the concentration of dye and the time of illumination. An aliquot of each sample was assayed for enzymatic activity and the remainder treated with 5% trichloroacetic acid, and finally with an acetone-ether mixture. The final residues were suspended in 5.7 N HCl and hydrolyzed for 32 hrs at 110°C in sealed evacuated tubes. The content of histidine residues was determined on the short column of the amino acid analyzer.

and the amount of protection against photooxidation has been found (Figure 5). It may, therefore, be concluded that the histidine residues in the complex were protected from photooxidation, presumably because they were present in the protonated state even at pH 7.9. This proton could have originated from a hydroxyl group of the substrate which would have exchanged with tritium-labelled water under the conditions of

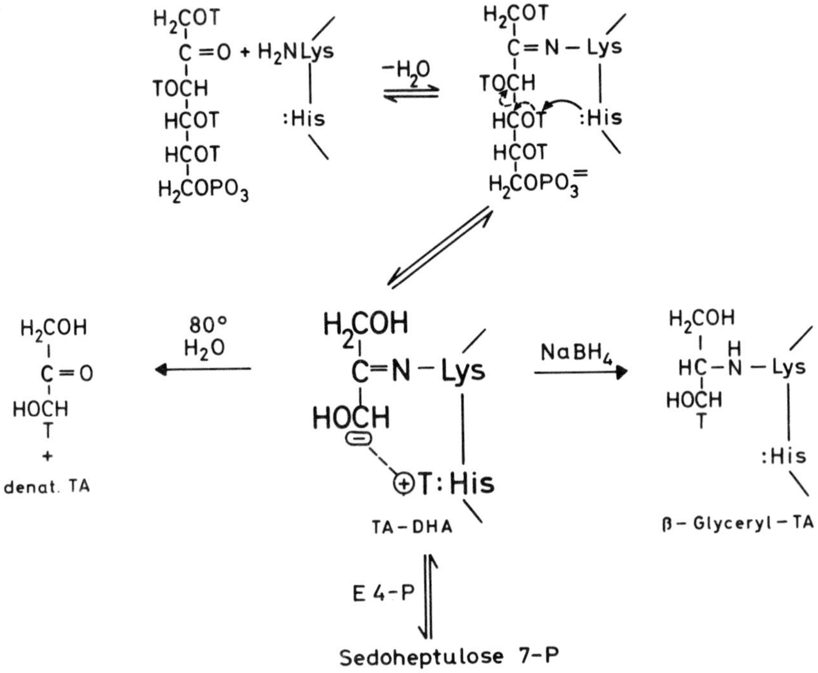

Figure 4. Scheme of the experimental procedures of the tritium labelling experiments and the mechanism of TA-DHA complex formation. The experimental conditions have been described previously (15).

the reaction. According to the reaction mechanism illustrated schematically in Figure 4, the complex could exist as a carbanion with a reactive carbon in the C-3 position of dihydroxyacetone and a protonated histidine forming a carbanion-imidazolium dipole. This proton on the imidazole nitrogen is then available for the following aldol concensation when E-4-P is added to the complex.

The following facts must be explained. This tritium on the imidazole nitrogen does not exchange, or exchanges very slowly, with water. Furthermore, there are two identical subunits in transaldolase, but only one active site for fructose-6-phosphate and only one critical histidine, acting as a base catalyst.

160

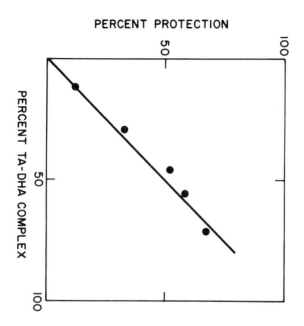

Figure 5. Linear relationship between the amount of complex present and the extent of protection against photooxidation. Experimental conditions have been described previously (15).

A possible explanation is that the two histidines at the active site are closely linked by a hydrogen bond (Figure 7). Both subunits are, therefore, contributing to one active site in transaldolase. Only one nucleophilic group is present at physiological pH to accept the proton from the C-4 hydroxyl group of F-6-P. This would explain why both histidines in the complex are protected against photooxidation when only one could be labelled with tritium.

Since a hydrophobic region around the active site in the transaldolase-dihydroxyacetone complex must be assumed in our model in order to prevent the exchange of tritium on the imidazole nitrogen, binding studies with 1-anilino-8-sulfonic acid (ANS) as a fluorochrome were carried out. This probe has been reported (17,18) to interact with non-polar regions of proteins causing an enhancement of fluorescence intensity

and a shift of the emission peak of ANS to shorter wavelengths. The results of these experiments are shown in Figure 6. When transaldolase was titrated with ANS, the fluorescence intensity increased to a saturation level. At 22°C
an equilibrium constant with a dissociation (K) of 2.2 x
10^{-5} M could be calculated for the transaldolase-ANS complex.
The number (n) of ANS binding sites per transaldolase molecule, determined according to the method described by Weber
and Young (17), has been found to be 2.2 (± 0.2). When
F-6-P was added in excess, the fluorescence intensity decreased almost completely, indicating that the substrate was
able to displace all the dye bound to the enzyme. When E-4-
P was added to the same mixture, the fluorescence intensity
returned to the original high level (19). This finding indicates that F-6-P displaces ANS either directly from the active
center or by causing a conformational change leading to a
decrease of hydrophobicity. The fact that addition of E-4-P,
which removes the triose from the complex and regenerates
free transaldolase, causes an increase of fluorescence intensity again supports the conclusion that one mole of ANS is
bound to each of both subunits close to the active site.
Addition of substrate F-6-P could cause a conformational
change, bringing the two subunits closer together and forming
a single active center (Figure 7) to which one mole of F-6-P
and one mole of dihydroxyacetone are bound after aldol cleavage has occurred (Figure 7).

This model would explain why one mole of F-6-P displaces
two moles of ANS bound to transaldolase.

Another interpretation of these results is that transaldolase binds two moles of F-6-P but is able to cleave only
one F-6-P because only one histidine with a nucleophilic
group is available to act as a base catalyst. Thus, two
moles of ANS can be displaced by two moles of F-6-P when
only mole of dihydroxyacetone is bound to transaldolase
forming the Schiff base intermediate. Whether or not the detailed interpretation is correct, the results obtained so far
indicate clearly that histidine is directly involved in
aldol cleavage, acting as a base catalyst.

References

1. Horecker, B.L., P.Z. Smyrniotis, H.H. Hiatt, and P.A.
    Marks. J. Biol. Chem., 212, 827 (1955).

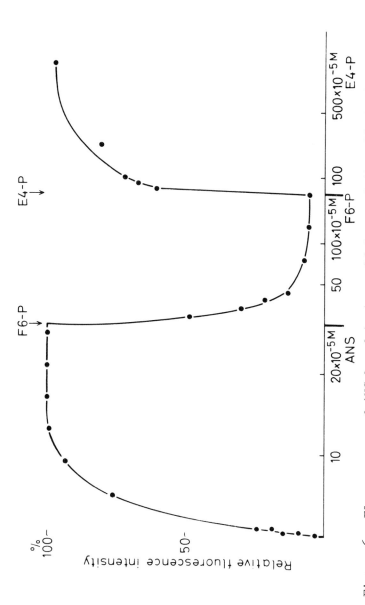

Figure 6. Fluorescence of ANS bound to transaldolase and the effect of fructose-6-phosphate (F-6-P) and erythrose-4-phosphate (E-4-P) on the fluorescence intensity. Conditions: Transaldolase, 7.3×10^{-6} M, dissolved in 0.05 M Na-phosphate buffer, pH 7.6. Test volume, 3 ml. Small amounts of ANS, F-6-P and E-4-P were titrated to the enzyme solution consecutively and the fluorescence intensity was recorded in an Eppendorf Photometer (excitation 313 + 366 nm; emission 420-3000 nm).

163

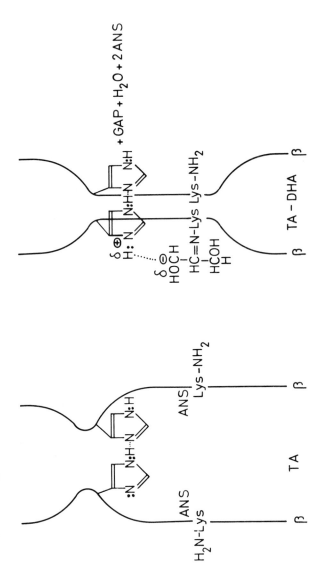

SUBUNIT STRUCTURE OF TRANSALDOLASE
AND MECHANISM OF TA-DHA COMPLEX FORMATION
(Binding of ANS and displacement by substrate F6-P)

Figure 7. Scheme of the subunit structure of transaldolase and the mechanism of TA-DHA complex formation. Displacement of ANS from the enzyme by substrate F-6-P during Schiff base formation and aldol cleavage.

2. Venkataraman, R., and E. Racker. J. Biol. Chem., <u>236</u>, 1883 (1961).

3. Horecker, B.L., S. Pontremoli, C. Ricci, and T. Cheng. Proc. Nat. Acad. Sci. US, <u>47</u>, 1940 (1961).

4. Horecker, B.L., T. Cheng, and S. Pontremoli. J. Biol. Chem., <u>238</u>, 3428 (1963).

5. Horecker, B.L., P.T. Rowley, E. Grazi, T. Cheng, and O. Tchola. Biochem. Z., <u>338</u>, 36 (1963).

6. Brand, K., and B.L. Horecker. Arch. Biochem. Biophys., <u>123</u>, 312 (1968).

7. Brand, K., and B.L. Horecker. Ztschr. Anal. Chem., <u>243</u>, 640 (1968).

8. Tsolas, O. Ph.D. Dissertation, Albert Einstein College of Medicine, New York, 1967.

9. Tsolas, O., Sia, C.L., and Horecker, B.L. Arch. Biochem. Biophys., <u>136</u>, 303 (1970).

10. Westhead, E.W. Biochemistry, <u>4</u>, 2139 (1965).

11. Hoffee, P., C.Y. Lai, E.L. Pugh, and B.L. Horecker. Proc. Nat. Acad. Sci. U.S., <u>57</u>, 107 (1967).

12. Martinez-Canion, M., C. Turano, F. Riva, and P. Fasella. J. Biol. Chem., <u>242</u>, 1426 (1967).

13. Chatterjee, G.C., and E.A. Noltmann. Europ. J. Biochem. <u>2</u>, 9 (1967).

14. Ray, W.J., Jr., and D.E. Koshland, Jr. J. Biol. Chem., <u>237</u>, 2493 (1962).

15. Brand, K., O. Tsolas, and B.L. Horecker. Arch. Biochem. Biophys., <u>130</u>, 521 (1969).

16. Horecker, B.L. In Pentose Metabolism in Bacteria, Wiley, New York, 1962, p. 81.

17. Weber, G., and L.B. Young. J. Biol. Chem., <u>239</u>, 1414 (1964).

18. Stryer, L. J. Mol. Biol., <u>13</u>, 482 (1965).

19. Brand, K. Unpublished experiments.

THE SYNTHESIS AND PROPERTIES OF A NEW CLASS OF CHROMOPHORIC ORGANOMERCURIALS (1)

C.H. McMurray* and D.R. Trentham

Molecular Enzymology Laboratory, Department of Biochemistry
University of Bristol, University Walk, Bristol, BS8 1TD

Abstract

With the recent advances in the study of biological systems at the molecular level, there is an increasing requirement for reagents which will react with specific sites on biological macromolecules. It is an advantage if a measurable signal occurs concomitant with the reaction, and if the magnitude of the signal is sensitive to the local environment.

The organomercurials described here are readily synthesised and are easy to handle. They have the following potential uses: chromophoric probes for thiol groups in proteins and other thiols, kinetic studies of reactivity of thiol groups, "reporter groups" (2) - indicators of perturbations in the biological systems to which they are attached, tertiary structure determination of biological macromolecules, estimation of cations, and biological activity.

Synthesis and Structural Determination of the Organomercurials

Four organomercurials have been synthesised.[+] 2-Chloromercuri-4-nitrophenol (I), 2-chloromercuri-4,6-dinitrophenol (II), 4-chloromercuri-2-nitrophenol (III), and 2,6-dichloromercuri-4-nitrophenol (IV).

The compounds were all synthesised by the mercuration of the parent nitrophenol.

The structures of the chloromercurials were established by their elemental analyses and U.V., visible and PMR spectra.

*Present address: Department of Chemistry, Harvard University, Cambridge, Mass., U.S.A.

[+]Available from Whatman Biochemicals Limited, Springfield Mill, Maidstone, Kent, England.

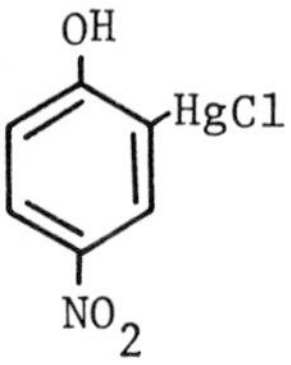

OH   OH   OH   OH

HgCl   ClHg NO_2 NO_2 ClHg HgCl

NO_2 NO_2 HgCl NO_2

I 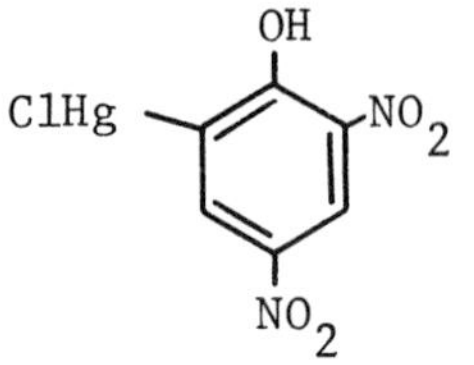   II 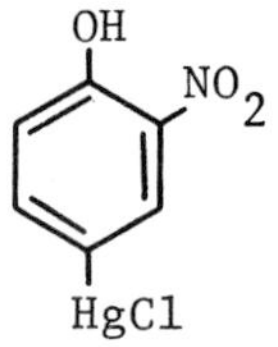   III 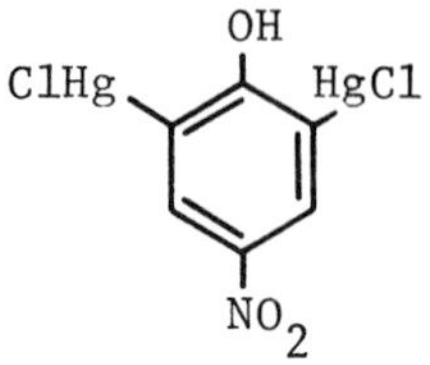   IV

The site of mercuration in the aromatic nucleus was evident from the PMR spectra and from the property of 2-nitrophenol to direct ortho and para and 4-nitrophenol ortho to the hydroxyl group.

Properties of the Organomercurials

Nitrophenolate ions have visible absorption bands so that the extinction of the chloromercurinitrophenols at any pH will be determined by the extent of ionisation of the phenol (Table I). The perturbation of the pK of the nitro-phenol and the resultant chromophoric change provide the main basis for the uses of the organomercurials outlined below. For this reason, these properties are discussed in some de-tail.

TABLE I

Extinction Coefficients at the Absorption Maxima and pK Values of Phenolic Hydroxyl Groups of Organomercurials

	ε	λ_{max}, $m\mu$	Added Ligand None	TGA	EDTA
2-Chloromercuri-4-nitrophenol (I)	1.74×10^4	405	6.5	7.1	8.85
2-Chloromercuri-4,6-dinitrophenol (II)	1.57×10^4 1.13×10^4	371 410 (shoulder)	4.0	5.0	6.3
4-Chloromercuri-2-nitrophenol (III)	4.1×10^3	416	6.8	7.3	7.9
2,6-dichloromercuri-4-nitrophenol (IV)	1.74×10^4	410	6.3	7.05	10.1

Organomercurials bind very tightly to thiols. The reaction of a thiol and a mercurinitrophenol might be expected
to introduce a pK change in the phenol and thus a chromophoric
change at certain pH values. This is the case to a certain
extent (Figures 1 and 2). However a much more dramatic pK
change is observed if a weakly bound ligand which interacts
with the hydroxyl group of the phenol is displaced (Figures
1 and 2, Table I). This reaction is illustrated in equation
1

$$\underset{\text{NO}_2}{\underset{\text{OH}}{\bigcirc}} \overset{\text{L}}{\underset{}{\text{Hg}}} + \text{RSH} \longrightarrow \underset{\text{NO}_2}{\underset{\text{OH}}{\bigcirc}} \text{HgSR} + \text{LH} \quad (1)$$

where L represents a ligand and RSH a thiol. A ligand such
as EDTA (ethylene diamine tetraacetic acid) shows the largest

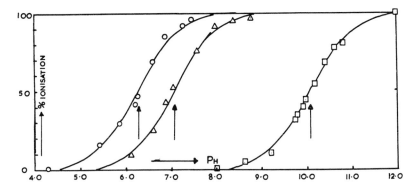

Figure 1. Ionisation constants of 2,6-dichloromercuri-4-nitro-
phenol. The variation of the phenol ionisation with pH was
measured in the presence of various ligands. The percent
ionisation of the phenol was deduced from the ratio of the absorbance at the measured pH to the plateau absorbance at high
pH values. The absorbance was measured at 410 mμ at 25° using
a cuvette with a 1 cm light path. The extinction of the un-
ionised phenol at 410 mμ is negligible. Legend: 0, 2,6-dichloro-
mercuri-4-nitrophenol (3.87×10^{-5} M); Δ, as 0 and thioglycollic acid (5×10^{-5} M); □, as 0 and EDTA (5 mM). The solutions
contained either sodium citrate (0.1 M) or triethanolamine
hydrochloride (0.1 M) adjusted to the measured pH with HCl or
NaOH. The solid lines are theoretical curves for species with
pK values indicated by the arrows.

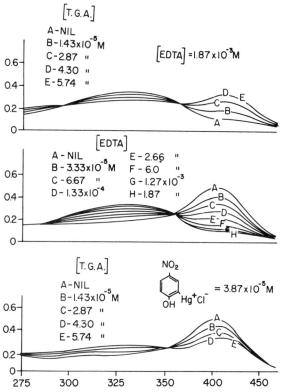

Figure 2. Absorption spectra of 2-chloromercuri-4-nitro-phenol showing the effect of added ligands. The solutions at 25° in a cuvette with a 1 cm light path contained 2-chloro-mercuri-4-nitrophenol (3.87 x 10⁻⁵ M), triethanolamine hydro-chloride (0.2 M), adjusted to pH 7.0 with NaOH. In addition, the solution in the top traces contained EDTA (1.87 mM) to which aliquots of thioglycollic acid (TGA) were added. Ali-quots of EDTA and TGA were added to the solutions in the cen-tre and bottom traces, respectively.

effect. EDTA has four carboxylate anions. Some of these anions probably interact with the hydroxyl group and prevent its ionisation.

The pK perturbations of the phenol are greatest when the mercury atom is ortho to the hydroxyl group. Of the mono-mercurials, the largest spectral changes ($\varepsilon \sim 10^4$ at $\lambda = 410$ mμ) in the pH range 3.3 to 6.5 are observed with 2-chloro-mercuri-4,6-dinitrophenol, and in the pH range 6.5 to 9 with

2-chloromercuri-4-nitrophenol. The pK values of the organomercurials associated with various ligands are summarised in
Table I.

The anions of all the organomercurials are water soluble, though the more substituted phenols (II, IV) are only
sparingly so (1 mg/ml).

Uses of the Organomercurials

The potential uses of the organomercurials are listed
below together with examples.

1. Chromophoric probes for thiol groups in proteins and other
   thiols.

The spectra (Figure 2) indicate that an extinction change
occurs when a thiol reacts with 2-chloromercuri-4-nitrophenol.
Hence the mercurial is a probe for thiol groups on proteins
and other biological macromolecules. This is illustrated by
the titration of 2-chloromercuri-4-nitrophenol with GPDH
isolated from lobster muscle* (curve A, Figure 3) and with the
same enzyme in which the active centre "essential" thiol
group has been selectively carboxymethylated (curve B, Figure
3). Activity of the enzyme is lost during the titration of
the first equivalent of mercurial at the active site, even in
the presence of 1 mM NAD$^+$ (4). It is apparent from curve A
that the extinction of the nitrophenol group differs on the
two thiols. This is a reflection of the different local environment around each nitrophenol. The extinction differences
therefore provide a probe to study the local environment of
the thiol groups. The variable extinction of the mercurinitrophenols on the protein does not prevent the stoichiometry of
the reaction with the thiol group being measured. The extinction may either be measured using titration curves (Figure
3) or the mercurial can be removed quantitatively from the
protein using an excess of thiol such as 2-mercaptoethanol.
The absorbance of the mercurinitrophenol 2-mercaptoethanol
complex of known extinction may be measured.

2. Kinetic studies of reactivity of thiol groups.

The thiol groups on the surface of a biological macromolecule will have different reactivities. The reactivity
will depend on such factors as the steric interaction of

*Abbreviations: GPDH, D-glyceraldehyde-3-phosphate dehydro-
genase (EC 1.2.1.12); CM-GPDH, GPDH carboxymethylated at the
active site cysteine 148 (3); TGA, thioglycollic acid.

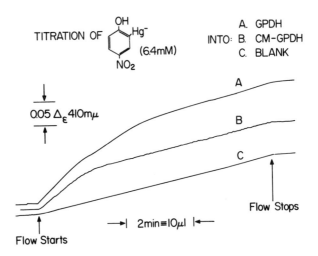

Figure 3. Spectrophotometric record of a titration of 2-chloromercuri-4-nitrophenol at constant rate against: Curve A, native GPDH; Curve B, CM-GPDH; and Curve C, buffer solution. Traces were recorded at 410 mμ and 4°. The observation chamber had a 1 cm light path. The reaction cell contained: Curve A, GPDH (4.82×10^{-8} moles of sites); Curve B, carboxymethyl GPDH (4.82×10^{-8} moles of sites); Curve C, blank. The solvent (5.0 ml) contained NAD^+ (5×10^{-5} M), EDTA (10^{-3} M), NaCl (10^{-2} M), triethanolamine (10^{-1} M) adjusted at 20° to pH 7.9 with HCl. 2-Chloromercuri-4-nitrophenol (6.4×10^{-3} M) was added to the stirred solution at a constant flow rate of 5 μl per min as indicated.

The extent of mercuration of the enzyme may be determined. 2.02 ± 0.1 mercurials bind per subunit of native GPDH and 0.97 ± 0.05 mercurials bind per subunit of CM-GPDH.

neighboring groups, the local solvent environment created by these groups and the role of neighboring acid or base catalysts. This reactivity may be investigated by measuring the rate of the reaction of the thiol groups with the chromophoric organomercurials. Thus, while two thiol groups on GPDH react with rate constants of 220 sec^{-1} and 2.2 sec^{-1} under specified conditions (Figure 4a), the thiol group of the carboxymethylated enzyme reacts with a rate constant of 2.0 sec^{-1} (Figure 4b). This probably reflects the greater reactivity of a thiol group at or near the active site.

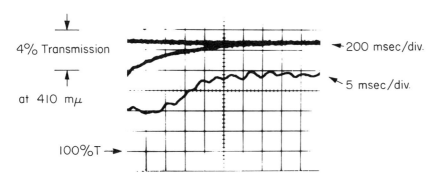

Figure 4a. Spectrophotometric record of the reaction of 2-chloromercuri-4-nitrophenol and native GPDH. Stopped flow traces were recorded at 410 mμ and 20° using a quartz iodine light source and grating monochromator. The observation chamber had a 1 cm light path. The reaction mixture contained GPDH (3.01 x 10^{-6} moles of sites), 2-chloromercuri-4-nitro-phenol (6.5 x 10^{-5} M), EDTA (10^{-3} M), NAD$^+$ (2 x 10^{-5} M), triethanolamine (10^{-1} M) adjusted to pH 7.9 with HCl. The reaction was initiated by rapid mixing of the mercurial with GPDH. The rate of the bottom trace is 5 msec/div and of the upper two traces 200 msec/div. The top trace was triggered about 5 sec after initiating the reaction and indicates end point stability. The plateau at the start of the bottom trace represents the constant absorbance of the solution during rapid flow of the solutions. The flow stopped about 10 msec after the oscilloscope trace was triggered.

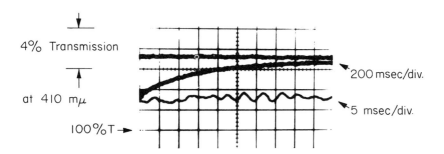

Figure 4b. As in Figure 4a except that CM-GPDH replaced GPDH.

3. "Reporter groups."

If a molecule that is bound to a biological macromolecule has the property of signalling changes that occur in another part of the macromolecule, it is termed a "reporter group" (2). The organomercurials have this property because chromphoric changes are associated with the ionisation state of the nitrophenols and the extent of ionisation is dependent on the environment. Since the thiol bound mercurinitrophenols have pK values around 7 (Table I), the maximum chromophoric signals following a pK perturbation will occur in the most suitable pH range for biological studies. For example, a conformation change of a protein is likely to induce a local solvent environment change around the organomercurial, thus perturbing its pK. The pK change will be signalled by a chromophoric change. Since the proton transfers associated with a pK change are extremely rapid, the kinetics of the slower processes, such as the conformation change, may be monitored. These points are illustrated by pK measurements of the nitrophenol residue of the CM-GPDH and the native enzyme, each treated with 0.8 mole of mercurial (I) per mole of subunit (Table II).

TABLE II

pK Values of 2-Chloromercuri-4-Nitrophenol Bound to
GPDH, CM-GPDH and Thioglycollic Acid

	Solution A	Solution B
GPDH	8.55	7.8
CM-GPDH	8.05	8.05
Thioglycollic Acid	7.7	7.7

The pK values were determined spectrophotometrically at 4°. The enzyme subunit or thioglycollic acid concentration was 5.0×10^{-5} M and organomercurial concentration was 4.0×10^{-5} M. Solution A contained EDTA (1 mM), pyrophosphate (0.1 M), and phosphate (0.1 M) adjusted to I 0.6 with NaCl. Solution B contained EDTA (1 mM), ethylene diamine (0.2 M) adjusted to I 0.6 with NaCl. The enzyme solutions contained NAD^+ (1 mM).

The pK of the nitrophenol residue near the active site is perturbed by pyrophosphate (Figure 5). An analysis indicates that there are at least two pyrophosphate binding sites near the active site. The perturbation by pyrophosphate of the pK is unlikely to be caused by mercurial migration since the perturbation also occurs when two moles of mercurial are bound per mole of subunit of native enzyme.

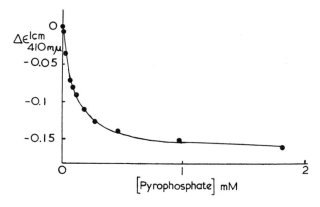

Figure 5. Spectrophotometric titration at 4° of pyrophosphate into native GPDH (5 x 10^{-5} M in subunits) treated with 2-chloro-mercuri-4-nitrophenol (4 x 10^{-5} M). The solution also contained triethanolamine hydrochloride adjusted to pH 8.6 with NH$_3$, EDTA (1 mM) and NAD$^+$ (1 mM).

4. **The tertiary structure determination of biological macro-molecules.**

To determine the tertiary structure of a large biological molecule such as an enzyme by X-ray diffraction analysis, it is essential to synthesise heavy atom derivatives of the molecule isomorphous with a parent molecule. This synthesis has proved difficult in the past because of the lack of satisfactory methods of achieving the heavy atom incorporation in stoichiometric amounts under well-defined conditions (5,6). That 2-chloromercuri-4-nitrophenol may be incorporated under defined conditions to the enzyme GPDH is clear (Figure 3). Moreover, analysis of the heavy atom content of samples of crystalline enzymes is readily achieved with all four mercurials since their absorption bands (λ_{max} 410 mµ) are separate from protein absorption bands.

175

Large crystals of CM-GPDH containing one mole of 2-chloromercuri-4-nitrophenol per mole of subunit have been grown. These crystals are isomorphous with those of the native enzyme.

5. Estimation of cations.

2-Chloromercuri-4-nitrophenol may be used as a chromophoric reagent for the colorimetric titration of certain cations. Use is made of the tighter binding of a ligand to the cation over the mercurial coupled with the extinction change associated with the binding of the ligand to the mercurial. For example, calcium may be estimated (Figure 6, equation 2).

$$\left[\begin{array}{c} \text{OH} \\ \text{Hg} \\ \text{NO}_2 \end{array} \cdot \text{EDTA} \right]^- + Ca^{++} \rightleftharpoons \begin{array}{c} \text{OH} \\ \text{Hg}^+ \\ \text{NO}_2 \end{array} + [Ca \cdot EDTA] \quad (2)$$

6. Biological Activity.

The mercurials being water soluble and reactive towards thiols may have biological activity. The relatively high solubility of 2-chloromercuri-4,6-dinitrophenol in methanol indicates that this compound is particularly interesting because of possible membrane permeability.

In conclusion, it may be noted that the organomercurials combine all the properties of other reversible thiol agents available to the biochemist, notably DTNB (5,5'-dithiobis-2-nitrobenzoic acid) and p-chloromercuribenzoic acid. However they have striking advantages over these materials in that they reflect further properties of the biological macromolecule and they are proving of great use to the X-ray crystallographer.

Acknowledgment

We are grateful to Dr. H. Gutfreund and Dr. H.C. Watson for helpful discussions. We thank Mrs. T.G. Vickers for technical assistance and the Medical Research Council and the Science Research Council for financial support.

176

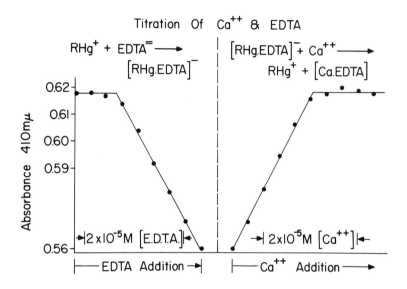

Figure 6. Spectrophotometric titration of calcium against ethylene diamine tetraacetic acid (EDTA). The left hand titration curve shows the absorbance changes of a solution of 2-chloromercuri-4-nitrophenol (3.9×10^{-5} M), triethanolamine hydrochloride (0.1 M) adjusted to pH 7.55 with NaOH when treated with 1 µl aliquots of EDTA (10.0 mM). The plateau at the start of the titration is because initially the EDTA binds cation impurities in the solution (such as Ca^{++} in the water) preferentially to the mercurial. The right hand titration curve shows the absorbance changes of the same solution which contains the added EDTA when treated with 1 µl aliquots of $CaCl_2$ (10.0 mM). The plateau at the end of the titration occurs when the calcium ions have completely removed EDTA from the mercurial.

References

1. McMurray, C.H. and D.R. Trentham. Biochem. J., $\underline{115}$, 913 (1969).

2. Burr, M., and D.E. Koshland. Proc. Nat. Acad. Sci., U.S., $\underline{52}$, 1017 (1964).

3. Davidson, B.E., M. Sajgò, H.F. Noller, and J.I. Harris. Nature, London, $\underline{216}$, 1181 (1967).

4. Trentham, D.R. Biochem. J., $\underline{109}$, 603 (1968).

5. Watson, H.C., and L.J. Banaszak. Nature, London, $\underline{204}$, 918 (1964).

6. Wassarman, P.M., and H.C. Watson. Abstr. 5th Meet. Fed. Durop. Biochem. Soc., Prague, 1968, p.193.

PART 2

STRUCTURAL INTERACTIONS IN LIGAND BINDING IN HEMOPROTEINS

CRYSTALLOGRAPHIC STUDIES OF LIGAND BINDING

B. P. Schoenborn

Department of Biology, Brookhaven National Laboratory
Upton, New York

At the 1966 Johnson Foundation Colloquium, Watson and Chance (1) presented some crystallographic data comparing the cyanide, hydroxide, and fluoride derivatives of myoglobin with each other. They concluded that no change in the iron position is evident, and noted that the difference maps of the cyanide and hydroxide derivatives were rather similar, suggesting that these derivatives exhibit similar conformational changes. Some of these changes were indicative of a porphyrin shift towards the iron. In the case of the cyanide derivative, which is a very low spin compound, in contrast to the mixed spin hyroxide derivative, such a change would support the notion that a low spin compound should exhibit a planar heme. This dilemma led to the full three-dimensional investigations of the hydroxide (2) and the cyanide (3) derivatives.

I should like to present here some of the conclusions derived from the analysis of the hydroxide derivative. Unfortunately, I will not be able to compare these findings with those of the cyanide compound which have not yet been published by Peter Bretscher. Nor will I be able to compare these findings with the oxymyoglobin work which is now being completed by Chris Nobbs.

Sperm whale myoglobin was crystallized from an 80% solution of ammonium sulfate at pH 6.9. After completion of crystallization, the pH was adjusted to 9.4 with NaOH. Crystals prepared in this manner proved to be unstable under X-rays. Fortunately, exposure to xenon stabilized these crystals and permitted data collection. A three-dimensional difference Fourier map was then calculated, suing 2,600 independent reflections and the standard metmyoglobin phases. The negative peaks at the gold and mercury sites that were first observed by Dickerson, Kendrew and Strandberg (4) are again

181

present. An inspection of the difference electron density map reveals a number of features depicted in Figures 1, 2, and 3.

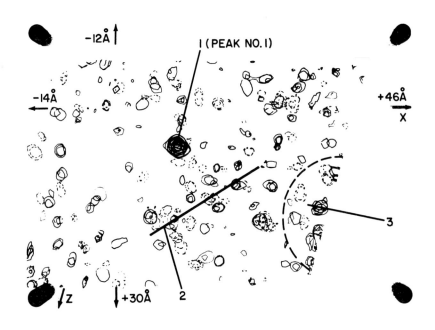

Figure 1. Difference electron density map, with superimposed y section from 0.7 to 0.9. 1, xenon peak #1; 2, indicated direction of E helix; 3, region of GH corner of the symmetry-related molecule.

Some of these features clearly indicate molecular rearrangements, while others are suggestive of some changes for which no definite new positions can be determined. The well defined alterations are listed in Tables I and II, and were refined by a full matrix least-squares analysis, keeping the rest of the molecule invariant. The other features, indicating changes whose exact nature could not be determined, are listed in Table III.

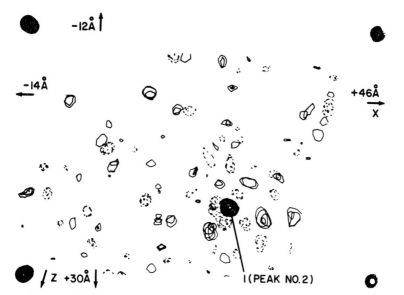

Figure 2. Difference electron density map with superimposed y section 0.35 to 0.45 depicting the xenon peak, #2, and the surrounding GH corner

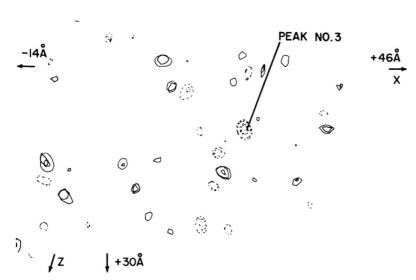

Figure 3. Difference electron density map with superimposed y section 0.0 to 0.1 depicting the sulfate peak, #3.

183

TABLE I

Groups Used in the Least Squares Refinement, with Molecular
Reorientation Given as Rotations along Specified Bonds

Group	Alteration
Histidine B5	$+20°$ $±8°$ ($C\alpha$-$C\beta$); $-15°$ $±8°$ ($C\beta$-$C\gamma$)
C1	$+25°$ $±9°$ ($C\alpha$-$C\beta$)
FG3	$-20°$ $±6°$ ($C\alpha$-$C\beta$)
GH1	$62°$ $±10°$ ($C\alpha$-$C\beta$); $+50°$ $±20°$ ($C\beta$-$C\gamma$)
G14	uncertain; change, $15°$ $±15°$
G17	uncertain; change, $10°$ $±10°$
Helical segment, E10 - E18	Twist, $+10°$ $±3°$, along axis of helix Kink, $+7°$ $±3°$; raises end of helix in y

TABLE II

Additional Features with Refined Fractional Coordinates*

Peak	eA^{-3} from Fourier	x	y	z	Weight in Electrons+
1	0.89	0.175 (0.177)	0.868 (0.864)	0.163 (0.168)	+49
2	0.89	0.401	0.400	0.522	+26
3	-0.34	0.369	0.056	0.272	-21

*The fractional coordinates for peak #1 as found in the two-dimensional analysis in xenon metmyoglobin are given in (5)
+From least squares.

TABLE III

Groups with Minor Alerations, as Indicated in Difference Map

Group	Possible Change
Glutamic acid A2	Ion on NH could be missing
Glutamic acid A4	Ion between N_ϵ of tryptophan A5 and
Tryptophan A5	carboxyl group of A4
Lysine A14	Shift of side chain
Glutamic acid A16	Ion on carboxyl with H-bond to lysine
Lysine E20	E20 broken, with side chain shift
Histidine EF4	Movement of ring to +x, +y
Histidine G14	Indicated ring and peptide shift to -z
Valine G15	Peptide shift to -z
Leucine G16	Peptide shift to -z
Histidine G17	Peptide and ring shift to -z
Serine G18	Peptide shift to -z

In summary, the following observations have been made: no change in the heme has been found that would alter its planarity. The first xenon site is identical to the site found in the metmyoglobin xenon compound (5). The second xenon site is new and lies at a point in the structure where two histidines did change their positions. At this high pH of 9.1,the histidines would have lost their hydrogen bonds and are therefore free to move. No changes are observed in the distal or heme-linked histidines. A negative peak close to N_ϵ of histidine E7 is observed which is similar to that found by Stryer (6) in azide myoglobin and by Nobbs (7) in deoxy-myoglobin. A small twist is observed in the E helix but no obvious links to functional changes are apparent. Most of the other observed perturbations are probably due to the changed ionic surrounding and might be evidence of the onset of alkaline denaturation.

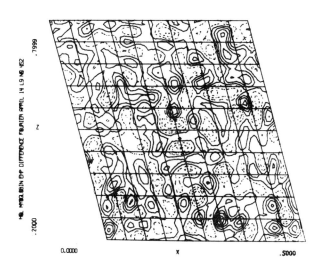

Figure 4. Difference electron density projection map (h0ℓ) of the ethyl hydrogen peroxide myoglobin derivative.

At this point I would like to introduce some results on the ethyl hydrogen peroxide derivative shown in Figure 4. The derivative was prepared by mounting the crystals in quartz capillaries and then soaking them in mother liquor containing 3 millimoles of ethyl hydrogen peroxide at pH 6.8 for 5-10 min at 20°C. The tubes were then sealed and cooled to 3°C during their X-ray exposures (4-8 hr). The resulting h0ℓ difference map is shown in Figure 4. The map is obviously too complicated to interpret but it should be noted that it shows considerable agreement with the hydroxide h0ℓ projection. The positive peak near the iron position could easily correspond to the ethyl group. The negative (sulfate) peak present in azide and deoxymyoglobin is not observed. It is, however, of little value to speculate further about the possible changes until more data is available.

References

1. Watson, H.C., and Chance, B. Hemes and Hemoproteins, (B. Chance, R. Estabrook, and T. Yonetani, eds.), Academic Press, New York, 1966.

2. Schoenborn, B.P. J. Mol. Biol., in press (1969).

3. Bretscher, R. Private communication.

4. Dickerson, R.E., Kendrew, J.C., and Strandberg, E. Acta Cryst., 14, 1188 (1961).

5. Schoenborn, B.P., Watson, H.C., and Kendrew, J.C. Nature, 207, 28, (1965).

6. Stryer, L., Kendrew, J.C., and Watson, H.C. J. Mol. Biol., 8, 96 (1964).

7. Nobbs, C., Watson, H.C., and Kendrew, J.C. Nature, 209, 339 (1966).

ALTERATIONS IN THE PROJECTED ELECTRON DENSITY OF GLYCERA HEMOGLOBIN ACCOMPANYING CHANGES IN LIGAND STATE*

Eduardo A. Padlan[+] and Warner E. Love

Thomas C. Jenkins Department of Biophysics
Johns Hopkins University, Baltimore, Maryland 21218

A single-heme, single-chain hemoglobin from the marine annelid worm, Glycera dibranchiata, has been isolated and crystallized. The crystals belong to the monoclinic space group P 2_1 and have a single molecule of 18,000 Daltons in the crystallographic asymmetric unit. The structure was first solved in projection at 3 Å resolution, (1) and then in three dimensions at 5.5 Å resolution (2). The method of multiple heavy atom isomorphous replacement was used.

To a remarkable degree this molecule resembles the α chain of horse hemoglobin. In addition, the plane of the heme fortuitously lies parallel to the 2-fold screw axis of the space group, and the heme is therefore seen edge-on in the centrosymmetric projection along b. Thus it becomes possible to obtain experimental data relevant to the prediction of Hoard et al. (3) that the iron in low-spin complexes would lie coplanar with the heme, while the iron in high-spin comlexes would tend to lie out of the plane. In order to employ the difference Fourier technique (4-6), the various derivatives must be closely isomorphous. The lattice constants for carboxy-, deoxy, acid met-, azide met-, and cyanide methemoglobin given in Table I show that this condition has been met.

The h0ℓ reflections for the five derivatives were collected, as well as an independent set of h0ℓ reflections from another cyanide methemoglobin crystal.

*This research was supported by USPHS AM 02528.

** Present address: Physics Department, University of the
                    Philippines, Quezon City, Philippines.

TABLE I

Lattice Parameters of Various Hemoglobin Derivatives

Ligand Form	a*	b*	c*	α**
Hb	61.22	32.55	41.06	109.55
HbCO	61.40	32.80	41.01	109.16
Hi	61.32	32.72	40.94	--
HiCN	61.43	32.85	40.91	109.23
HiN$_3$	61.18	32.74	40.92	109.07

*In Ångstrom units; the standard deviations are 0.04, 0.02, and 0.03 Å for a, b, and c respectively.
**In degrees; the standard deviation is 0.02°.

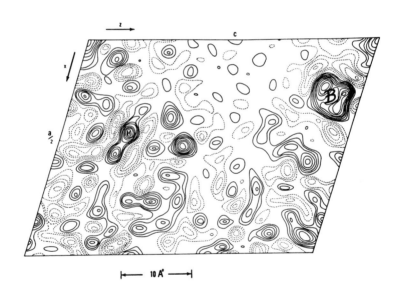

Figure 1. Fourier synthesis for Glycera cyanide methemoglobin prjected along the b axis. The heme, edge-on to the left of center, is marked H. The large peak marked B is the B helix, also edge-on. The F$_{ooo}$ term was left out. Contours at uniform arbitrary intervals.

The coefficients for the syntheses were

$$m \cdot s(F_p) \cdot (F_D - F_p)$$

where m is the combined figure of merit (7), $s(F_p)$ is the sign of the coefficient (actually the sign of the cyanide methemoglobin compound, which was the structure solved by isomorphous replacement) and $(F_D - F_p)$ is the difference in structure amplitudes of the two derivatives being compared. Since the changes in electron density accompanying the alterations in ligand state under consideration here are very small, it can safely be assumed that the cyanide methemoglobin signs will be those for all the derivatives. Figure 1 shows the b̲ axis projection for cyanide methemoglobin.

To evaluate the significance of the features in the difference syntheses, coefficients were formed by differencing the two independent sets of data for cyanide methemoglobin. This synthesis is shown in Figure 2. The root-mean-square density was taken as a measure of the error and the difference maps are all contoured at twice this interval.

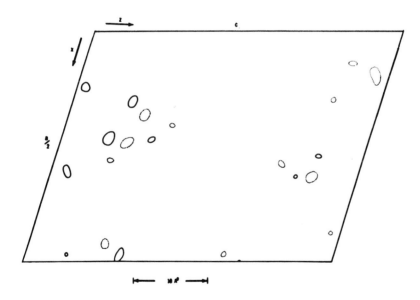

Figure 2. Fourier difference synthesis of two independent sets of data for Glycera cyanide met hemoglobin.

189

Figure 3 shows a comparison of azide methemoglobin (+) with deoxy hemoglobin (-), and the peak shows the binding side of the heme to face the end-on projection of the B helix, just as in the case of sperm whale myoglobin.

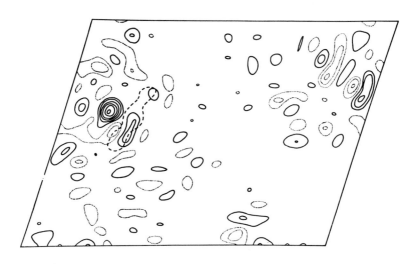

Figure 3. Fourier difference synthesis of azide met minus deoxy hemoglobin (Glycera)

Spin state alterations were postulated by Hoard et al. (3) and further by Caughey (8) to be accompanied by motion of the iron relative to the porphyrin. The weaker ionic bonds in the high spin states were proposed to be so long that the iron was out of the heme plane, while in the covalent strong (short) bond case, the iron was thought to lie in the heme plane.

These proposals have been examined relative to difference Fourier syntheses prepared for pairs of spin-states as follows: carboxy hemoglobin vs. cyanide methemoglobin (low-low); azide methemoglobin , carboxy hemoglobin, and cyanide methemoglobin, each vs. deoxy hemoglobin (low-high), and acid methemoglobin vs. cyanide methemoglobin (high-low), and acid methemoglobin vs. deoxy hemoglobin (high-high).

Except for a single peak at the binding site containing two contour lines, Figure 2 in its featurelessness would much resemble the difference map of carboxy hemoglobin vs cyanide methemoglobin. This low-low comparison accords well with the prediction of Hoard et al. (3). Figure 3 exemplifies the three comparisons, cyanide met, azide met, and carbony hemoglobin, all vs. deoxy hemoglobin. There is a positive peak at the ligand site and there are areas elongated parallel to the heme of more or less positive and negative difference density which would result from shifts of the porphyrin relative to the rest of the molecule. The change from cyanide methemoglobin to acid methemoglobin is from low to high spin. The difference between liganded H_2O and CN is 4 electrons, which must be at or below the level of significance. Figure 4 shows this synthesis. The heme moves away from the F helix when cyanide is replaced by water in methemoglobin.

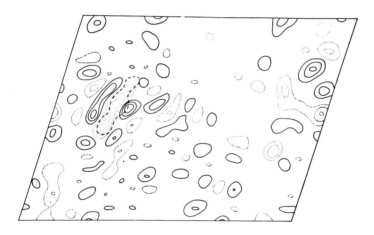

Figure 4. Fourier difference synthesis of acid methemoglobin vs cyanide methemoglobin (Glycera).

Figure 5 shows acid methemoglobin vs. deoxy hemoglobin. In agreement with Nobbs et al. (5) the binding site of acid methemoglobin is much more occupied than it is in deoxy hemoglobin. However, there are clearly other significant alterations in electron density accompanying this transition.

These preliminary results show that there are significant alterations in the structure of Glycera hemoglobin which accompany alterations in ligand state. The degree of change correlates somewhat with the degree of change of spin state

191

in accordance with Hoard et al. (3). Only when the structure
has been solved at high resolution in three dimensions will
it be possible to understand these structural alterations
in atomic terms.

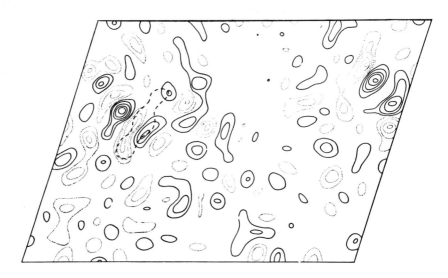

Figure 5. Fourier difference synthesis of Glycera acid methemoglobin vs. deoxy hemoglobin.

References

1. Padlan, E. A. Ph.D thesis, Johns Hopkins University, 1968.

2. Padlan, E. A. and W. E. Love. Nature, 220, 376 (1968).

3. Hoard, J. L., M. J. Hamor, T. A. Hamor, and W. S. Caughey.
   J. Am. Chem. Soc., 87, 2312 (1965)

4. Stryer, L., J. C. Kendrew, and H. C. Watson. J. Mol. Biol.
   8, 96 (1964).

5. Nobbs, C. L., H. C. Watson, and J. C. Kendrew. Nature,
   209, 339 (1966).

6. Watson, H. C. and B. Chance. In Hemes and Hemoproteins
   (B. Chance, R. W. Estabrook, and T. Yonetani, Eds.) Academic Press, New York, 1966. p. 149

7. Blow, D. M., and F. C. Crick. Acta Crystal., 12, 794 (1959

8. W. S. Cuaghey. Ann. Rev. Biochem., 1967, p. 611.

DISCUSSION

Mildvan: I wonder if Dr. Love might comment on why the shifting heme was not seen in sperm whale myoglobin?

Love: I cannot comment as I do not know.

Moss: When you say that the distance out of the plane depends on the spin state, don't you mean only that in the high spin iron state more of the iron orbitals which are occupied happen to be of the σ type rather than the π type, so that in the high spin form there is more energetically unfavorable overlap of electron density with the porphyrin nitrogen π orbitals? The correlation is really electronic; it is the spatial symmetry of the orbitals which are occupied with determines if the iron will "fit" the plane, and the "magnetic" correlation with spin is only coincidental.

Love: The prediction was made by Hoard and his co-workers that there would be a structural correlation with spin state (1). The higher the spin state, the more the iron would rise up out of the plane of the heme. I do not know about orbitals, but we can see motion of the heme relative to the rest of the structure.*

Chance: Dr. Schoenborn, can you yet determine the excess electron density at the heme site to be equivalent to the difference between C_2H_5OOH and HOH?

Schoenborn: It is not possible in such a projection difference map irrevocably to assign any peak to a specific group. However, in the region of the heme we do observe a large positive peak which is consistent with an ethyl group. There is little indication of a second oxygen atom; it is assumed that one oxygen is at the site occupied by water in metmyoglobin. In addition, there is evidence that the phenylalanine CD1 is shifted. However, I would like to repeat at this point that this projection difference map is too complicated to interpret

*Some further discussion of spin-state structure and spin-state correlations are to be found on p. 321.

193

precisely; the analysis presented is consistent with the data, but not final proof.

Chance: My point is simply that the absence of a change of unit cell, as is observed in one type of lamprey crystals, or the presence of a 10% change of unit cell, as occurs in the other type of lamprey crystals, is so small, compared with the complete change of unit cell parameters observed in the oxy-deoxy hemoglobin transition by Perutz (2) in mammalian material, that the observation of cooperativity in the lamprey crystals with minimal structural changes is something that must be considered in mechanisms for the control of reactivity in hemoglobin in general.

Love: I would call your attention to Dr. Perutz's paper (2) where he states that the deoxy to oxy (actually met) change of quaternary structure is accompaied by a change of 7% in b. The other lattice constants remain "unchanged." This was with horse hemoglobin; some other deoxy crystals crack upon exposure to oxygen.

References

1. Hoard, J.L., M.J. Hamor and W.S. Caughey. J. Amer. Chem. Soc., 87, 2312 (1965).

2. Perutz, M.E. et al. Nature, 203, 687 (1964).

HIGH RESOLUTION PROTON MAGNETIC RESONANCE
STUDIES OF MYOGLOBIN*

R.G. Shulman, K. Wüthrich, and J. Peisach+

Bell Telephone Laboratories, Incorporated,
Murray Hill, New Jersey

In paramagnetic heme compounds, the unpaired electrons of the iron are partially delocalized to the porphyrin ring (1) In cyanoporphyrin iron (III) compounds, the unpaired electron is in the π-orbitals of the heme group. Unpaired electron density is transferred from the π-orbitals of the ring carbon atoms to the protons of the heme group (Figure 1) by spin polarization and hyperconjugation. The resulting contact shifts, measured as frequency differences, $\Delta\nu_c$, are given by (2):

$$\Delta\nu_c = A \ \frac{\gamma_e}{\gamma_H} \ S(S+1) \ \frac{\nu}{3kT}$$

where γ_e and γ_H are the gyromagnetic ratios of the electron and proton, respectively; A is the contact interaction constant in cycles per second; T, the absolute temperature; S, the total electronic spin; ν, the NMR spectrometer frequency; and k the Boltzmann constant. From the measured shift of a heme proton resonance one can calculate a value for A which in turn is proportional to the spin density in the π-orbital of the ring carbon atom next to the observed protons. Pseudo-contact interactions which might also contribute to the observed hyperfine shifts appear to be small compared to contact coupling in low spin ferric heme compounds (3).

*The portion of this investigation at the Albert Einstein College of Medicine was supported in part by PHS GM 10959 and 1-K3-GM-31, 156. This is communication No. 172 from the Joan and Lester Avnet Institute of Molecular Biology.

+Permanent address: Departments of Pharmacology and Molecular Biology, Albert Einstein College of Medicine, Yeshiva University, Bronx, New York, 10461.

Figure 1 shows the NMR spectrum at 220 MHz of porpoise cyanoferrimyoglobin (MbCN) (4). The poorly resolved resonances of about 900 protons are found in the central region of the spectrum between 0 and -9 ppm. Much better resolved resonances are found at higher and lower fields. We have been able to show from their temperature dependence that most of the shifted resonances come from the heme protons and these have been

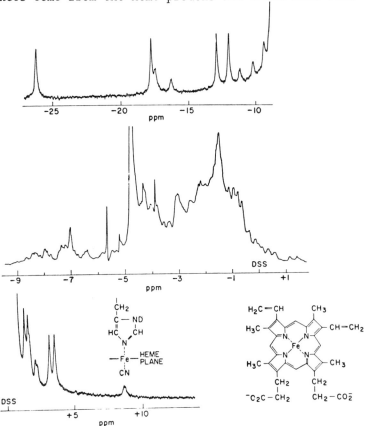

Figure 1. Proton NMR spectrum at 220 MHz of cyanoferrimyo-globin. Different horizontal and vertical scales were used for the spectral regions from from 1 to -9 ppm, and +1 to +10 and -10 to -30 ppm. The sharp lines between -3.8 and -6 ppm correspond to the resonance of HDO and its first and second spinning side bands. The structure of protoheme IX and of the axial ligands of the heme iron in cyanoferrimyoglobin are shown at the bottom.

shifted by hyperfine interactions. For example, four low field resonances of intensities corresponding to three protons each are from the ring methyl groups (Figure 1). In order to identify other resonances of the heme groups, we have studied (5) in collaboration with Drs. E. Antonini and M. Brunori, the heme resonances in ferrimyoglobins reconstituted with porphyrins other than protoporphyrin IX.

A comparison with the deuteroporphyrin IX derivative (Deut MbCN) is shown in Figure 2. Many of the resonances are very similar in both compounds. Note, however, that the two resonances with the intensities of one proton each, labelled "f" in the MbCN spectrum do not appear in the Deut MbCN spectrum. Hence they arise from the CH protons of the vinyl groups at the 2 and 4 positions of the porphyrin which in Deut MbCN have been replaced by hydrogen atoms. The 2,4-hydrogen resonances are seen at very high fields in the Deut MbCN spectrum. Further identifications of the proton resonances are given in Figure 2. From their intensities and from comparison with the NMR spectrum of Deut MbCN one cannot distinguish between the resonances of several heme substituents which are of the same kind, for example the four ring methyl groups. Recently such distinctions have been made from theoretical calculations of the heme cyanide wave functions, (6) in which the electron distributions were adjusted to fit the observed spin densities.

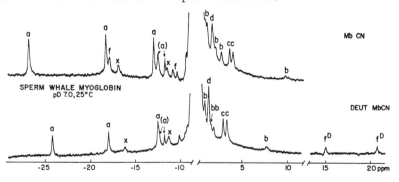

Figure 2. A comparison of the 220 MHz NMR spectra taken at 25° of native sperm whale cyanoferrimyoglobin (MbCN) and reconstituted cyanodeuteroferrimyoglobin (Deut MbCN). The assignments of the resonances to heme protons are: (a) ring methyls; (b) mesoprotons; (c,d) methylene protons of propionic acid groups; (f) –CH protons of the vinyl groups; (f^D) 2,4-protons in Deut MbCN; (x) tentatively assigned to the imidazole protons of the proximal histidine.

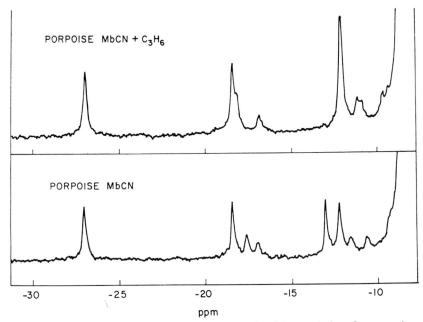

Figure 3. NMR spectrum in the low field region of porpoise cyanoferrimyoglobin (lower trace) and its cyclopropane derivative (upper trace).

Experimental support for these theoretical assignments comes from NMR studies (7) of the MbCN complex with cyclopropane whose spectrum is shown in Figure 3. Only some of the low field resonances of MbCN are affected by the addition of cyclopropane. In porpoise MbCN the methyl resonance at -2880 Hz at 26°C is shifted upfield by ca. 200 cycles/sec by the addition of cyclopropane so that it coincides with the other methyl resonances at -2695 Hz. This resonance which is moved by the presence of cyclopropane was assigned from the theoretical analysis, to the ring methyl which X-ray crystallography had shown (8) to be nearest to the cyclopropane site in crystalline myoglobin. After these assignments of the heme proton resonances the NMR data yield spin densities on the ring carbon atoms around the porphyrin ring which delineate the electronic wave functions containing the unpaired electron.

In addition to hyperfine shifted resonances, the NMR spectra of heme proteins also include resolved ring current shifted resonances, which, as Dr. Phillips mentioned yesterday, are structure sensitive because they depend upon the relative positions of the protons observed and the aromatic rings. In heme proteins the largest ring current shifts will usually come from the heme group. Unless there are conformational changes with temperature, ring current shifts are temperature independent, and can thus be distinguished from hyperfine shifts (1).

An interesting comparison of ring current shifted resonances, as they reflect conformational changes in a protein, is seen (9) with deoxymyoglobin and oxymyoglobin. In Figure 4 the upfield regions of the NMR spectra of Mb and MbO_2 are compared. In oxymyoglobin, which is diamagnetic, all the shifted resonances must come from ring currents. The weak, highest field resonance at +624 Hz has the intensity of one proton, and is independent of temperature. The resonance at +125 Hz in Mb has the intensity of three protons and is temperature independent. The two nearby resonances at slightly lower fields in Figure 4 are temperature dependent, and thus most likely come from hyperfine shifts. There are no resonances of corresponding intensity near +125 Hz in oxymyoglobin.

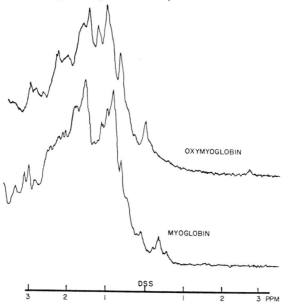

Figure 4. A comparison of the upfield region of the 220 MHz NMR spectra of sperm whale myoglobin (upper trace) and deoxy myoglobin (lower trace).

199

This implies that there are conformational changes upon oxygenation of myoglobin which affect these ring current shifted resonances.

Figure 5 compares the aromatic region of Mb and MbO_2. There is a resonance at -6.20 ppm in the spectrum of Mb which does not appear in the MbO_2 spectrum. This resonance has been assigned to the two protons in the meta-positions of phenylalanine CD1. In going from Mb to MbO_2, this resonance is shifted by at least 0.2 ppm.

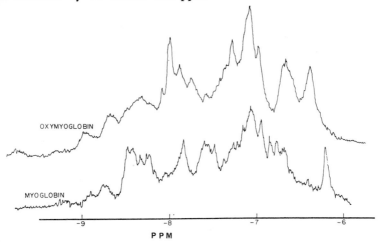

Figure 5. A comparison of the aromatic region of sperm whale oxymyoglobin (upper trace) and deoxymyoglobin (lower trace).

In order to calibrate this shift, we have plotted in Figure 6 the ring current shifts vs. positions observed by Caughey (10) and Kenny (11) in model systems of porphyrins and similar compounds. Using Kendrew and Watson's coordinates, we have determined that a change of the ring current shift of 1 ppm corresponds to a change in the relative positions of the phenylalanine ring and the heme group of approximately 1 A. Therefore, a change in the ring current shift of 0.2 ppm, which we have observed, corresponds to a displacement of approximately 0.2 Å of the phenylalanine CD1, if one assumes that the two rings are parallel. If the rings do not remain parallel, the displacement would have to be somewhat larger than 0.2 Å to yield a shift of the proton NMR of 0.2 ppm. Hence high resolution NMR shows that there are conformational changes upon oxygenation of myoglobin.

200

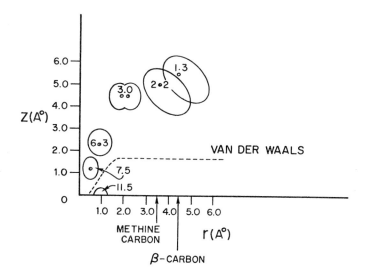

Figure 6. A plot of the ring-current shifts measured (10,11) in porphyrin and phthalocyanine complexes where the positions can be deduced from the chemical formulas. The estimated uncertainties in the positions are indicated by the ellipses and are larger than the errors introduced in deriving the ring current shifts from the measured resonances. The coordinates z and r refer to the distances perpendicular to the plane of the ring and radially outward in the plane of the ring, respectively. The van der Waals thickness of the porphyrin π-electron system is indicated by the dashed line.

References

1. K. Wüthrich, R. G. Shulman and J. Peisach, Proc. Nat. Acad. Sci. (U.S.) 60, 373 (1968).

2. N. Bloembergen, J. Chem. Phys., 27, 595 (1957).

3. K. Wüthrich, R. G. Shulman, B. J. Wyluda and W. S. Caughey Proc. Nat. Acad. Sci. U.S., 62, 636 (1969).

4. K. Wüthrich, R.G. Shulman, T. Yamane, B.J. Wyluda, T. Hugli and F.R.N. Gurd. J. Biol. Chem., 245, 1947 (1970).

5. R. G. Shulman, K. Wüthrich, T. Yamane, E. Antonini and M. Brunori. Proc. Nat. Acad. Sci. U.S., 63, 623 (1969).

6. R. G. Shulman, S. H. Glarum and M. Karplus. J. Mol. Biol., in press (1970).

7. R. G. Shulman, J. Peisach and B. J. Wyluda. J. Mol. Biol., 48, 517 (1970).

8. B. P. Schoenborn, Nature 214, 1120 (1967).

9. R. G. Shulman, K. Wüthrich, T. Yamane and W. E. Blumberg (to be published).

10. W. S. Caughey and P. K. Iber, J. Org. Chem., 28, 269 (1963).

11. J. N. Esposito, J. E. Lloyd and M. E. Kenney, Inorg. Chem., 5, 1979 (1966).

12. J. C. Kendrew and H. C. Watson (private communication).

DISCUSSION

Koenig: Why do you say that the average between the two states is thermal rather than quantum mechanical, and how fast is the exchange?

Shulman: We get a beautiful fit to the experiment by taking a linear combination of the squares of the coefficients for each atom, which can be a thermal mixture. For a quantum mechanica mixture, you must take the squares of the sum of the coefficients, rather than the sum of the squares. The former does not fit. The experiments indicate that the exchange between the two states is faster than 10^5 sec^{-1}

Chance: Can NMR distinguish motion of the iron from twisting of the heme plane?

Shulman: That is hard to do at present. I think that as we come to know more about the effect of axial ligands, all this can be correlated with the EPR, so that we can use all that data as well in studying the effect of different axial ligands

Phillips: Would you comment on the differences in the NMR characteristics of high-spin and low-spin heme systems?

Shulman: The spectra look very different. However, instead of just showing you that one spectrum looks very different

from another, I tried to confine my attention to those differences that we are absolutely sure come from conformational
changes, because as you go from one spin state to another
there are changes in magnetic moments of the Fe which give
different dipolar effects upon the protons. In the spectra
and the data I presented, I have tried to emphasize the parts
which are due only to conformational changes. They are extensive, and in the particular case of oxy to deoxy myoglobin,
one is the phenylalanine CD1 moving by 0.2 Å.

Phillips: How certain are you that there are no pseudo-
contact contributions to the contact shifts that you observe
in the paramagnetic heme system?

Shulman:* Although at the time this paper was presented we
had no evidence for appreciable contributions of pseudo
contact interactions to the shifts, subsequent studies of
MbCN in H_2O (1) showed too many shifted resonances to be explained by contact shifts alone. The contributions of the
pseudo contact interactions to the observed shifts have been
evaluated in the analysis of the data (2).

Kretsinger: Dr. Shulman noted the general similarity and detailed differences between the porpoise and the sperm whale
myoglobin NMR spectra. Crystallographic studies also show
this similarity. About ten years ago, Kraut (personal communication) showed that the space group and unit cell dimensions of the two were identical, i.e., they crystallized isomorphous to one another. I collected the three-dimensional
data to 2.0 Å resolution and calculated the porpoise Fourier
synthesis, using the sperm whale phases and porpoise intensities -- $\rho(x) = \sum_h F_{h(por)} \cdot \cos[hx + \phi_{h(s.w.)}]$

I judge the electron density map to be accurate, since I
detected changes in amino acid sequence, such as gly - ala,
which were subsequently confirmed. Bradshaw, Gurd, and I
completed the amino acid sequence determination, and found
differences, nearly all at the surfaces of the molecule.

To summarize the X-ray studies, in no place did the main
chain shift more than 0.4 Å; in some places it did move a
significant amount, some 0.2 Å.

*Added in proof November 18, 1970.

203

The sperm whale and porpoise myoglobin structures are now being refined by Diamond, using a program which fits standard bond lengths and angles to the calculated electron density.

References

1. Sheard, B., T. Yamane and R.G. Shulman. J. Mol. Biol., 53, 35 (1970).

2. Shulman, R.G., S.H. Glarum and M. Karplus. J. Mol. Biol., in press.

3. Bradshaw, R.H., R.H. Kretsinger, and F.N.R. Gurd. J. Biol. Chem., 244, 2159 (1969).

NUCLEAR RELAXATION STUDIES OF DISSOLVED AND CRYSTALLINE METHEMOPROTEINS[*]

A.S. Mildvan, N.M. Rumen and B. Chance

Johnson Research Foundation, University of Pennsylvania
and Department of Dermatology,
Medical College of Georgia, Augusta, Georgia

We have been interested in the mechanism of ligand exchange reactions on dissolved and crystalline hemoproteins
(1,2). As mentioned before (3), kinetic and structural information can be obtained by measuring the relaxation rates
of ligands. We have measured the relaxation rate of water
protons (at 24.3 MHz) and ^{17}O (at 8.1 MHz) and of the fluoride ligand (at 56.4 MHz) as a function of temperature in
solutions and crystalline suspensions of various methemoproteins. The following problems have been considered.

The Nature of the Sixth Ligand on Acid Methemoproteins.
To determine whether the sixth ligand on acid methemoproteins
was water or a strongly hydrogen bonded hydroxyl group (1,2),
we compared the effects of various acid and alkali methemoproteins on the proton relaxation rate of water at 24.3 MHz.
From the exchange limited region we concluded that protons
exchange into the coordination sphere of acid methemoproteins
(seal, sperm whale metmyoglobin, lamprey methemoglobin) at a
rate of $\sim 10^4$ sec^{-1} (1,2). The alkaline forms exchange protons 56-78% slower in solution and 49-59% slower in crystalline suspension without a significant change in the energy
of activation (2) (Figure 1).

This finding is consistent with fewer protons available

[*]
This work was supported in part by U.S. Public Health Service
Grants Am-13351, GM-12246, CA-06927, and FR-05539, by National Science Foundation Grant GB-8579 and by an appropriation
from the Commonwealth of Pennsylvania. This work was done
during the tenure (A.S.M.) of an Established Investigatorship from the American Heart Association.

[**]Present address, Institute for Cancer Research, Fox Chase,
Philadelphia, 19111.

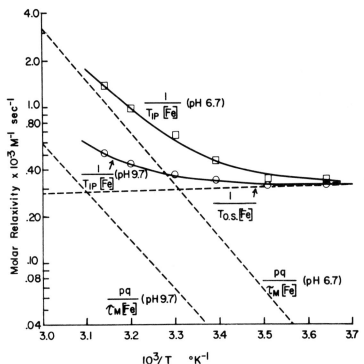

Figure 1. Arrhenius plot of the affect of temperature on the longitudinal molar relaxivity of water protons in solutions of acid metmyoglobin and alkali metmyoglobin from seal. The data are fit by the summation of an outer sphere contribution $(1/T_{o.s.} [Fe])$ and an exchange contribution $(pq/\tau_m [Fe])$. The energies of activation in Kilocalories per mole are: $1/T$ o.s. [Fe] = 0.4; pq/τ_m [Fe] (acid form): 15.2 ± 1.8; pq/τ_m [Fe] (alkaline form): 14.6 ± 0.7.

for exchange in the alkaline forms and supports the ligand structure proposed by Stryer et al.(4) in which the acid form coordinates a water molecule and the alkaline form coordinates a hydroxyl ion.

Nature of the Chemical Rate Process Measured by Proton Relaxation. We believe this process to be the escape (k_{off}) of protons for several reasons:

$$Fe-O\overset{H}{\underset{H}{\diagdown}} \underset{\longleftarrow}{\overset{k_{off}}{\rightleftharpoons}} Fe-O_{\diagdown H} + H^+$$

a) The reverse process (k_{on}) is probably diffusion controlled. Below 12° the relaxation rate (Figure 1) has become slower than outer sphere relaxation ($1/T_{o.s.}$). The latter process is a measure of the effect of the Fe on the relaxation rate of protons which do not gain access to the coordination sphere. From $1/T_{o.s.}$ we can calculate the distance of closest approach to Fe of the water protons (5). Using a steric factor equal to the heme area/surface area of the molecule and the correlation time of $1 \times 10^{-10} < \tau_s < 2 \times 10^{-10}$ sec, we obtain a distance of 2.8 to 3.5 A. Since a coordinated water is 2.8 A thick, these results indicate that protons can diffuse through a channel to the second coordination sphere. Hence, k_{on} is probably very rapid, and therefore does not limit the rate of proton exchange.

b) Below pH 6.7 the proton exchange rate is independent of proton concentration. The absence of acid catalysis in this process argues against it being the exchange or escape of entire water molecules.

The rate constant for seal metmyoglobin (2.4×10^4 sec^{-1} at 25°) as determined by relaxivity agrees, within a factor of two, with the rate constant required by Ilgenfritz (6) to fit his kinetic data using a totally different technique, field jump.

c) No effect of 0.5 to 6.3 mM acid metmyoglobin on the ^{17}O resonance of water (natural abundance and 1.06% enriched) was observed at temperatures up to 40° (2), suggesting a slow exchange of entire water molecules. However, our experimental variability in a large number of line width determinations of the broad ^{17}O resonance corresponds to a transverse relaxation rate $< 10^5$ sec^{-1} while our measured proton exchange rates are 10^4 sec^{-1}. Hence, the results of the experiments using $H_2^{17}O$ are inconclusive. Therefore the fluoride resonance experiments were carried out.

The Stepwise Nature of the Acid to Alkali Conversion.
The van't Hoff plot of Figure 2 indicates that the pK of seal metmyoglobin as detected by nuclear relaxation (for the conversion of the slowly exchanging to the rapidly exchanging form) is independent of temperature while the spectroscopic pK for the conversion of the brown to red form is dependent on the temperature (ΔH = 9 Kcal/mole). This suggests that different processes are being observed by the two techniques and a minimal scheme of the following type is required to

accommodate the observations (2,8):

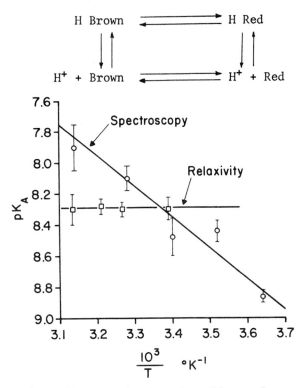

Figure 2. Van't Hoff plot showing the effect of temperature on the pK of the conversion of the acid to alkaline forms of seal metmyoglobin as determined by optical spectroscopy (ΔH = 9 K cal/mole) and by relaxivity (ΔH < 1.5 K cal/mole.)

The Fluoride Complex. The Arrhenius plot in Figure 3 compares the proton relaxation rate of solutions of acid met-myoglobin, the fluoride complex, and the azide complex. Because of its magnitude and temperature dependence, the relaxivity of the fluoride complex appears to be in the region of rapid proton exchange, i.e. $1/pT_{1p} = 1/T_{1M}$.

If this is the case, then protons exchange into the environment of iron in the fluoride complex at a rate which is greater than 2.4×10^5 sec^{-1}. Similar observations have been made by Fabry and Koenig for the fluoride complex of methemoglobin (personal communication).

208

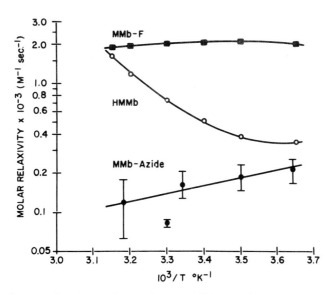

Figure 3. Arrhenius plot of the effect of temperature in the
longitudinal molar relaxivity of the protons of water in solutions of acid metmyoglobin (HMMb), the fluoride complex
(MMb-F), and the azide complex (MMb-Azide).

The distance between this proton and the iron in the
fluoride complex (calculated from the Solomon Bloembergen
equation assuming $1.0 \times 10^{-10} < \tau_c < 2.0^{-10}$ sec) is $2.9 \pm$
0.1 A. This value is in agreement with the value expected
for the distance of a proton which is hydrogen bonded to the
fluoride ligand (2.85 - 3.35 A) but is somewhat greater than
the value expected for a proton covalently bound to the fluoride ligand (2.68 - 2.74 A), as determined by crystallographic
studies on small complexes.

We have recently found the effect of metmyoglobin-F on
the transverse relaxation rate of water $(1/T_{2p})$ to be identical to its effect on $(1/T_{1p})$ over the same range of temperatures (G.L. Cottam and A.S. Mildvan, unpublished observations).
Hence, the hyperfine contribution to $1/T_{2p}$ is negligible as
would be expected for a hydrogen bond. Hopefully, Dr. Schoenborn's neutron diffraction data on metmyoglobin may provide
us with this distance on the same complex in the crystalline
state.

A study of the relaxation rates of the fluoride ion in
solutions and crystalline suspensions of the metmyoglobin

fluoride complex by ^{19}F NMR was made (2) (Figure 4). With the exception of the crystalline alkaline form, metmyoglobin broadened the NMR line width of the fluoride ion, and increased the RF power necessary to saturate the resonance. Chance and Rumen found the alkali crystals to be inert to substitution by fluoride ion (8).

The paramagnetic contributions to the transverse $(1/pT_{2p})$ and longitudinal $(1/pT_{1p})$ relaxation rates of fluoride were of the same order of magnitude $(3.6 \pm 1.9 \times 10^3$ sec^{-1} at 26°C) and increased with temperature, suggesting that they reflected the rate of exchange of fluoride into the coordination sphere of iron.

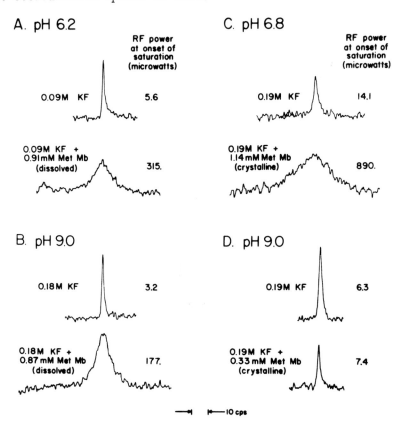

Figure 4. Effect of dissolved (A,B) and crystalline (C,D) metmyoglobin on the ^{19}F NMR spectrum of the fluoride ion at low (A,C) and high (B,D) pH.

210

Moreover, $1/pT_{2p}$ doubled when the concentration of fluoride was doubled, suggesting that fluoride displacement by fluoride on iron proceeded by way of an S_N2 mechanism with a rate constant of $3.5 \times 10^4 M^{-1} sec^{-1}$. An S_N2 mechanism might involve a Griffiths type septa-coordinate transition state (9) for fluoride. Regardless of the mechanism, fluoride exchange into the coordination sphere of iron (at 0.1M [F]) is two orders of magnitude slower than proton exchange.

The Azide Complex. The effect of the low spin metmyoglobin-azide complex on the proton relaxation rate of water decreases with increasing temperature, but its magnitude is small (Figure 3). Such behavior is consistent with either rapid proton exchange or, possibly, with outer sphere relaxation. Outer sphere relaxation is unlikely because of our observation that the magnitude of $1/T_{2p}$ is approximately twice that of $1/T_{1p}$.

Hence, as with fluoride we may calculate that protons exchange into a position which is 3.1 ± 0.1 Å from the iron at a rate greater than $1.7 \times 10^4 sec^{-1}$. The crystallographic distance from the iron to the proton, calculated from the published locations of the azide and water ligands in sperm whale metmyoglobin (4,10) is 2.7 Å if the proton is covalently bonded to the α nitrogen of the azide ligand and 3.6 Å if it is hydrogen bonded. Hence, our distance calculations and the inequality of $1/T_{1p}$ and $1/T_{2p}$ suggest covalent bonding of a proton to the azide ligand in agreement with Beetlestone's conclusions from thermodynamic data (11).

Effect of Crystallization On the Structure and Reactivity of Hemoproteins. In a study of several methemoproteins (seal and horse metmyoglobin and horse and lamprey methemoglobin) in the dissolved and crystalline state, (2) it was observed that crystallization produces small reductions on the effect of the iron on the proton relaxation rate of water.

Decreases of the outer sphere contribution to the relaxation rate (19-49%) suggest small increases (6-15%) in the distance of closest approach of water protons to iron (1). Decreases in the water proton exchange (or escape) rate by 30-70% occur despite a lower energy of activation for this process. (Figure 5). Hence, crystallization introduces an entropy or statistical barrier to proton escape (1,2). The slower proton exchange observed in dissolved alkaline hemoproteins is also observed in crystalline suspensions (Figure 5), consistent with fewer protons in the coordination sphere of iron at high pH.

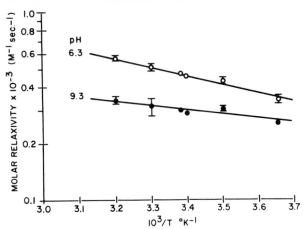

Figure 5. Arrhenius plot of the effect of crystalline suspensions of seal metmyoglobin on the longitudianal molar relaxivity of the protons of water. The energies of activation, in kilocalories per mole are at pH 6.3, 1.8 ± 0.4 and at pH 9.3, 1.3 ± 0.3.

References

1. Maricic, S., A. Ravilly, and A.S. Mildvan. in Hemes and Hemoproteins (B. Chance, R. Estabrook, and T. Yonetani, eds.) Academic Press, New York, 1966, p. 157.

2. Mildvan, A.S., N. Rumen, and B. Chance. Fed. Proc., 27, 525 (1968) and Abs. 156th American Chem. Soc. Meeting, Biol., 32 (1968).

3. Mildvan, A.S. This symposium.

4. Stryer, L., J.C. Kendrew, and H. Watson. J. Mol. Biology, 8, 96 (1964).

5. Luz, Z. and S. Meiboom. J. Chem. Phys., 40, 2686 (1964).

6. Ilgenfritz, G. This symposium.

7. Pauling, L. The Nature of the Chemical Bond, (third edition), Cornell University Press, Ithaca, N.Y., 1960, p. 460.

8. Chance, B. and N.M Rumen. Abs. VII Int'l. Congress of Biochemistry, A 85 (1967).

9. Griffith, J.S. Proc. Roy. Soc. A 235, 23 (1956).

10. Kendrew, J.C., R.E. Dickerson, B.E. Strandberg, R.G. Har and D.R. Davies. Nature, 185, 422 (1960).

11. Beetlestone, J. This symposium, p. 267

DISCUSSION

Koenig: Dr. Cohn has presented equations due to Bloembergen and Solomon and ideas due to Luz and Meiboom (1) and she considered them adequate to explain the effect of paramagnetic probes on the enhancement of the relaxation rate of solvent protons in protein solutions. You are taking the same view. These equations predict a very specific functional form for the frequency dependence of the relaxation enhancement. As you know, we have been investigating the frequency dependence and temperature dependence of this effect for some time, over the Larmor frequency range of 10 kHz to 50 MHz. Along with several colleagues (T. Fabry, M. Fabry, J. Kim, and W. Schillinger) we have studied iron-containing transferrin, cobalt-substituted carbonic anhydrase, and various hemoglobin derivatives in great detail. Though the complete results will be reported elsewhere, the following generalization holds in all cases we have studied: over the approximate temperature range from 0° to 30°, and at the frequencies that you have used, the major part of the observed relaxation enhancement cannot be reconciled with the Solomon-Bloembergen-Luz-Meiboom view that you and she have invoked.

Mildvan: Have you tried seal metmyoglobin?

Koenig: Not yet

References

1. Luz, Z. and S. Meiboom. J. Chem. Phys., 40, 2686 (1964).

213

A UNIFIED THEORY FOR LOW SPIN FORMS OF ALL FERRIC HEME PROTEINS AS STUDIED BY EPR*

W.E. Blumberg and J. Peisach

Bell Telephone Laboratories, Inc.
Murry Hill, New Jersey 07974
and
Departments of Pharmacology and Molecular Biology
Albert Einstein College of Medicine, Yeshiva University
Bronx, New York 10461

All heme proteins either occur naturally as, or can be converted to, ferric low spin forms amenable to EPR spectroscopy. Ferric low spin forms of heme proteins are compounds in which the heme iron has six covalently bonded ligand atoms, four from the porphyrin (as in all heme proteins), and two, designated as Z ligands, which are either supplied by amino acids of the protein, or by exogenous ligating groups. Optical spectra in the visible region of all these compounds resemble one another, and their identification by optical criteria, especially in samples containing more than one low spin compound, becomes exceedingly difficult. EPR spectroscopy, as we shall see later, differentiates among the different low spin compounds, and in fact, conveys information concerning the geometry and strength of the chemical bonds of the non-porphyrin Z ligands.

The forte of the EPR method in the studies of low spin ferric heme compounds extends only to the immediate ligand environment of the iron, and the technique conveys no information concerning subtle changes of the protein not immediatly affecting the ligand field of the iron. Substitutions

*First presented at the Third International Congress of Magnetic Resonance in Biology, Airlie House, Warrenton, Virginia, October 14-18, 1968. The portion of this investigation at the Albert Einstein College of Medicine was supported in part by NIH GM-10959 and 1-K3-GM-31156 to J.P. This is communication no. 168 from the Joan and Lester Avnet Institute of Molecular Biology.

of one Z ligand for another, accompanying gross changes in protein configuration, are readily sensed by EPR, and we shall show how different changes in protein structure, which in turn affect the nature of the iron ligands, can be distinguished one from another. Furthermore, since the number of possible ligand atoms which are endogenous to the protein and which can bind iron is limited, EPR can be used to identify all of them as each has a particular spectroscopic signature. For both exogenous as well as endogenous ligands, EPR is useful in ascribing structure to low spin ferric heme compounds.

All ferric low spin compounds have spin = 1/2, and the interaction of the spin with an external magnetic field can be completely described by three g values. In the case of the ferric heme proteins, these three g values are all different, indicating that the heme of these compounds has less than axial symmetry. Figure 1 shows that for a frozen sample of a heme protein in solution, where the molecules are randomly oriented, the three g values can be directly determined from three features of the EPR absorption derivative spectrum: an end maximum, an end minimum, and a midpoint crossing (1). Since the EPR spectrum is a broadening convolution over a second rank tensor interaction, a knowledge of the three g values and the broadening function of any one of the features is sufficient to define the whole spectrum, and thus the total number of low spin heme iron atoms in the sample. One can then easily quantitate the concentration of low spin compounds in a sample even if the

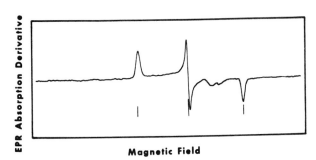

Figure 1. EPR spectrum of isolated ferric alpha chains of hemoglobin, pH 8.7. The three g values are 2.56, 2.18 and 1.88 (after ref. 2). The vertical lines show where the three g values are measured on the low spin EPR spectrum.

216

sample is heterogeneous and contains a mixture of compounds, each with its own set of three g values.

For a pure ferric low spin compound, the wave function describing the spin is constituted solely of a combination of the three ε d orbitals (3). In such a case, the three g values are inter-related in such a way that any two determine the third. This can be used to check on the assignment of an EPR spectrum or spin state to a specific compound. For these compounds, the g values are determined completely by a formalism involving two crystal field components. These can be specified arbitrarily as an element of tetragonal symmetry and one of orthohombic symmetry. The g values can be used to determine the ratios of the coefficients of these elements (Δ and V, respectively) to the spin orbit coupling energy, λ. There is a unique correspondence between sets of three g values and sets of two dimensionless ratios, Δ/λ and V/λ. Depending on the choice of an axis system, there may be several numerically different but geometrically equivalent pairs of these ratios. We have adopted one self-consistent set of axes for all the compounds studied. The ratio Δ/λ, called the tetragonality (Figure 2), depends primarily on the charge of the iron atom. This in turn is determined by the sum of the electronegativities of the two Z ligands, the electron donation by the porphyrin ligands remaining constant. The greater the electron donation to the iron atom, the smaller the tetragonality, Δ/λ. The ratio V/λ, called the rhombicity, is purely a function of the geometrical arrangement of the ligand atoms and the π bonds between them and the iron atom. Thus, the rhombicity is an indication of whether different low spin compounds have similar geometry or not.

In Figure 2 we present the results of the analysis of EPR data for a series of low spin derivatives or hemichromes of hemoglobin. It can be seen that the low spin forms can be classified by the types P, O, H, B, and C, and each type is set off with its own characteristic range of crystal field parameters, all derived from sets of three g values measured by EPR. We have not included analyses based on only two g values, as these are somewhat less accurate.

Hemoglobin A, when isolated and also in the red cell, exists in the diamagnetic oxy-form. Oxidation by ferricyanide converts it to the high spin ferric form, ferrihemoglobin [Hb$(d_{5/2}^2)$H$_2$O)] (6) which has a characteristic EPR absorption

217

with g values near 6, 6, and 2. The sixth ligand of iron in this compound is oxygen from a water molecule which occupies the position at the heme originally occupied by oxygen in oxyhemoglobin. At elevated pH, a proton is dissociated from the water molecule, yielding a negatively charged oxygen atom of hydroxide ion at the heme. This proton dissociation

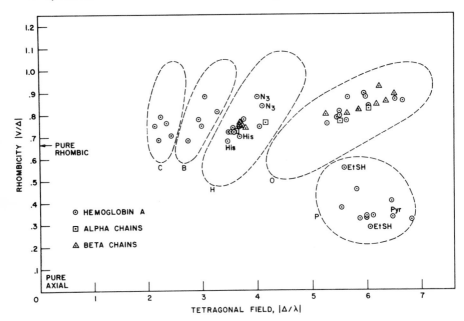

Figure 2. Crystal field parameters for ferric low spin compounds of hemoglobin A and its isolated constituent chains. Hemoglobin A was isolated from human red cell hemolysates without the use of toluene. Isolated α-chains were prepared from oxyhemoglobin by the method of Bucci and Fronticelli (4) as modified by Parkhurst, Gibson and Geracci (5). β-chains were obtained as oxyhemoglobin H. Oxidation to the ferric form was performed with ferricyanide on a Biogel P-2 column at pH 7.

Unless otherwise indicated, all points in this figure are for compounds formed from ligands endogenous to the hemoglobin molecule. Low spin compounds can be formed using the following reagents referred to in the figure: EtSH, mercaptoethanol; Pyr, pyridine; His, histidine; N_3, azide.

The five areas enclosed by the dashed lines define the regions where parameters for the five different compounds may be expected to lie.

effectively increases the strength of the cubic part of the ligand field of the heme iron, changing the spin state of iron from high spin to low spin. This low spin form, designated as the O type $[Hb(d_{1/2}^5) - OH]$ can be prepared from ferrihemoglobin A tetramers, from ferric α-chains or from ferric β-chains. In all instances, the ligand arrangement of the heme iron is approximately the same, and there is no suggestion from our data that differences in protein moieties affect the electronic structure of these compounds as measured by EPR. This is in contrast to our findings (2) that there is a difference between the electronic structure of the heme of ferric α-chains compared to ferrihemoglobin A. Since the hydroxide ion in the O-type low spin form occupies the same position in space as did the oxygen in oxyhemoglobin, no conformational change of the protein need take place to form this paramagnetic compound. Thus the geometry of the iron and its five nitrogen atom ligands remains essentially undistorted from that of the native protein.

It is well known that azide binds to ferrihemoglobin, yielding a low spin compound $[Hb(d_{1/2}^5) - {}^-N_3]$. Histidine also binds to ferrihemoglobin, yielding a compound in which the electronic environment of the heme is very similar. This is not too surprising, since in both cases the new sixth ligand replacing oxygen of water is an imino nitrogen atom. We designate these compounds collectively as the H-type low spin form. A similar derivative of hemoglobin can be produced with an endogenous sixth ligand contributed by the protein. This can be prepared as a spontaneously formed compound from ferrihemoglobin A or, more quickly, from ferrihemoglobin H, a β-chain tetramer (7). It is believed that this compound involves a nitrogen atom of the distal histidine (E7) of hemoglobin, which, in the native protein, is at a distance too far removed to bind to the iron (8). Through a process of hydrogen bond breaking, the tertiary structure of the protein is sufficiently altered so that now this distal histidine can bind to the iron to form a dihistidine low spin form. The compound formed from ferrihemoglobin H is no different from that formed from isolated ferric alpha chains or from ferrihemoglobin A. Here again, EPR is insensitive to differences in protein structure. As the members of the class of H-type low spin compounds have the same rhombicity as the O-type, this suggests that the five original nitrogen ligands of the heme remain the same and determine the symmetry of the iron ligand field in the same way.

Yet another low spin form, designated as the C-type, can also be observed in low concentrations from ferrihemoglobin A under appropriate conditions. This compound, as a majority component low spin form (9), can be obtained from HbRiverdale (10), a mutant hemoglobin where an amino acid substitution of an arginine has been made for glycine in the β-chains at position 24, far removed from the heme. The effect of this substitution is to weaken the stabilizing forces between the B and E helical regions, decreasing the rigidity as well as the stability of the molecule. The EPR parameters for the C-type low spin form are the same as those for cytochrome c (11) in which the fifth and sixth ligands of iron are a nitrogen atom of histidine and a sulfur atom of methionine, respectively (12). It is therefore assumed that the ligand environment of the iron in the C-type low spin form of hemoglobin is the same as that for cytochrome c. In this case also, the five nitrogenous ligands are the same as in the H- and O-type low spin forms.

Another low spin form, B-type, which can be prepared as a majority species from hemoglobin incubated with salicylate (13), has the same EPR parameters reported for cytochrome b_2 (14) and b_5 (15). In the case of the B-type, the sixth ligand remains unidentified at present, but from the analysis of EPR data, two facts can be deduced:

1) The five original nitrogen ligand atoms are the same as in the O-, H-, and C-types.

2) The sixth ligand has an electronegativity lying between that of a nitrogen atom of histidine and the sulfur atom of methionine.

The last low spin form that can be prepared from endogenous ligands of the protein is the P-type. This material can be made by incubating ferrihemoglobin A in 5 M urea, or by incubation of the protein at very high pH. A compound having identical EPR spectra can be made by adding a thiol, such as mercaptoethanol, to ferrihemoglobin or by adding a very large excess of pyridine. In the latter case, the pyridine does not bind to the heme but instead raises the pH of the solution. Pyridine, buffered at pH 7, is not effective in converting ferrihemoglobin to a low spin form. We believe that a necessary Z ligand for the P-type low spin form is a mercaptide sulfur atom. In the case of the compound prepared with urea or at high pH, it is believed that gross denaturation of the tertiary structure of the protein must

220

take place so that a cysteine on the proximal side of the heme can now be brought close enough for bonding to iron. Evidence that a sulfur atom is a ligand in the P-type low spin form of hemoglobin is based upon the following observations:

1) The same compound can be prepared by addition of a thiol.

2) The P-type low spin form cannot be formed with sperm whale myoglobin, a protein that does not contain cysteine.

3) Mercurial-treated hemoglobin forms the P-type low spin form at a much slower rate.

As can be seen in Figure 2, the rhombicity of the P-type form is different from all the others. This is not surprising in view of the fact that a sulfur atom has replaced a nitrogen atom on the proximal side of the heme. As sulfur can make δ bonds with iron, this imposes a different geometry on the iron ligand field.

The sum of the electronegativities of the two Z ligands in the P-type low spin compound is approximately the same as that of hydroxide oxygen plus imidazole nitrogen, as in the O-type low spin compounds, giving a tetragonal field of 6λ. As one would expect the electronegativity of a mercaptide group to be greater than that of either hydroxide oxygen or imino nitrogen, it is concluded that the P-type low spin compound does not simultaneously involve both a mercaptide sulfur atom and either of these two other ligand atoms. It is likely that a water molecule completes the coordination sphere of the iron in the P-type form. Further aspects of this problem, involving the addition of thiols to heme proteins, will be discussed elsewhere.

The clinical implications of the formation of precipitated low spin forms (Heinz bodies) in erythrocytes of patients suffering from α-thalassemia has been studied by EPR of erythrocyte ghosts, and has been discussed elsewhere (16). Unlike optical spectroscopy, where the presence of low spin forms is often masked by the presence of oxyhemoglobin, EPR can be used very effectively to study small quantities of paramagnetic material in the presence of a diamagnetic heme protein.

Although each of these five low spin forms of hemoglobin occupies a small spread in the two crystal field parameters, one can take an average value of each of them over the distribution characteristically typical of the group. The g values of the EPR spectra corresponding to these characteristic crystal fields are given in Table I.

TABLE I

Characteristic g Values for the Endogenous Low Spin Compounds of Hemoglobin

Low-spin Type	g_1	g_2	g_3
O	2.55	2.17	1.85
H	2.80	2.26	1.67
C	3.15	2.25	1.25
B	2.95	2.26	1.47
P	2.41	2.25	1.93

A similar analysis of the g values of low spin compounds of various heme proteins which have been measured in this laboratory or elsewhere is shown in Figure 3.

The points for hemoglobin cyanide (Hb-CN) and cytochrome c cyanide (C-CN) are almost identical, indicating that the same ligand environment of the heme exists in both compounds. These points are far outside any of the dashed regions, since cyanide is so different from any of the endogenous ligands.

As mentioned before, the points for Hb$_{Riverdale}$ and cytochrome c are so close that it is concluded that identical environments for iron must exist in both compounds.

The points for cytochromes b$_2$ and b$_5$ in the pH ranges in which the proteins are reported to be native, fall in the B-type class, also signifying an identical ligand arrangement. Cytochrome c at pH 13 also shows a minority species falling into this class (point not shown; cf ref. 28).

The azide compounds of the catalases, cytochrome a$_3$, myoglobin (point not shown, cf ref. 29) and Hb M$_{Boston}$ (HbM$_B$) lie in the same region as hemoglobin-azide (Figure 2). Cytochromes b$_2$ and b$_5$ at pH 12 fall in the same class, suggesting that at this pH these compounds are dihistidine low spin forms.

Cytochrome c peroxidase, myoglobin, cytochrome a$_3$, and probably chloroperoxidase can form hydroxides which resemble those formed by hemoglobin. Cytochrome c at pH 13 also shows

TRUTH DIAGRAM NO. I

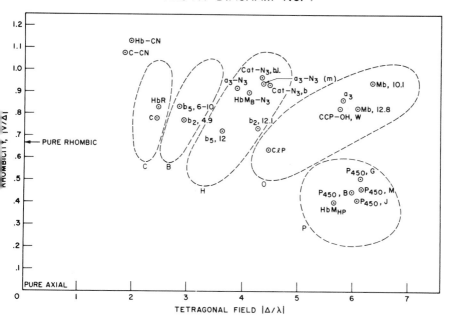

Figure 3. Crystal field parameters for ferric low spin forms of various heme proteins. The dashed lines represent the same regions as in Figure 2.

The abbreviations used are: P-450, J, P-450, M, rabbit liver microsomal cytochrome P-450 (17,18); P-450, B, rat liver microsomal P-450 (19); P-450, G, bacterial cytochrome P-450 (20); HbM$_{HP}$, Hemoglobin M$_{Hyde Park}$ (21), ClP, chloroperoxidase (22); CCP-OH, W, cytochrome c peroxidase (23); Mb, 12.8, sperm whale myoglobin, pH 12.8 (24); Mb, 10.1, sperm whale myoglobin, pH 10.1 (24); a$_3$, cytochrome a_3 (25); b$_5$, 12, cytochrome b_5, pH 12.1 (15); b$_2$, 12.1, cytochrome b_2, pH 12.1 (14); HbM$_B$-N$_3$, Hemoglobin M$_{Boston}$ azide (21); a$_3$-N$_3$, cytochrome a_3 azide (26); Cat-N$_3$, b, horse erythrocyte catalase azide (27); a$_3$-N$_3$ (m), cytochrome a_3 azide, minority component (26); Cat-N$_3$, b.l., beef liver catalase azide (28); b$_2$, 4.9, cytochrome b_2, pH 4.9 (14); b$_5$, 6-10, cytochrome b_5, pH 6 to 10 (15); c, cytochrome c (11); Hb$_R$, Hemoglobin$_{Riverdale}$ (9); C-CN, cytochrome c cyanide (28); Hb-CN, ferrihemoglobin cyanide (28).

The analysis for ClP, C-CN, and Hb-CN are based on two g values, while all other points are based on three g values.

223

a similar hydroxide form (point not shown; cf ref. 28).

The various preparations of cytochrome P-450 all fall in the same class as the mercaptide low spin form of hemoglobin (Fig. 2) indicating that mercaptide sulfur is a ligand for this cytochrome. The only low spin compound reported for Hemoglobin M$_{Hyde Park}$ (HbM$_{HP}$) lies in this group. Both HbM$_{HP}$ and HbM$_B$ differ from hemoglobin A in that there is a substitution of tyrosine for histidine in the abnormal β-chain, proximal to the heme in the former and distal to the heme in the latter. Obviously, neither protein can produce a dihistidine (H-type) low spin form with endogenous ligands. HbM$_B$, however, can form an H-type low spin form when an exogenous imine nitrogen ligand, such as from azide, is added. This compound is the same as is formed in the case of azide added to ferrihemoglobin A. For HbM$_{HP}$ to form a low spin compound, however, the tyrosine proximal to the heme must be replaced by a more electronegative ligand. By far the most likely replacement is the sulfur of cysteine (β 93) which is rather close to the heme (8).

Analyses of the EPR spectra of the low spin ferric forms of several heme proteins reveal compounds which are dissimilar to those produced by hemoglobin and the other heme proteins described in Figures 2 and 3. These results are shown in Figure 4. The addition of ethyl isocyanide to rabbit liver cytochrome P-450 slightly modifies the optical and EPR spectra for this heme protein. However, the analysis of EPR data for the isocyanide treated material shows that the crystal field parameters still lie in the region of the P-type low spin forms, with mercaptide as the ligand.

The g values for horse erythrocyte catalase cyanide yield a point on Figure 4 lying in the region of the H-type low spin form, and far removed from the Hb-CN and C-CN forms (Figure 3). This shows that either the catalase cyanide compound is in no way similar to those of hemoglobin and cytochrome c or that the material examined in this experiment was a dihistidine (H-type) low spin form.

The analysis of data for the hydroxide forms of three peroxidases - cytochrome c, horseradish, and Japanese radish yield points lying one region to the left of the hydroxide forms of hemoglobin (O-type, Figure 2). As the hydroxide oxygen must have the same electronegativity in all heme hydroxide compounds, the proximal ligand of the peroxidase must be

more electronegative than the proximal histidine of hemoglobin. Likewise, the point shown for Japanese radish peroxidase azide (and also for horseradish peroxidase azide, cf
ref. 28) is to the left of the azides of Figures 2 and 3,
also consistent with this hypothesis.

The data points for ferric horseradish peroxidase cyanide (HRP-CN) are displaced to the right in Figure 4 from
the position assumed in Figure 3 by the cyanide derivatives
of hemoglobin and cytochrome c. Thus, the iron-ligand structure of this cyanide compound is not the same as that of
hemoglobin or cytochrome c cyanide.

TRUTH DIAGRAM NO. 2

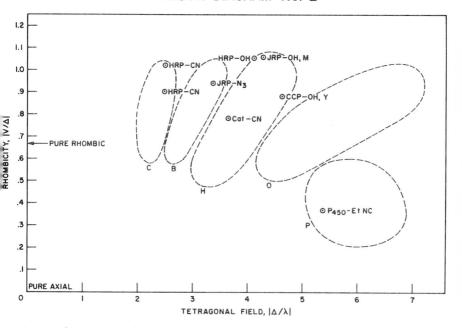

Figure 4. Crystal field parameters for ferric low spin forms
of various heme proteins which do not conform to the contours
of Figure 2: P_{450}-EtNC, rat liver cytochrome P-450 treated
with ethyl isocyanide (17); CCP-OH,Y, cytochrome c peroxidase hydroxide (30);Cat-CN, horse erythrocyte catalase cyanide (27); JRP-OH,M, Japanese radish peroxidase hydroxide
(31); HRP-OH, horseradish peroxidase hydroxide (32); JRP-N_3,
Japanese radish peroxidase azide (31); HRP-CN (upper) horse
radish peroxidase cyanide (28); HRP-CN (19234) horseradish
peroxidase cyanide (32).

Furthermore, the formation of HRP-CN from ferric horse-radish peroxidase and cyanide ion does not represent a simple addition of cyanide ion to the iron atom with a ligand environment as it exists in the native protein. Instead, the Z ligand contributed by the protein must be less electronega-tive than the Z ligand from the protein in both hemoglobin and cytochrome c cyanides.

Thus, we see that EPR is a useful tool in describing the nature of the immediate ligand field environment of heme iron in low spin compounds but is insensitive to subtle changes of the configuration of the protein moiety or even substitutions of one protein moiety for another. The number of low spin compunds that can be prepared from a heme protein with endo-genous ligands is a small number, as the number of suitable endogenous heme ligands is also a small number. A study of the interconversion of these low spin compounds provides a method for examining gross conformational changes accompany-ing heme protein reactions.

Acknowledgment

The authors would like to thank Dr. Eliezer Rachmilewitz for stimulating discussions, Dr. Helen Ranney for generous samples of hemoglobin H and Hemoglobin$_{Riverdale}$, and Miss Rhoda Oltzik for generous samples of hemoglobin A and for technical assistance.

References

1. Kneubühl, F.K.   J. Chem. Phys., $\underline{33}$, 1074 (1960).

2. Peisach, J., Blumberg, W.E., Wittenberg, B.A., Witten-berg, J.B., and Kampa, L.   Proc. Natl. Acad. Sci. U.S., $\underline{63}$, 934 (1969).

3. Griffith, J.S.   Nature, $\underline{180}$, 30 (1957).

4. Bucci, E., and Fronticelli, C.   J. Biol. Chem., $\underline{240}$, PC 551 (1965).

5. Parkhurst, L.J., Gibson, Q.H., and Geracci, G.   Personal communication.

6. Peisach, J., Blumberg, W.E., Wittenberg, B.A., and Wit-tenberg, J.B.   J. Biol. Chem., $\underline{243}$, 1871 (1968).

7. Rachmilewitz, E.A.   Second Symposium on Cooley's Anemia, Ann. N.Y. Acad. Sci., $\underline{165}$, 171 (1969).

8. Perutz, M.F., Muirhead, H., Cox, J.M., and Goaman, C.G. Nature, 219, 139 (1968).

9. Ranney, H., Peisach, J., and Blumberg, W.E. Unpublished observations.

10. Ranney, H., Jacobs, A.S., Udem, L., and Zalusky, R. Biochem. Biophys. Res. Commun., 33, 1044 (1968).

11. Salmeen, I., and Palmer, G. J. Chem. Phys., 48, 2049 (1968).

12. Dickerson, R.E., Kopka, M.L., Weinzierl, J.E., Varnum, J.C., Eisenberg, D., and Margolaish, E. In Structure and Function of Cytochromes (K. Okunuki, M.D. Kamen, and I Sekuzu, eds.), University Park Press, Baltimore, 1968, p. 225.

13. Rachmilewitz, E.A., Blumberg, W.E., and Peisach, J. Unpublished observations.

14. Watari, H., Groudinsky, O., and Labeyrie, F. Biochim. Biophys. Acta, 131, 592 (1967).

15. Bois-Poltoratsky, R., and Ehrenberg, A. European J. Biochem., 2, 361 (1967).

16. Rachmilewitz, E.A., Peisach, J., Bradley, T.B. and Blumberg, W.E. Nature, 222, 248 (1969).

17. Ichikawa, Y., and Yamano, T. In Recent Developments of Magnetic Resonance in Biological System (S. Fujiwara and L.H. Piette, eds.), Hirokawa Publishing Co., Tokyo, 1968, p. 108.

18. Miyake, Y., Gaylor, J.L., and Mason, H.S. J. Biol. Chem., 243, 5788 (1968).

19. Klein, M., Blumberg, W.E., and Peisach, J. Unpublished observations.

20. Gunsalus, I.C., Katagiri, M., Ganguli, B., and Yu, C.A. Presented at the Symposium on Membrane Function and Electron Transfer to Oxygen. Miami, January 20-24, 1969.

21. Watari, H., Hayashi, A., Morimoto, H., and Kotani, M. In Recent Developments of Magnetic Resonance in Biological System (S. Fujiwara and L.H. Piette, eds.), Hirokawa Publishing Co., Tokyo, 1968, p. 128.

22. Palmer, G., and Hager, L.P. Private communication.

23. Wittenberg, B.A., Kampa, L., Wittenberg, J.B., Blumberg, W.E., and Peisach, J. J. Biol. Chem., 243, 1863 (1968).

24. Gurd, F.R.N., Falk, K.-E., Malmström, B.G., and Vänngärd, T. J. Biol. Chem., 242, 5724 (1967).

25. van Gelder, B.F., Orme-Johnson, W.H., Hansen, R.E., and Beinert, H. Proc. Natl. Acad. Sci., U.S., 58, 1073 (1967).

26. Beinert, H. Personal communication.

27. Torii, I., and Ogura, Y. In Recent Developments of Magnetic Resonance in Biological System (S. Fujiwara and L.H. Piette, eds.), Hirokawa Publishing Co., Tokyo, 1968, p. 101.

28. Peisach, J., and Blumberg, W.E. Unpublished observations.

29. Gibson, J.F., and Ingram, D.J.E. Nature, 180, 29 (1957).

30. Yonetani, T., and Schleyer, H. J. Biol. Chem., 242, 3926 (1967).

31. Morita, Y., and Mason, H.S. J. Biol. Chem., 240, 2654 (1965).

32. Blumberg, W.E., Peisach, J., Wittenberg, B.A., and Wittenberg, J.B. J. Biol. Chem., 243, 1854 (1968).

DISCUSSION

Caughey: I should like to ask Dr. Blumberg to comment on the interconversion among normal hemoglobins; that is, how can one induce the Blumberg shifts he has just described?

Blumberg: I think Dr. Caughey wants to know how we get from one low-spin compound to another. In general, that is not so easy. It is much easier to start over with a new sample of either oxy- or ferrihemoglobin. In hemoglobin, there is a natural pathway through these compounds to get from oxyhemoglobin to garbage. It is: oxyhemoglobin to high-spin ferric hemoglobin, to the hydroxide form, to the dihistidine form, and into one or the other of the cytochrome forms, b or c, and then to the mercaptide form. When the reaction has proceeded that far, you throw the sample away because you can

228

never go back again. In certain places along the line, you can go back. Different heme proteins have different equilibria in regard to these ligands, for example, cytochrome P_{450} exists in the native form in the form that hemoglobin is in when it is garbage. The various cytochromes also behave in this way. From cytochrome c, for example, one can make other compounds: cytochrome c hydroxide, cytochrome c b-type, etc. by suitable recipes.

Riggs: I would just like to know whether you think that these shifts of helixes in hemoglobin really occur,because the F-helix would have to rotate as much as 180° as well as shift up and down. This would put many polar groups into contact with the heme and many hydrophobic groups would be oriented outwards to the medium. This would surely change the solubility. Since heme exchange occurs, it would seem that the heme could come off and become bound elsewhere. Heme migration is rendered likely by the high denaturing pH used. If one converts hemoglobin to "garbage", to use Dr. Blumberg's expression, it is difficult to see how the measurements should be interpreted. If the heme becomes detached, it would have nothing to do with the conformation of the "garbage" from which it came. Indeed, in the completely denatured protein "conformation" loses its meaning. The finding of gross changes may be real, but is it significant?

Blumberg: Now you know the meaning of gross conformational changes, as I mentioned in the first sentences of my talk. These changes do happen, and I am not indicating at all that the changes are subtle. They are indeed gross.

Stryer: I wonder if you could tell us how, in your diagrams, you measured your geometry axis?

Blumberg: The geometry axis in these diagrams is a certain combination of crystal field parameters; it is the ratio of the rhombic part to the tetragonal part. These are crystal field coefficients A_2^0 and A_2^2.

DEPARTURES FROM AXIAL SYMMETRY AS MEASURED BY EPR FOR HIGH SPIN FERRIC HEME PROTEINS*

J. Peisach[+] and W.E. Blumberg

Departments of Pharmacology and Molecular Biology
Albert Einstein College of Medicine
Yeshiva University, Bronx, New York 10461
and
Bell Telephone Laboratories
Murray Hill, New Jersey 07974

All heme proteins either occur naturally or, through chemical modification, can be converted to the high spin ferric form (Heme $d^5_{5/2}$)(1). The catalases (2), peroxidases (3-6), bacterial sulfite reductase (7) and bacterial tryptophan pyrrolase (8) are in this form as they are usually isolated. Myoglobin and hemoglobin can be converted to this form by a one-electron oxidation. The cytochromes, in general, may be converted to the high spin ferric form by replacement of a covalent bond to the iron. In other heme proteins, such as cytochrome c oxidase, the high spin ferric form may appear transiently during function of the protein (9). In this spin state, all of these proteins are amenable to study by EPR spectroscopy.

The EPR spectrum of heme iron in either polycrystalline materials or frozen solutions has an EPR absorption extending from g = 6 to g = 2 (Figure 1). In EPR spectra, taken in the absorption derivative form, the only prominent features of the spectrum are two excursions at either end of the absorption envelope, the one at g = 6 being the stronger and typically diagnostic for high spin ferric heme. Since the EPR signal

*First presented at the Third International Congress on Magnetic Resonance in Biology, Airlie House, Warrenton, Virginia, October 14-18, 1968. The portion of this investigation done at the Albert Einstein College of Medicine was supported in part by PHS GM 10959. This is communication no. 170 from the Joan and Lester Avnet Institute of Molecular Biology.
+Research Career Development Awardee (1-K3-GM-31,156).

231

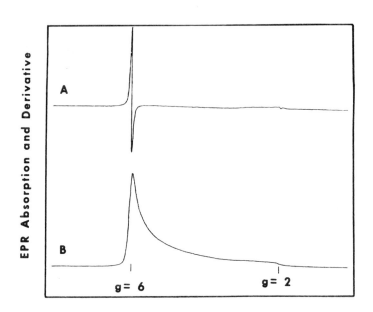

Magnetic Field

Figure 1. EPR absorption (lower) and absorption derivative
(upper) of hemin in dimethyl formamide.

arises from absorption between energy levels that can be described by an effective spin = 1/2, the unpaired electrons of
the ferric heme are only sensitive to two types of symmetry
elements in the environment, tetragonal and rhombic. It is
the large value of tetragonal field which is responsible for
the spread of the absorption between $\underline{g}$ = 6 and $\underline{g}$ = 2. The
rhombic component, when present, modifies the apppearance of
the EPR spectrum in the region of $\underline{g}$ = 6 so that now the single
absorption derivative is split into two components. The
splitting at $\underline{g}$ = 6 is directly related to departures from
tetragonality of the heme by

$$E/D = \Delta g/48$$

where E is the coefficient of the rhombic operator, D is the
coefficient of the axial operator in the spin Hamiltonian, and

232

Δg is the absolute difference in $\underline{g}$ values between the two components near $\underline{g}$ = 6. Since E/D has a maximum value of 1/3 (10), one may express the rhombicity as a percentage of the way between a completely tetragonal field and a completely rhombic field. Thus:

$$R = (\Delta g/16) \times 100\%$$

where R is the percent rhombicity.

This paper describes the EPR characteristics of various high spin ferric heme proteins where the rhombicity, as determined from EPR parameters, reflects differences in the protein environment of these molecules. For the simplest cases, ferrihemoglobin and ferrimyoglobin, the EPR is described as a single absorption derivative near $\underline{g}$ = 6 (Figure 2A). The small rhombicity that may be observed in single crystals of these proteins is not resolved in the randomly oriented molecules that occur in frozen solution. Thus the signal near $\underline{g}$ = 6 shows no splitting. Careful analysis of the spectrum shows that the effective $\underline{g}$ value at X-band is 5.92

Separation of oxyhemoglobin into isolated constituent chains can be chemically effected. Oxidation of oxy-α chains to the ferric form with ferricyanide yields, among other pro-

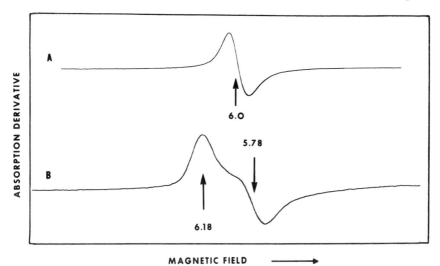

Figure 2. Low field portion of the EPR of the high spin forms of ferric hemoglobin A (A) and isolated ferric α chains (B). After reference (11).

233

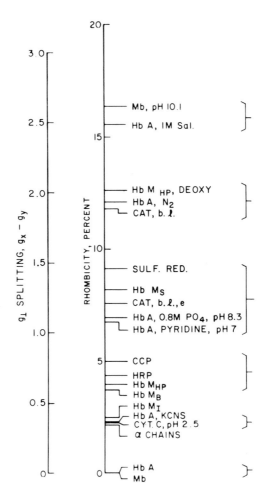

<u>Figure 3.</u> Values of the g splitting near g = 6 and computed percent rhombicity for high spin ferric heme proteins: Mb, myoglobin; HbA, hemoglobin A; Cyt. c, cytochrome c, pH 2.5 (18); HbM$_I$, hemoglobin M$_{Iwate}$; HbM$_B$, hemoglobin M$_{Boston}$; HbM$_{HP}$, hemoglobin M$_{Hyde\ Park}$ (15); HRP, horseradish peroxidase (5); CPP, cytochrome c peroxidase (6); Cat, b.l., beef liver catalase (18); Cat, e, erythrocyte catalase (2); HbA,N$_2$, hemoglobin A over which nitrogen gas is blown for one hour (18); HbM$_{HP}$, deoxy, deoxygenated hemoglobin M$_{Hyde\ Park}$ (15); HbA, 1M sal, ferrihemoglobin A incubated for one hour in 1 M salicylate.

234

ducts, a high spin ferric form in which the signal near g = 6 is split into two components with g values of 6.18 and 5.78 (Figure 2B)(11), giving a rhombicity of 2.3%. This departure from tetragonality is not an artifact of the chain separation since ferricyanide oxidation of a preparation of oxy-α chains added to an equivalent amount of oxy-β chains, as hemoglobin H, yields a material with an EPR signal identical to that of ferrihemoglobin A. Thus the departure from the tetragonality of the heme of the α chain depends on its association with the β chain in the tetramer of hemoglobin A.

A mixed state hemoglobin was prepared in which the α chains were in the ferric form and the β chains in the oxy form. The EPR of this material was due to the paramagnetism of the heme of the α chains only, since the oxygenated β heme was diamagnetic. Here, the EPR of this material was no different from that of ferrihemoglobin A (12). Furthermore, deoxygenation of the β chains did not affect the EPR signal from the α chain (12). Thus we conclude that although the heme symmetry of the α chain is sensitive to the presence of the β chain, it is sensitive neither to the state of oxygenation nor to the state of oxidation of the β chain. These experiments place some restriction on the mechanism of subunit interaction in hemoglobin A. Changes in one subunit upon oxygenation do not produce forces at the hemes of neighboring subunits sufficient to change the heme symmetry (13).

Various modifications of ferrihemoglobin can be prepared in which the symmetry of the heme has been altered while remaining attached to the globin (Figure 3). For example, the addition of a large excess of KCNS to ferrihemoglobin yields a product with both optical and magnetic properties altered from that of the parent protein. The EPR of this material resembles that of ferric alpha chains, with a comparable departure from tetragonality.

Even greater departures from axiality in ferrihemoglobin can be accomplished by changing the physical **environment of the protein, such as incubating in 0.8 M phosphate** buffer or in pyridine solution buffered at pH 7 or blowing nitrogen gas across the surface of a protein solution. The departure from tetragonality is greater in these cases than with thiocyanate treatment. Although these materials cannot be prepared in 100% yield as the thiocyanate-treated material can be, they represent discrete compounds. The interaction of the heme with the protein is different than from that in ferrihemoglobin A.

Yet further departure from axiality can be observed for minority components in solutions of ferrimyoglobin at pH 10.1 or ferrihemoglobin in 1 M salicylate, pH 7.0. In the former case, the major constituent of the sample is the hydroxy low spin form and in the latter case, the B-type low spin form (14). The chemical nature of these minority high spin derivatives remains unknown. It is likely either that their formation is obligatory along the pathway of formation of the low spin forms or that they represent the products of side reactions.

The hemoglobins M represent a class of abnormal hemoglobins in which either the α or β chains naturally occur in the nonfunctional ferric state as a result of an amino acid substitution near the heme. The non-substituted chain is normal and does not contribute to the EPR. Of the four listed Hemoglobins M (Table I), a tyrosine is substituted for histidine in the abnormal chain at a position lying close to the heme. In both hemoglobin $M_{Hyde\ Park}$ and hemoglobin M_{Boston}, where the β chains are abnormal, the percent of rhombicity of the hemes is approximately the same (15). For the α chain substituted abnormal hemoglobins, hemoglobin M_{Iwate} (16) and hemoglobin $M_{Saskatoon}$ (17), the departure from tetragonality depends on whether the amino acid substitution is proximal to, as in the former case, or distal to the heme. Thus we see that altering the primary structure of the protein specifically alters the symmetry of the heme.

In the case of hemoglobin $M_{Hyde\ Park}$, where the amino acid substitution of the β chain is at the point where the protein binds to the heme, the departure from tetragonality is dependent on the state of oxygenation of the normal α chain (15). Reversibly deoxygenating the molecule increases the rhombicity of the hemes of the adjacent abnormal chains by a factor of 3. This phenomenon demonstrates that it is possible for a conformational change occurring in one subunit to

TABLE I

Location of the Tyrosine Substituted for Histidine
in the Hemoglobins M

Location	Substituted Chains	
	α	β
proximal	Saskatoon	Hyde Park
distal	Iwate	Boston

236

produce an effect on the heme symmetry of an adjoining subunit. This same phenomenon is not observed for the other
three hemoglobins M (15).

Although cytochrome c is usually studied in its native
low spin form, it is possible to convert this protein reversibly to a high spin derivative. Below pH 4, the methionine
sulfur atom ligated to the heme iron is very likely removed
and replaced by oxygen from water. The heme symmetry is almost identical to that found in isolated ferric α chains and
ferric hemoglobin treated with KCNS. At pH below 1, the EPR
of cytochrome c is the same as for non-protein bound heme,
but the heme is still attached to protein through the thioester linkages, as even this material is almost completely re-
naturable when returned to a pH greater than 4.

When isolated, horseradish peroxidase (HRP), cytochrome
c peroxidase (CCP), sulfite reductase and catalase are in the
high spin ferric form and can exhibit EPR spectra with split
signals near g = 6. The heme symmetries of the non-axial
forms of HRP and CCP are almost the same. This observation
and the analysis of EPR spectra of low spin forms of both
proteins suggests that the protein constituents in the vicinity of the heme are very similar, although heme-linked reactions in these proteins differ significantly. It also suggests that the environment of the hemes in the peroxidases is
unlike that of hemoglobin and myoglobin.

In the case of catalase and sulfite reductase, the heme
symmetry is even more rhombic, reflecting differences in the
heme environments from those proteins previously discussed.
The EPR signal of erythrocyte catalase can actually be used
as a signature for its presence in intact red cells (18).

In Figure 3, there is a summary of all data available
for high spin forms of ferric heme proteins. For the most
part, the symmetry of the heme can be described as axial with
departures toward rhombicity varying up to 17%. One can see
that the symmetries fall in clusters set off in the figure as
brackets. For some of these proteins, such as in the upper-
most cluster, it would not be unreasonable to propose that the
structure in the vicinity of the heme is the same for all
members of the cluster. For some of the other clusters, there
is insufficient data to draw the same conclusion. It is clear,
however, that the symmetry of the heme of any heme protein depends on the structure of the protein moiety at all levels,
primary, secondary and tertiary, in contradistinction to low

spin heme protein compounds, where the symmetry of the heme is primarily governed by the immediate ligand atoms of the heme iron.

Acknowledgment

We would like to thank Drs. Jonathan B. Wittenberg and Beatrice A. Wittenberg for stimulating discussions and Lidija Kampa and Rhoda Oltzik for technical assistance.

References

1. Peisach, J., Blumberg, W.E., Wittenberg, B.A. and Wittenberg, J.B. J. Biol. Chem., 243, 1871 (1968).

2. Torii, K. and Ogura, Y. J. Biochem. (Tokyo), 64, 171 (1968).

3. Morita, Y. and Mason, H.S. J. Biol. Chem., 240, 2654 (1965).

4. Yonetani, T. and Schleyer, H. J. Biol. Chem., 242, 3926 (1967).

5. Blumberg, W.E., Peisach, J., Wittenberg, B.A. and Wittenberg, J.B. J. Biol. Chem., 243, 1854 (1968).

6. Wittenberg, B.A., Kampa, L., Wittenberg, J.B., Blumberg, W.E. and Peisach, J. J. Biol. Chem., 243, 1863 (1968).

7. Siegel, L.M. and Kamin, H. In Flavins and Flavoproteins (K. Yagi, ed.), University Park Press, 1968, p. 15.

8. Feigelson, P., Koike, K. and Poillon, W.N. Fed. Proc., 27, 588 (1968).

9. van Gelder, B.V., Orme-Johnson, W.H., Hansen, R.E. and Beinert, H. Proc. Nat. Acad. Sci. (U.S.), 58, 1073 (1967).

10. Blumberg, W.E. In Magnetic Resonance in Biological Systems (A. Ehrenberg, B.G. Malmström and T. Vänngård, eds.), Pergamon Press, Oxford, 1967, p. 119.

11. Peisach, J., Blumberg, W.E., Wittenberg, B.A., Wittenberg, J.B. and Kampa, L. Proc. Nat. Acad. Sci. (U.S., in press.

12. Peisach, J., Ogawa, S. and Blumberg, W.E. Unpublished observations.

13. Shulman, R.G., Ogawa, S., Wuthrich, K., Yamane, T., Peisach, J. and Blumberg, W.E.   This volume, p.195

14. Blumberg, W.E. and Peisach, J.   This volume, p.215

15. Watari, H., Hayashi, A., Morimoto, H. and Kotani, M. In Recent Developments of Magnetic Resonance in Biological System (S. Fujiwara and L.H. Piette, eds.), Hirokawa Publishing Co., Tokyo, 1968, p. 128.

16. Hayashi, A., Shimizu, A., Yamamura, Y. and Watari, H. Biochim. Biophys. Acta, 102, 626 (1965).

17. Watari, H., Hwang, K.-J., Kimura, K. and Murase, K. Biochim. Biophys. Acta, 120, 131 (1966).

18. Peisach, J. and Blumberg, W.E.   Unpublished observations.

SPIN-LABELED HEMOGLOBIN IN SOLUTION, POLYCRYSTALLINE SUSPENSIONS AND SINGLE CRYSTALS*

R.T. Ogata[+] and H.M. McConnell

Stauffer Laboratory for Physical Chemistry
Stanford, California 94305

In their early spin-label studies of hemoglobin in solution, Ogawa and McConnell noted that the resonance spectra of spin-labelled horse oxy- and methemoglobin were significantly different, implying differences in the protein conformation of these derivatives (1). This observation was somewhat surprising, in view of the results of the single crystal X-ray studies of Perutz and Mathew, where it was found that these two derivatives must have very similar secondary and tertiary structures and identical quaternary structures (2). In an effort to determine whether the differences in hemoglobin conformation detected by the spin-label in hemoglobin solutions was carried over into crystalline hemoglobin, a study of the resonance spectra of spin-labelled hemoglobin in single crystals and in polycrystalline suspensions was undertaken. It was felt that a study of this kind would aid in the interpretation of certain features of the resonance spectrum of spin-labelled hemoglobin (3).

In this study, two spin labels, N-(1-oxyl-2,2,5,5-tetramethyl-3-pyrrolidinyl) iodoacetamide (I) and N-(1-oxyl-2,2,6,6-tetramethyl-4-piperidinyl) iodoacetamide (II) were used in labelling the β-93 cysteine residues of hemoglobin.

*Supported by the ONR Contract No. 225(88) and benefitted from facilities made available by the Advanced Research Projects Agency through the Center for Materials Research at Stanford University.
+Supported by an NIH Predoctoral Fellowship.

241

Results

Because of the near cylindrical symmetry of the nuclear hyperfine term in the spin Hamiltonian, and its large aniso- tropy, the principal axis for the large nuclear hyperfine splitting is found experimentally rather easily in single crystals of spin-labeled hemoglobin. This principal axis is approximately in the direction of the $2p\pi$ -orbital of the ring nitrogen and was found by systematic searches in which the applied field was rotated relative to the crystal axes.

In all hemoglobin derivatives studied - carbonmonoxy, acid met, met azide, and met fluoride -two main spin-label conformational states were found. To within an experimental error of less than $\pm$ 10°, the orientation of the labels in these states was identical for all four hemoglobin derivatives and, remarkably, also identical for both labels I and II. A difference was detected, however, between the spectra of the low-spin derivatives - carbonmonoxy and met azide - and those of the high-spin derivatives - acid met and met fluoride. This

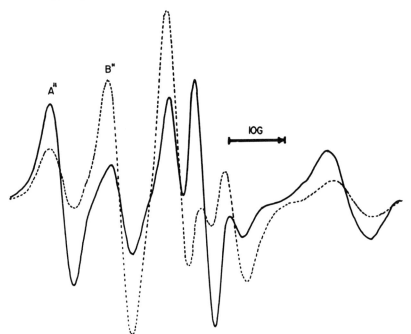

Figure 1. Resonance spectra of single crystals of methemo- globin azide (———) and methemoglobin (- - -) labelled with II. The applied field is perpendicular to b, and makes an angle of 45° ±3° with a and an angle of 24° ±3° with c

was a difference in the relative populations of the two main spin-label conformational states. As can be seen in Figure 1, the resonance spectrum of met azide hemoglobin (the spectrum of carbonmonoxy is identical) shows a much larger proportion of labels in the state labeled A" than does the spectrum of acid methemoglobin (the spectrum of met fluoride is identical to that of acid met). It is felt that these states are in equilibrium, for their relative populations are dependent in a sensitive way on the ionic composition of the solvent, as well as on the ligand bound to the iron atoms. The spectra of the high and low-spin derivatives become more nearly equal as the crystal is allowed to dry slowly. A similar effect is observed on increasing the ionic strength of a solution or polycrystalline suspension of hemoglobin as seen in Figures 2 and 3.

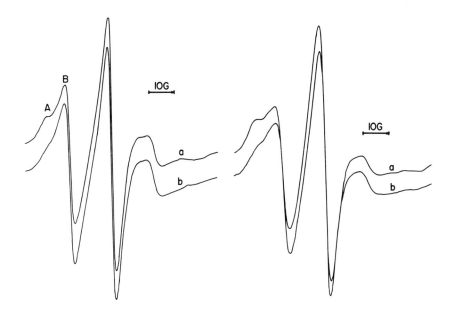

Figure 2. Solution (0.01 M anion) resonance spectra of hemoglobin labelled with II: a) carbonmonoxy- b) methemoglobin

Figure 3. High buffer concentration (2.5 M anion) solution resonance spectra of hemoglobin labelled with II a) carbonmonoxy- b) methemoglobin

It should be noted that the resonance spectrum of spin-labeled deoxyhemoglobin in solution shows no evidence of this two-state equilibrium (3-5). The spectra of deoxy and met-hemoglobin, though similar in low ionic strength solutions, are quite different at high ionic strengths. In a solution of high ionic strength, the methemoglobin resonance spectrum clearly shows the existence of two conformational states (see Figure 3). No such evidence for two states is seen in the spectrum of deoxyhemoglobin under similar conditions (6). Thus the spectrum of deoxyhemoglobin is uniquely different from the resonance spectra of the four derivatives studied here. This finding is consistent with the well-known fact that the structure of deoxyhemoglobin in crystals is uniquely different from the structure of the other hemoglobin derivative (2,7,8).

In this study, a comparison was also made between the local protein structure in the vicinity of β93 for hemoglobin in solution and in crystals. For a given ionic composition of the solvent, the passage from solution to crystals is accompanied by an increase in the degree of immobilization of the labels and very probably also a change in the relative populations of the two conformational states. It was concluded however, that there can be no large change in protein structure, since in both cases there is a delicately balanced equilibrium between the two isomeric spin-label states.

These results show that the local protein conformation in the vicinity of the label at a β93 cysteine depends on the spin state of the label of the iron atoms. However, the high- and low-spin structures must be very similar, since the principal axis orientations for both I and II were found to be very nearly the same for all four derivatives. Evidently the equilibria between the isomeric states are sensitive to very small changes in protein conformation.

This study establishes that the differences in the solution spectra of labeled carbonmonoxy and methemoglobin persist in the single crystal spectra. Because the change of protein structure associated with this difference must be small, these results are not inconsistent with the previously mentioned conclusion of Perutz and Mathews (2) that the met, oxy, and met azide derivatives have very similar structures.

References

1. Ogawa, S. and H.M. McConnell. Proc. Natl. Acad. Sci. U.S., 58, 19 (1967).

2. Perutz, M.F. and S.F. Mathews. J. Mol. Biol., 21, 199 (1966).

3. A more complete report of this study is given in (a) McConnell, H.M. and C.L. Hamilton, Proc. Natl. Acad. Sci. U.S., 60, 776 (1968) and (b) McConnell, H.M., W. Deal, and R.T. Ogata, Biochemistry, 8, 2580 (1969).

4. Ogawa, S., H.M. McConnell, and A. Horwitz. Proc. Natl. Acad. Sci. U.S., 61, 401 (1968).

5. McConnell, H.M., S. Ogawa, and A. Horwitz. Nature, 220, 787 (1968).

6. Ogawa, S., A. Horwitz, and H.M. McConnell. Unpublished results.

7. Muirhead, H., J.M. Cox, L. Mazzarella, and M.F. Perutz. J. Mol. Biol., 28, 117 (1967).

8. Bolton, W., J.M. Cox, and M.F. Perutz. J. Mol. Biol., 33, 283 (1966).

MAGNETIC SUSCEPTIBILITY OF HEMOPROTEINS

Akira Tasaki*

Johnson Research Foundation, School of Medicine
University of Pennsylvania, Philadelphia, Pennsylvania

Introduction

In the energy level scheme of the iron ion in hemoproteins, it is well known that the difference in energy level between the high and low spin states is very small. Since similar findings have been reported for many kinds of hemoproteins, it is generally believed that all complex iron compounds are similar in this respect. Therefore, it is important to investigate how various energy levels in hemoproteins are realized and how energy level splittings of the central iron ion in the hemoprotein can be evaluated.

Let us consider the energy levels of a $3d^n$ incomplete shell. We put n-electrons into the five d-orbits, taking interelectronic Coulomb repulsion into account. Each of the calculated energy levels is characterized by spin angular momentum S and orbital angular momentum L. In a free iron ion, the energy difference between the ^6S (S = 5/2) of $3d^5$ and the lowest doublet ^2T (S = 1/2) is about 6 eV (40,000 cm^{-1}) and the lowest quintet ^4G (S = 3/2) is higher than the ^6S state by about 32,000 cm^{-1}. The detail of this $3d^5$ configuration is now well understood. As reported by Sugano and Tanabe [1], in the crystal field of the surrounding atoms, the energy levels of both the high spin and low spin states decrease with increasing of crystalline field strength. It is emphasized here that in hemoproteins, the strength of the cubic crystalline fields is almost equal to the value at which these two energy levels cross each other. The discrepancy between the crystalline field strength and the crossing point of the two energy levels is as small as 0.5%. Thus, the large

*Present address: Osaka University, Osaka, Japan.

energy difference of these two states is reduced down to about 200 cm^{-1} which is very close to the thermal energy at room temperature. Therefore, the high and the low spin state coexist at room temperature.

Low spin compounds are frequently encountered in inorganic compounds. However, the coexistence of high and low spin states at room temperature has never been reported in the literature. It is improbable that this accurate energy equalization between the high and the low spin state is merely accidental. Instead, it is considered highly probable that an important relation exists between this particular electronic state and the biological activity of this protein.

Figure 1 illustrates the energy level scheme of a high spin ferric hemoprotein, Mb (Fe^{3+}). It shows how the differences in the energy levels can be detected. In the cubic ligand field, the high spin state and the low spin state are called $^6A_1$ and $^2T_2$, respectively. On account of spin-orbit interaction, the lowest high spin state ($^6A_1$) is split into three Kramer's doublets. This splitting can be described by spin Hamiltonian DS_z^2, where D includes the spin-orbit coupling constant and the energy separation between $^6A_1$ (S = 5/2) and $^4T_1$ (S = 3/2). The energy splitting due to this spin-orbit interaction is about 10 cm^{-1}. Therefore, at temperatures lower than 15°K, the electron population of the state $S_z = \pm 5/2$ reduces to a level below that of $S_z = \pm 3/2$ or $S_z = \pm 1/2$ and the effective number of magneton (n_{eff}) of the iron ion in this state becomes smaller than the value in the high spin state ($n_{eff} = \sqrt{35}$).

In a high temperature range, the temperature dependence of magnetic susceptibility obeys Curie's law. Below 20°K, however, the slope of the ($\chi - 1/T$) curve gradually decreases and the curve bends at about 15°K. In other words, it is possible to evaluate the value of D from the temperature dependence of magnetic susceptibility in the range of temperature below 20°K. This measurement was suggested by Prof. M. Kotani (2) and was carried out by the author in 1967 (2). Finally, these three Kramer's doublets can be separated by the applied magnetic field and an EPR absorption signal is expected to appear on transition between the separated lowest doublet ($S_z = \pm 1/2$).

In Figure 1, it can be seen that the EPR technique and magnetic susceptibility measurements at low temperatures

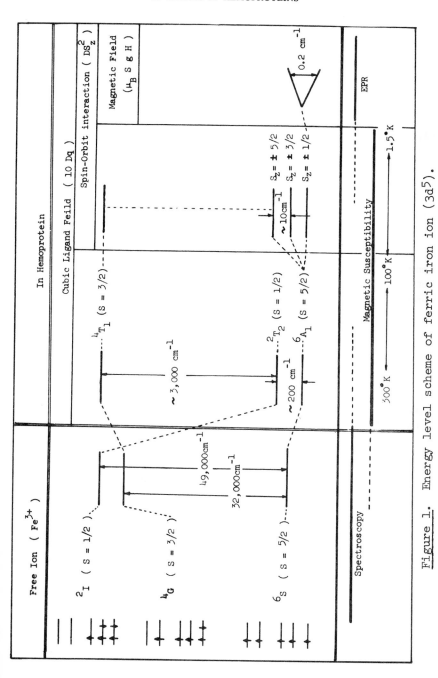

Figure 1. Energy level scheme of ferric iron ion ($3d^5$).

(1.5 ∿ 100°K) are two complementary methods which are required for determination of the fine structure of the energy levels. It should be noted that, in inorganic complex compounds, the calculated value of n_{eff} obtained by simple evaluation of $g_j \sqrt{J(J+1)}$ or $2\sqrt{S(S+1)}$ frequently disagree with the observed value. A precise quantum mechanical theory of the paramagnetic susceptibility was advanced by Van Vleck (4) and actual calculation of the effective number of magnetons on the complex compounds of $3d^n$ system (5) and on the hemoproteins (2) were carried out by Prof. M. Kotani. The observed magnetic susceptibility of some hemoproteins is smaller than that expected from the spin state and this small magnetic susceptibility has been often (and erroneously) attributed to the existence of an intermediate spin state (S = 3/2). However, from the recent magnetic susceptibility measurements at very low temperature, it has become certain that the small value of the susceptibility is not due to the intermediate spin state but is due to mixing of low spin compounds, since the temperature at which an abrupt change in the slope of the (χ - 1/T) curve occurs is characteristic of its spin state.

Recent magnetic susceptibility studies made by Prof. Kotani's group at Osaka University can be divided into the following three categories.

(1) Magnetic susceptibility measurements down to 1.5°K. Mb (Fe^{3+}) at various pH, Hb (Fe^{3+}), catalase, Japanese radish peroxidase (carried out by A. Tasaki and J. Otsuka).

(2) Magnetization measurements at 4.2°K using a superconduction magnet. Mg (Fe^{3+}), Hb (Fe^{3+}), Hb (Fe^{2+}) (A. Tasaki, J. Otsuka and Y. Nakano).

(3) Thermal equilibrium analysis of high and low spin states. Susceptibility measurements down to 77°K (T. Iizuka) and theoretical interpretation (J. Otsuka and A. Tasaki).

In the present paper, the results of these studies will be briefly discussed.

Magnetic Susceptibility Measurements down to 1.5°K.

(1) Mb (Fe^{3+}) at various pH's. It has been reported previously that ferrous myoglobin in a solution of pH 6 is purely of the high spin type. The magnetic susceptibility of this compound can be fully explained by virtue of a spin Hamiltonian DS_z^2 with $D = 10$ cm^{-1}. The solid lines in Figure 2 show the result of numerical evaluation of the paramagnetic susceptibility formula given by Van Vleck (4). The dots in

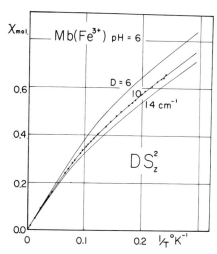

Figure 2. Magnetic susceptibility of Mb (Fe^{3+}), pH 6. The solid lines represent values calculated on the basis of Van Vleck's formula by the use of an electronic computer. The dots represent the observed values.

the figure represent the observed values. Agreement between the experimental values and the calculated values is very satisfactory. Thus, it is concluded that this compound is purely of the high spin type and also that the fine structure of the iron ion can properly be described by the above mentioned spin Hamiltonian.

The experimental susceptibility data of ferric myoglobin in solutions at pH 7.5, 8.5, 9.3 and 9.7 are shown in Figure 3. As has been pointed out by Schoffa et al. (6), the magnetic susceptibility of alkaline myoglobin is smaller than that of high spin compounds. Beetlestone et al. (7) attributed these intermediate values of susceptiblity to thermal equilibrium between high and low spin states. In the present work, however, measurements are carried out at very low temperatures; therefore, mixing of a high energy state is negligible (0.7% at 77°K). Yet, experiments carried out under

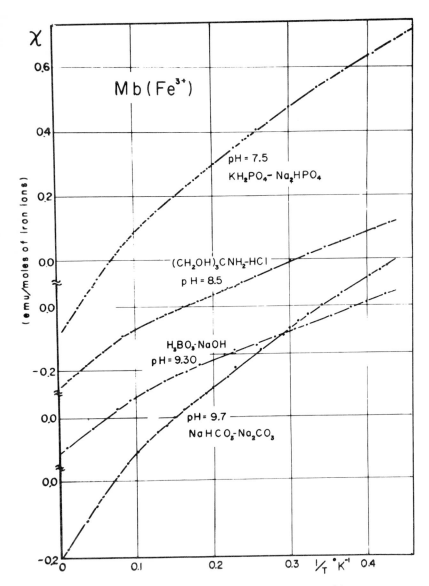

Figure 3. Magnetic susceptibilities of Mb (Fe^{3+}) at various pH's plotted against the reciprocal of the temperature ($^{\circ}$K).

these conditions yield a value intermediate between that of a high spin compound and that of a low spin compound. Furthermore, the temperature at which the $(\chi - 1/T)$ curve bends is the same as that at pH 6. From these results, it can be concluded that a portion of the iron ions still maintains its high spin and that the remaining portion is in the low spin state whose $(\chi - 1/T)$ relationship is given by an approximately straight line within the observed range of temperature. It is assumed that acid myoglobin $Mb(H_2O)$ is of the high spin type and the alkaline $Mb(OH)$ is of the low spin type and also that the fraction of the acid type and the fraction of alkaline type are a and (1 - a), respectively. The observed magnetic susceptibility can then be represented by

$$\chi_{obs} = a \chi_{high} + (1 - a) \chi_{low} \qquad (1)$$

The thermal equilibrium effect is expected to occur in each of the two compounds at high temperatures; however, at low temperatures where these magnetic susceptibility measurements were carried out, thermal agitation is so small that the mixing of higher energy state can safely be neglected. The $(\chi - 1/T)$ curves constructed on the basis of this simple mixture model are in good agreement with the experimental values. As a consequence of these low temperature measurements, it has become possible to determine the fraction a as a function of pH without being disturbed by the thermal effect of mixing. The (ln a - pH) curve obtained in the present study suggests the existence of a high order chemical reaction or of a phase transition. A further study of this point will be carried out in the near future.

(2) _Ferric hemoglobin_. The magnetic susceptibility of Hb (Fe^{3+}) at pH 6 is considerably smaller, as reported in the previous paper (3), than that expected from the spin value S = 5/2. During the course of repeated measurements, the value of the magnetic susceptibility was found to vary from time to time in one and the sample preparation, although the bending point of the $(\chi - 1/T)$ curve remained the same as that of Mb (Fe^{3+}) at pH 6. It is assumed, therefore, that a portion of the iron ion of Hb (Fe^{3+}), pH 6, has a low spin and that equation 1 can be applied to this compound. In Figure 4, the curve calculated for a = 0.29 in equation (1) is shown together with experimental data. Agreement between the observed values and the calculated curves is extremely good; however, we are still perplexed by the large amount of low

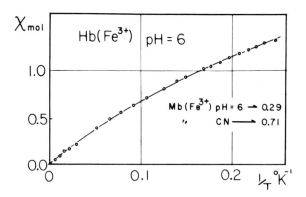

Figure 4. Magnetic susceptibility of a mixture of hemoproteins of the high and low spin types. The typical material for the high spin type is Mb (Fe^{3+}) pH 6 and that of the low spin type is Mb (Fe^{3+}) CN. The ratio of the two components was taken as 0.29 (high spin): 0.71 (low spin). The dots represent the observed values of Hb (Fe^{3+}) pH 6.

spin type compound which is required to explain these data. To explain present experimental results, other possibilities, such as the existence of an intermediate spin state, were examined theoretically; but, as mentioned in the previous section, the present behavior of the ($\chi - 1/T$) curve cannot be explained simply by assuming an intermediate spin state. Thus, it may be concluded that, under our experimental conditions, Hb (Fe^{3+}) pH 6 contains both high and low spin components. The ground state of energy level of the high spin compound is expressed by the spin Hamiltonian DS_z^2 (D = 10). It is an interesting, yet puzzling, problem that a low spin signal has not been detected by the EPR technique. A plan is now under consideration to conduct an EPR study at liquid helium temperature once again.*

(3) Deoxy ferrous hemoglobin. Since the EPR technique cannot be applied to deoxy ferrous hemoglobin, a magnetic

*Recently, the low spin signal which was not detected at 77°K was observed below 20°K.

susceptibility measurement is required for determination of the electronic structure of the central ferrous ion of this compound. As in the case of ferric hemoglobin, the magnetic susceptibility of the compound was found to vary from preparation to preparation. By taking many possible electronic configurations into consideration, Otsuka (7) calculated the $(\chi - 1/T)$ curves for the compound based on his previous theoretical analysis of electronic states. He arrived at the conclusion that this compound also contains a low spin component and that the energy level of the high spin component is explainable in terms of the spin Hamiltonian DS_z^2 (D = 4 cm^{-1}) as shown in Figure 5.

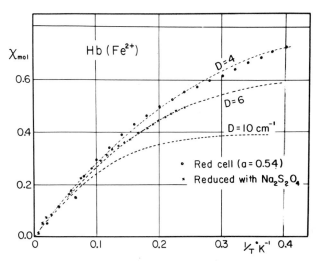

Figure 5. Magnetic susceptibility of deoxy Hb (Fe^{2+}). ●, the experimental data for the high spin component calculated from equation 1 by assuming $\underline{a}$ = 0.54. x, the experimental data of the sample which was obtained by reduction of Hb (Fe^{3+}) with $Na_2S_2O_4$. —, the results of calculation for various values of D.

Magnetization Measurements by Superconducting Magnet

It is well known that the magnetic property of a paramagnetic material cannot be represented by simple magnetic susceptibility constant when it is measured under a strong magnetic field combined with a low temperature where the term

255

$\mu_B H/kT$ cannot be considered far smaller than unity. With the magnetic field strength produced by a superconducting magnet* fixed at a level of 50 KOe, the term $\mu_B H/k$ corresponds to 5°K and consequently higher terms in Lagivin's function (4) have to be taken into consideration. The strength of the magnetic field at which paramagnetic saturation takes place varies to a considerable extent with the energy state and the degree of spin orbit coupling of the paramagnetic ion. In other words, it is a function of the g-value and of the D and E values of the spin Hamiltonian (5). It is therefore possible to obtain certain information concerning the electronic structure of the ion from the saturation effect. We assume the following spin Hamiltonian:

$$\mathcal{H} = DS_z^2 + E\,(S_x^2 - S_y^2) + \mu_B\,\vec{H}\cdot\tilde{g}\cdot\vec{S} \tag{2}$$

and calculate the $(\chi - 1/T)$ and $(M - H)$ curves for various values of D and E by the use of an electronic computer; and by comparing the experimental data with the calculated values, we determine the values of D and E of the compound. In Figure 6, the experimental data for red cell Hb (Fe^{2+}) deoxy are shown together with various calculated curves. As was expected from the previous results indicating the temperature dependence of the magnetic susceptibility, the experimental data agree with the calculated curve for $D = 4$ cm^{-1}. It is to be noted here that the present result for $D = 4$ cm^{-1} suggests the possibility of detecting its EPR absorption under a very high magnetic field which may be obtainable by the use of a superconducting magnet.

Thermal Equilibrium (Enthalpy-Entropy Compensation)

Lumry reported the enthalpy-entropy compensation effect in protein solutions (11); Iizuka, Kotani and Yonetani (8,9) reported the same effect on myoglobin, hemoglobin and cytochrome c peroxidase (CCP) based on their magnetic susceptibility measurements. The phenomenon reported by Lumry is that in chemical equilibrium involving protein, an enthalpy change and an entropy change compensate each other so as to leave the net free energy of the system unchanged. Thus, if the chemical properties of the added solutes are altered by using a number of compounds forming a homologous series, thermal equilibrium

*Maximum field 60 KOe , 16 mm : Mitsubishi Electric Co., Japan.

256

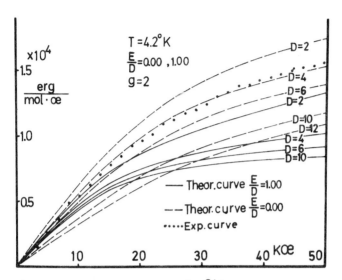

Figure 6. Magnetization of Hb (Fe^{2+}) plotted against the magnetic field strength. The calculated curves are based on the spin Hamiltonian of equation 2.

of the system under consideration is maintained in the same temperature range in spite of large changes of the enthalpy or of the potential barrier. Ordinarily, this phenomenon is attributed to the interaction between the protein molecule and water or to protein conformation changes. However, as reported by Iizuka and Kotani (6), the compensation effect has been observed in high and low spin equilibrium of hemoproteins which has taken place at temperatures below -50°C where water is completely frozen and conformation changes can-not·take place.

From the statistical point of view, the equilibrium constant K between two states is expressed by

$$K = \frac{1}{\gamma} \tag{3}$$

where ε is the energy difference between the two states and is the ratio of the degrees of freedom in the two states (i.e., entropy factor). In the case of high and low spin equilibrium in CCP and its derivatives, γ is 2.6×10^4 when the energy difference ε is equal to $+1800$ cm^{-1} and γ is as small as 5.6×10^{-4} when ε is -1200 cm^{-1}. These extreme values may be attributed to coupling of the spin state and various lattice vibrations of the large protein molecule.

To explain the experimental results for various compounds by this postulation, however, a large degree of freedom has to be attributed, sometimes to the high spin state and sometimes to the low spin state. Therefore, it seems unreasonable to ascribe this large value of γ to vibration of the protein molecule.

One possible explanation for this large value of γ is to attribute it to large temperature dependence of ε. In a large protein molecule, it is plausible that the energy difference between two states or potential barriers is affected by the state of the protein molecule, i.e., by conformation changes or interactions between the porphyrin and polypeptide chain. Now, it is assumed that ε is temperature dependent. By expanding ε as a function of temperature around T_1 and taking the first two terms of the series, ε may be written as

$$\varepsilon(T) = \varepsilon(T_1) + (T-T_1) \left(\frac{\partial \varepsilon}{\partial T}\right)_{T=T_1} \tag{4}$$

Substitution of equation (4) into equation (3) yields

$$\ln K = -\ln \gamma + \frac{\varepsilon(T_1)}{kT} - \frac{T_1 \varepsilon'(T_1)}{kT} + \frac{\varepsilon'(T_1)}{k} \tag{5}$$

where $\varepsilon'(T_1)$ represents $(\partial \varepsilon / \partial T)_{T=T_1}$. The second and the third terms in (5), representing the effect of the energy barrier, are enthalpy terms and the fourth term is a new apparent entropy term. It is clear that the absolute value of the apparent entropy term becomes large at the temperature where the conformational state of the protein changes extensively. If the first and second terms are smaller than the third term, these small terms may be neglected and equation (5) becomes

$$\ln K = \frac{T_1 \varepsilon'(T_1)}{kT} + \frac{\varepsilon'(T_1)}{k} \tag{5'}$$

This relation briefly describes entropy-enthalpy compensation. The conditions for realization of equilibrium of this type may be satisfied when the temperature dependency of energy barrier ε' is almost linear within the observed temperature range and the value of $\varepsilon(T_1)$ is not very large. These conditions seem plausible (see Figure 7). The enthalpy term calculated from equilibrium experiments does not correspond to the energy difference between these two states at T_1; but, as shown in Figure 7, it corresponds to the value extrapolated to absolute zero temperature.

Recently, by taking the interaction between the porphyrin and polypeptide chain of protein into consideration, a statistical mechanical evaluation of the entropy and enthalpy was carried out. It is true, as assumed, that weak interaction exists between the nearest bonds. Thus, the number of bonds are temperature dependent; and when the interaction between

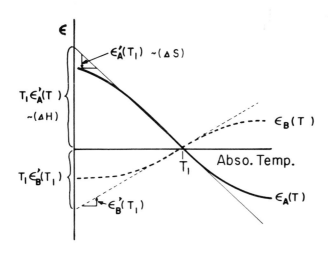

Figure 7. Illustration of the energy difference between the high and low spin states. $T_1(\partial\varepsilon/\partial T)_{T=T_1}$ and $(\partial\varepsilon/\partial T)_{T=T_1}$ correspond to the enthalpy and entropy term, respectively.

the nearest bonds is strong, the number changes cooperatively with temperature. The relative position of the porphyrin to the polypeptide varies with the change of number of
the bonds. The free energy of the system can be represented
by these interaction terms and the state of iron ion. Actual
calculation of the entropy and the enthalpy along this line
was made by Otsuka. He obtained results that were qualitative
satisfactory. The details of the calculation will be reported
elsewhere.

Summary

1. The energy levels of the 3d electrons of the central
Fe-ion in hemoproteins are discussed on the basis of the
crystal field theory.

2. The results of measurements of the magnetic susceptiblity of myoglobin in the range of temperature down to
1.5°K are described.

3. The magnetic susceptibility of hemoglobin under various experimental conditions in the range of temperature down
to 4.2°K is discussed.

4. The temperature dependence of the magnetic susceptiblity of hemoglobin is discussed, based on the experimental
results obtained with a superconducting magnet.

5. The problem of entropy-enthalpy compensation in
equilibrium involving complex compounds at low temperature is
briefly reviewed.

Acknowledgment

The author would like to express his sincere thanks to
Prof. M. Kotani for many valuable and helpful discussions.

References

1. Tanabe, Y. and S. Sugano. J. Phys. Soc., Japan, 13, 394 (1958).

2. Kotani, M. Rev. Mod. Phys., 35, 717 (1963).

3. Tasaki, A., Otsuka, J., and M. Kotani. Biochim. Biophys. Acta, 140, 284 (1967).

4. Van Vleck, J. *Theory of Electric and Magnetic Suscep-tibilities.* Oxford Univ. Press, London and New York, 1932

5. Kotani, M. Prog. Theoret. Phys. Suppl. $\underline{14}$, 1 (1960).

6. Schoffa, G., W. Scheller, O. Postau, and F. Jung. Acta Biol. Med. Ger., $\underline{3}$, 65 (1959).

7. George, P., J. Beetlestone, and J.S. Griffith. *Hematin Enzymes.* Pergamon Press, New York, 1961, p. 105.

8. Otsuka, J. J. Phys. Soc., Japan, $\underline{24}$, 885 (1968).

9. Iizuka, T., and M. Kotani. Biochim. Biophys. Acta, $\underline{197}$, 257 (1968).

10. Iizuka, T., M. Kotani, and T. Yonetani. Biochim. Bio-phys. Acta, $\underline{197}$, 257 (1968).

11. Lumry, R., and R. Biltonen. *Structure and Stability of Biological Marcromolecules* (S. Timacheff and G. Facam, eds.), Dekker, New York, 1969, p. 65.

Discussion

Shulman: Dr. Tasaki, the previous interpretation of your data had indicated that there are conformational changes taking place during the high to low spin transition. It is possible that $-25°C$ is an unique temperature, because the ice structure changes in that region, allowing these conformational changes, and that if you did this experiment in a glass such as ethyl-ene glycol–water, which softens at another temperature, you might find the spin changes occurring at a different tempera-ture? That would help to support the interpretation based upon conformational changes.

Tasaki: You are right.

Shulman: With different freezing points of the solvent, do you have the same T_0?

Tasaki: Iizuka tried this experiment, and I think he obtained almost the same T_0.

REVERSIBLE CHANGES AT THE IRON SITE
ON DEHYDRATING METMYOGLOBIN

Thomas Moss

IBM, Watson Laboratory
New York, New York

I want to show a quick piece of data, gathered in collaboration with Drs. Ehrenberg and Feher, using a magnetic susceptibility device designed by Dr. A. Redfield of Watson
Laboratory. I think it confirms accurately a suspicion that
many people have held for a long time--that whatever the
effects of dehydration on the conformation of a heme protein,
there can certainly be profound effects on the magnetic state
of the iron. We looked at the effects of temperature cycling
and drying on the magnetic susceptibility of masses of randomly oriented crystals of metmyoglobin. The real rationale
for this work was to see whether we could find an explanation
for the observation which many people have made that, in
doing single crystal heme protein EPR work, one loses EPR
intensity after temperature cycling or manipulating the crystals.

The data for the crystals in the mother liquor (Figure 1)
fit the theoretical curve for $D = 5$ cm^{-1} somewhat better
than that for $D = 10$ cm^{-1} (the data of Tasaki et al. (1) was
best fit by $D = 10$ cm^{-1}), but they are acceptably close to
previous results on metmyoglobin.

The susceptibility after temperature cycling, and after
simply removing the mother liquor, is only a little below
that of the original crystals -- the difference is only mar-
ginally greater than the experimental error which is about
as large as the symbols marking the experimental points.

When the crystals are dessicated over P_2O_5, however,
there is a very large drop in the susceptibility to a value
perfectly characteristic of a low spin state. The change
goes on to completion in a relatively short time, as illustrated by the identity of the curves for 12 and 72 hr dessi-

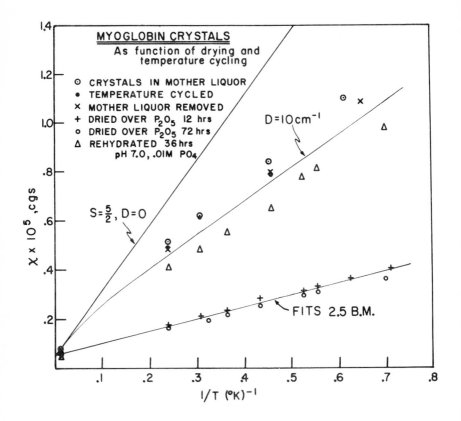

Figure 1. Magnetic susceptibility of myoglobin crystals as a function of temperature cycling and dehydration.

cation. One can also see a change to a hemichrome-like spectra in the optical absorption of the dried crystals. The magnetic change is largely reversible, since the susceptibility returns almost completely to the high spin value just by adding buffer to the dehydrated crystals (curve marked by Δ).

The main point for experimentalists is simply that one must really avoid any evaporation from the surfaces of crystals used in magnetic measurements. An additional speculative point to make about the data is to note that the 2.5 Bohr magnetons measured for the moment of the dehydrated material is just about what Tasaki and his colleagues (1) have measured for the susceptibility of myoglobin at pH 9.0. At the last of these Johnson Foundation Symposia (2), Dr. Caughey suggested that the real change which takes place on dehydrating metmyoglobin is <u>not</u> that a water is removed from the coordination sphere of the iron, but that, instead, a water is removed from the region of the distal nitrogen (Nδ) on the imidazole ring (histidine E7) located just beyond the bound water molecule. He supposed that this could cause a rearrangement of the imidazole double bonds such that one of the water protons would move much closer to the proximal imidazole nitrogen, leaving the iron coordinated to something resembling a hydroxyl ion—as we suppose myoglobin at pH 9 to be. The data does not prove this, of course; many low spin iron forms have moments near 2.5 Bohr magnetons. However, the coincidence is striking enough to make one think again about Dr. Caughey's suggestion.

References

1. Tasaki, A., J. Otsuka, and M. Kotani. Biochem. Biophys. Acta, <u>40</u>, 284 (1967)

2. Caughey, W.S. Hemes and Hemoproteins (B. Chance, R. Estabrook, and T. Yonetani, eds.), Academic Press, New York, 1966, pp. 276-277

STRUCTURE-REACTIVITY CORRELATIONS IN METHEMOGLOBIN REACTIONS

J. G. Beetlestone and D. H. Irvine

Department of Chemistry, University of Ibadan,
Ibadan, Nigeria

Two aspects of the structural control of reactivity in proteins will be discussed in this paper: Firstly, the effect of configurational changes in the protein on the thermodynamics of reactions of a protein with small molecules, and secondly, the effect of differences in the amino acid composition of the protein on the thermodynamics of such reactions.

The system we have examined over the past few years is the reaction of methemoglobins with various ligands (1-15) which, although of no physiological significance, is an excellent model system insofar as the structure of the protein is known to high resolution, the many available ligands provide means whereby the reaction may be varied in a known manner, there are many variants (both abnormal human and from different species of known amino acid composition), and data of high precision may be obtained with relatively simple techniques. Furthermore, this system allows a study to be made of the effect on reactivity of the intra-polypeptide configurational change that has been shown to occur in some hemoglobin reactions (16,17) in the absence of the much more dramatic inter-polypeptide configurational change (18) that is presumed to underlie the phenomenon known as heme-heme interaction in the reaction of hemoglobin with oxygen.

We have investigated primarily the pH dependence of the free energy and enthalpy of reactions in which the water molecule bonded to the sixth ligand coordination position in methemoglobin or metmyoglobin is replaced by another ligand, and in this paper we shall summarize the status of the main experimental observations, and a working hypothesis by which differences in reactivity may be correlated with differences in structure. The reaction may be written symbolically as

$$Hb\ OH_2 + L = HbL + H_2O$$

For these reactions, the Hill constant, n, is unity, implying that each of the heme groups has approximately the same affinity for a ligand and does not interact with the other hemes. We may therefore define an equilibrium constant, K_L:

$$K_L = \frac{[HbL]}{[HbOH_2][L]}$$

Standard free energies and enthalpies for this reaction have been determined by measuring spectrophotometrically K_L as a function of temperature and pH at ionic strength of 0.05 M. We first compare the behavior of fluoride, azide, and hydrosulfide ions as ligands. (The formation of the cyanide ion complex has also been studied for two hemoglobins and the behavior is the same as for azide ion [8]. However, since the precise determination of the formation constant is technically more difficult for cyanide than for azide ion, the latter has been the ligand of choice.) The data are presented in Figure 1 and may be conveniently discussed as follows:

1) The variation of $\Delta G°$ with pH, or between methemoglobins from different species, is much smaller than the corresponding changes in $\Delta H°$ which are accompanied by compensating variations in $\Delta S°$. This is conveniently demonstrated by plotting $T\Delta S°$ against $\Delta H°$ for either different pH's for the same methemoglobin, or for different methemoglobins at the same pH. As Figure 2 shows, a linear plot is obtained with a slope close to unity. Similar large compensating enthalpy and entropy changes have been reported for the successive formation constants for the reaction of hemoglobin with oxygen (19) and for the oxidation of hemoglobin to methemoglobin at different pH's (20). We envisage two possible ways in which such large compensating enthalpy and entropy changes could arise:

a) Configurational changes in the protein, which would give rise to enthalpy and entropy changes of the same sign which would compensate, at least to some extent.

b) Hydration changes which would not give rise to any contribution to the free energy change, since water in the hydration sheath is always in equilibrium with the solvent water but which, like a phase change, would be accompanied by exactly compensating enthalpy and entropy changes (21).

2) At no pH values do the $\Delta G°$'s for the formation of the fluoride complex of the methemoglobins and sperm whale met myoglobins differ by more than 250 cal/mole. Furthermore, if electrostatic effects arising from different net charges on

the methemoglobins are minimized by comparing $\Delta G°$'s at the isoelectric points, the maximum difference for any two methemoglobins is only 80 cal/mole, which is comparable with the experimental error (15). By contrast, $\Delta G°$'s for the formation
of the azide and hydrosulfide complexes differ by up to 1.2
and 1.8 kcal/mole, respectively. We attribute this difference
in behavior between fluoride ion on the one hand, and azide
and hydrosulfide ion on the other, to the conformational change
in the protein that is known to accompany the formation of

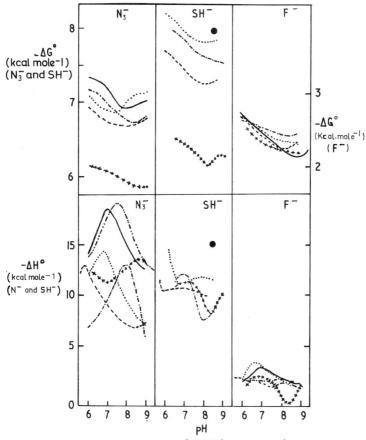

Figure 1. The variation of $\Delta G°$ (27°) and $\Delta H°$ with pH for the
formation of the fluoride, azide, and hydrosulfide complexes
of the following hemoglobins: human A, ——.and ; dog, ...;
guinea pig, ---; pigeon, -.-; human C, —··——; sperm whale
metmyoglobin, +++.

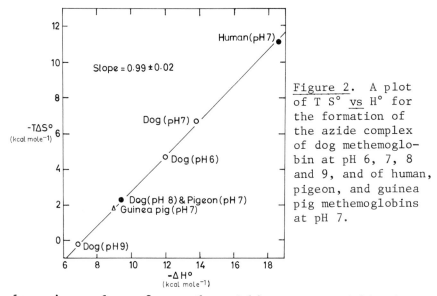

Figure 2. A plot of T S° vs H° for the formation of the azide complex of dog methemoglobin at pH 6, 7, 8 and 9, and of human, pigeon, and guinea pig methemoglobins at pH 7.

low-spin complexes from methemoglobin or metmyoglobin, but which is absent in the formation of high spin complexes (16, 17). The contribution of this configurational change to the free energy of complex formation will vary from one methemoglobin to another as a consequence of differences in amino acid composition. The question arises as to whether the configurational change is the same for all ligands forming low spin complexes. Figure 3 shows a plot of the average value between pH 6 and 9 of ΔG° for the hydrosulfide complex against that for the azide complex over the same pH range. This is good linear correlation, with a slope of 1.5. This result is open to three possible interpretations:

a) The configurational change accompanying ligand binding is spin-specific (<u>i.e.</u>, it is the same for all low-spin complexes) and the departure from unit slope of the plot in Figure 3 merely reflects the different percentages of the low-spin form in the azide and hydrosulfide complexes (22). Preliminary measurements of K_L for the cyanide complex of metmyoglobin and of other methemoglobins (23) would suggest, however, that this simple interpretation is not strictly valid, since although there is a correlation between the ΔG°'s for the formation of the cyanide and hydrosulfide complexes which are both entirely low spin, the slope is less than unity.

b) The configurational change is ligand-specific, but

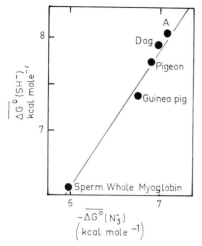

Figure 3. A plot of the average value between pH 6 and 9 of the standard free energy of formation of the hydrosulfide complex, $\Delta G°(SH^-)$ <u>vs</u>. the average value over the same pH range of the standard free energy of formation of the azide complex, $\Delta G°(N_3^-)$ for various hemoglobins.

for any two ligands the different contributions to the free energies of complex formation are the same for all the methemoglobins. This interpretation is consistent with the experimental observations, but would appear to have little theoretical justification. Furthermore the intermediate behavior exhibited by ligands such as the nitrite ion which form complexes which are mixtures of high and low spin forms would seem to suggest that the spin state is significant (24).

c) The configurational change is not so much spin specific or ligand specific but is associated with a particular structural feature that occurs concomitantly with low spin complex formation, namely the optimising of a hydrogen bond between the ligand bonded to the iron atom and the "internal" nitrogen of the distal imidazole. The extent to which a change in structure is required to optimise hydrogen bonding would be dependent both on the ligand and on the haemoprotein. We shall return to this question of hydrogen bonding later.

3) We now examine the variation of $\Delta H°$ with pH for fluoride complex formation (15). In the absence of a configurational change the simplest model for a methaemoglobin reacting with a small negative ion would be a dielectric cavity model, which indeed has been shown by us (1,7) to predict correctly the pH variation of the ionic strength dependence of the formation constants of methaemoglobin complexes. However such a model implies that free energy changes arising from electrostatic effects should be much larger than enthalpy changes, which is not the observed behaviour. We conclude therefore

that either there is a small configurational change not observable at the resolution of the structural observations or that there is a mechanism not involving a configurational change by which reaction of the fluoride ion at the iron atom inside the molecule can produce changes in hydration. While the former may occur we suggest that changes in hydration could also arise as a result of the loss of polarization on the distal imidazole ring. Schematically

In the methaemoglobin the polarized form would be stabilized by virtue of the formal positive charge carried by the iron atom in methaemoglobin. Loss of positive charge on the nitrogen atom which "sees" the solvent would give rise to changes in the hydration structure.

4) We compare now the pH variation of $\Delta H°$ for the formation of the fluoride and hydrosulphide complexes of sperm whale metmyoglobin. It may seem from Figure 1 that the pH variation of $\Delta H°$ is of the same form for both ligands and Figure 4 shows that after an arbitrarily chosen 9.1 kcal/mole has been subtracted from each of the values of the fluoride ion complex the two sets of data agree within experimental error. We conclude that for sperm whale metmyoglobin the configurational change which presumably accompanies the formation of the low spin hydrosulfide complex does not give rise to any pH dependent contribution to the enthalpy of complex formation.

5) If we now turn to the formation of the fluoride and hydrosulphide complexes of methaemoglobins, we see from Figure 1 that $\Delta H°$ as a function of pH is different for the two ligands. This, together with the data for metmyoglobin discussed under [4], suggests that the intra-chain configurational change gives rise to a pH dependent contribution to $\Delta H°$ by virtue of changes in inter-chain interactions, even though for these reactions the gross inter-chain configurational changes does not occur.

6) Inspection of Figure 1 shows that the behaviour of

272

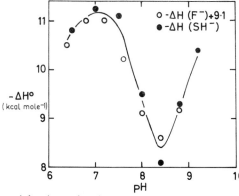

Figure 4. A plot of $\Delta H°$ vs. pH for the formation of the fluoride (o-o) and hydrosulfide (o-o) complexes of sperm whale metmyoglobin. The plotted values for the fluoride complex are the observed values minus an arbitrarily chosen value of 9.1 kcal/mole.

azide ion (and cyanide ion) is distinctly different from that for hydrosulphide ion in that (i) $-\Delta H°$ for the former systems varies by 6 to 7 kcal/mole^{-1} over the pH range compared with 3 to 4 kcal/mole^{-1} for hydrosulphide ion, and (ii) the plot of $-\Delta H°$ against pH shows a distinct maximum at the pH at which $-\Delta H°$ for the hydrosulphide ion reaction shows either a minimum or a distinct change in slope. Taken with the free energy data discussed earlier, these observations could be explained on the basis of the configurational change accompanying complex formation being different for azide and cyanide ions on the one hand and hydrosulphide ion on the other. Alternatively, it could mean that the reaction with azide and cyanide ion is accompanied by an additional mechanism which produces variable enthalpy and entropy changes, but which is absent in the hydrosulphide reaction. Any postulated additional mechanism for azide and cyanide ion must however conform to the following:

a) It must be common to methaemoglobins and sperm whale metmyoglobin; that is to say it must not be dependent on subunit interaction, since the variation of $\Delta H°$ with pH for azide and cyanide complexes is of the same form for both haemoproteins.

b) Its contribution to the free energy of formation of the complex must differ from one methaemoglobin to another, in a manner consistent with the correlation obtained in Figure 3.

A mechanism conforming to these restrictions is suggested by the work of Stryer et al. (25), in which they reported change in the electron density in the region of the haem ring when metmyoglobin reacts with azide ion in the crystal. They interpreted these observations in terms of the replacement of

273

the water molecule bonded to the iron atom by the azide ion,
and the transfer of a hydrogen atom from the "external" to
the "internal" nitrogen of the distal imidazole with the concomitant formation of a hydrogen bond with the ligand and the
loss of a sulphate ion from a position adjacent to the external nitrogen, the sulphate ion presumably being hydrogen bonded to this nitrogen in the crystal of metmyoglobin. Following this formulation we therefore write the formation of the
azide complex schematically as

The reorientation of the water molecule that is required in
order to maintain hydrogen bonding with the "external" nitrogen will disrupt the hydration structure in its vicinity and
hence give rise to compensating changes in enthalpy and entropy which will be pH dependent by virtue of the fact that
the hydration structure around the protein will be pH dependent. This mechanism clearly conforms to the two restrictions
above, and furthermore it is immediately apparent why it
should be absent for the hydrosulphide ion. In this case the
hydrogen bond with the distal imidazole would be formed with
the hydrogen atom on the ligand itself. Schematically

Hence for the hydrosulphide complex a hydrogen atom is present
on the "external" nitrogen atom on the distal imidazole and
the hydration changes which we have associated with the removal of this hydrogen atom will not accompany the formation of
this complex. The implication of this is that the enthalpy

274

of formation of any complex with a ligand which brings its own hydrogen atom into the complex should show behaviour similar to hydrosulphide rather than azide and cyanide ion. If we make the commonly made assumption that the spectrophotometrically observable ionization of methaemoglobin to its alkaline form can be identified with the formation of the hydroxyl ion complex then there should be a correlation between ΔH° for the·formation of the hydroxyl ion complex (4,5) and ΔH° for the formation of the hydrosulphide complex (15) but not between ΔH° for the formation of the hydroxyl ion complex and ΔH° for the formation of the azide complex (8,9,14). Figure 5 shows that this is indeed observed. Further we should expect $-\Delta H^\circ$ as a function of pH for the formation of the hydroxyl ion complex to be of the same form as that for the hydrosulphide complex rather than that for the azide complex. It is not possible to make the measurements for methaemoglobins, but this prediction has been verified for the ionization of sperm whale metmyoglobin (14).

It could be argued that the hydroxyl ion and the hydrosulphide ion are so closely similar that they would be expected to behave similarly whatever the nature of complex formation with methaemoglobin. We therefore determined the enthalpy of formation of a complex with methylamine; a ligand that forms a low spin complex and which carries its own hydrogen atom into the complex, but which differs from hydrosulphide ion not only in structure but also by virtue of being neutral rather than negatively charged. Figure 5 shows a plot of ΔH° for the formation of the methylamine complex (26) against ΔH° for the formation of the hydrosulphide and azide complexes for a series of methaemoglobins. As we would predict, there is a linear relation between ΔH° for the formation of the hydrosulphide and methylamine complexes.

7) In the discussion so far we have emphasized the differences between the behaviour of different ligands. We now turn our attention to a feature of the variation of ΔH° with pH that is common to all ligands. Inspection of Figure 1 shows that for each methaemoglobin there is a characteristic pH, which varies from 5.8 for guinea pig methaemoglobin to 8.6 for sperm whale metmyoglobin, at which the plot of ΔH° against pH for any ligand shows either a distinct maximum or minimum, or at least a dramatic change in slope. In passing we draw attention to the results of Hammes and Schimmel (27) on the binding of cytidine E' phosphate to ribonuclease for which the variation with pH of ΔH° is much greater than that for

275

ΔG° and for which, as shown in Figure 6, $-\Delta H^{\circ}$ as a function of pH passes through a maximum. This suggests that the phenomenon of a characteristic pH may not be restricted to haemoglobin.

Clearly any quantitative explanation of the variation of ΔH° with pH must involve a knowledge of the pH dependence of any configurational change and of the structure of the hydration sheath of the protein as a function of pH. Neither of

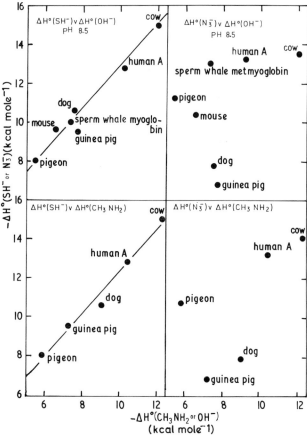

Figure 5. (1) Plots of ΔH° for the formation of the hydrosulphide complex against ΔH° for the formation constant of the hydroxyl complex and of the methylamine complex. (2) Plots of ΔH° for the formation constant of the azide complex against ΔH° for the formation of the hydroxyl complex and of the methylamine complex.

276

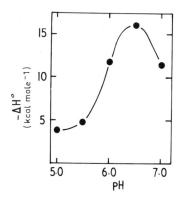

Figure 6. A plot of $\Delta H°$ for the binding of cytidine-3'-phosphate to ribonuclease vs. pH.

these is known but it is safe to assume that both will be dependent on the number, type, and orientation of polar groups on the surface of the molecule, and that the differing characteristic pH's of the methaemoglobins are reflections of changes in some of these factors. If this is the case, then there might be a correlation between the characteristic pH and the isoelectric point of a methaemoglobin.

Figure 7 shows that there is indeed a rough correlation of this type, but that the characteristic pH cannot be direc-

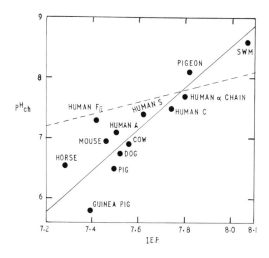

Figure 7. A plot of the characteristic pH, pH_{ch}, for a methaemoglobin vs. isoelectric point (23). The dashed line corresponds to pH_{ch} = I.E.P.

tly dependent on the value of the isoelectric point since the characteristic pH varies over a much wider range of pH than the isoelectric point, the characteristic pH being up to 0.5 unit above the isoelectric point for some methaemoglobins and up to 1.6 units below it for others. The correlation suggests however, that the characteristic pH and the isoelectric point may both be functions of the same property of the methaemoglobin.

An obvious common factor would be the relative proportion of different types of charged groups in the molecule and we have attempted to find such a correlation for those methaemoglobins where the precise amino acid composition is known. Table I shows the changes in amino acid composition relative to human haemoglobin A for four abnormal human haemoglobins together with their characteristic pH, pH_{ch}.

TABLE I

Hemoglobin	Amino Acid Changes from Human A		pH_{ch}
Human A			7.1
Human C	β6 Glu	Lys	7.5
Human S	β6 Glu	Val	7.4
Human D Ibadan	β87 Thr	Lys	7.4

We note that irrespective of the position in the molecule the gain of a lysine residue or the loss of a glutamic acid residue leads to an increase of pH_{ch}. Further, bearing in mind the experimental uncertainty in pH_{ch} of about $\pm$ 0.1 unit, the magnitude of the shift in pH_{ch} suggests that pH_{ch} may be independent of the position of the charged residues in the molecule. This leads us to suspect that pH_{ch} may be directly related to the difference between the numbers of negatively and positively charged groups irrespective of their position in the molecule. Since histidine residues will be predominantly in the neutral form in the region of pH under consideration this difference will be given by (Lys + Arg - Glu - Asp - 8) where the symbols represent the number of lysine, arginine, glutamic acid and aspartic acid residues in the haemoglobin tetramer. The number eight appears in order to take into account the two propionic acid side chains on each haem ring.

In order to test this hypothesis we have determined pH_{ch} for those haemoglobins where the precise amino acid composition is known from sequence studies, and Figure 8 shows a

278

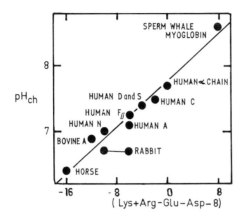

Figure 8. A plot of the characteristic pH, pH_{ch}, against the function, (Lys + Arg - Glu - Asp - 8) for a series of haemoglobins (9,12,14,23,29,30). The amino acid composition on which this plot was made were taken from reference 32 with the exception of that for horse haemoglobin which was that given by Perutz (28). For the purpose of this plot metmyoglobin has been treated as if it were a tetramer. The double point for rabbit corresponds to an uncertainty of four per tetramer in the number of glutamic acid residues.

plot of pH_{ch} for these haemoglobins against this function of the amino acid composition. In spite of the multiplicity of amino acid differences between the different haemoglobins the points all fall on a straight line within experimental error. A further point to note is that for haemoglobins where pH_{ch} is greater than the isoelectric point the amino acid function is positive but where pH_{ch} is less than the isoelectric point the function is negative. All of the charged side chains of amino acid residues are either on the surface of the haemoglobin molecule or point out into the solvent (28) and to account for this correlation we suggest that depending on their relative proportions the amino acids which carry charges in the pH range 6-9 cooperatively adopt different configurations above and below the characteristic pH, thereby producing a different hydration structure above and below this pH. This suggestion is supported by the observation that for the three methaemoglobins so far examined a dielectric cavity model for the protein successfully accounts for the pH dependence of the ionic strength variation of the formation constant for a complex of a methaemoglobin with a charged ligand at all pH's between 5.8 and 9 except at the characteristic pH of the met-haemoglobin (7,31). The dielectric cavity model involves the assumption of fixed charges and clearly if at any pH the charged groups are in transition from one configuration to another this assumption would not be valid since the orientation of the charged groups would be expected to be ionic strength dependent.

The importance of this characteristic pH for a haemoglobin has been emphasized in some recent experiments carried out in our laboratory. We have examined the Bohr effect and the reverse Bohr effect of a number of animal haemoglobins by the direct measurement of proton release upon oxygenation (23). The details of this work will be published elsewhere but briefly the finding is that the magnitude of the Bohr effect as measured by this method varies only slightly from one haemoglobin to another but the reverse Bohr effect differs markedly, and as Figure 9 shows, there is a good correlation between the characteristic pH of a haemoglobin and the magnitude of its reverse Bohr effect. Possible explanations for this correlation will be discussed elsewhere, but it is apparent that the characteristic pH has significance in the context of the reaction of haemoglobin with oxygen. It will be noted that several of the haemoglobins used in obtaining the correlation between pH_{ch} and the reverse Bohr effect are not the same as those used in obtaining the correlation between pH_{ch} and the function (Lys + Arg - Glu - Asp - 8). The reasons for this are (i) the limited availability of some of the haemoglobins used to obtain the latter correlation and (ii) the

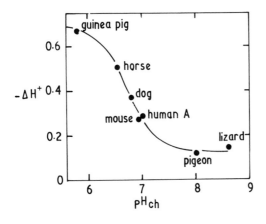

Figure 9. A plot of the magnitude of the reverse Bohr effect at pH 5.3 vs. pH_{ch} (21). The solid line is a smooth curve drawn through experimental points.

fact that the precise amino acid compositions are not known for some haemoglobins such as those of guinea pig and pigeon which have widely different pH_{ch}'s and reverse Bohr effects. Nevertheless taken together the two correlations in Figs. 8 and 9 imply that the functional properties of a haemoglobin depend on the relative proportions of charged amino acids irrespective of their position in the molecule. This implies that mutations giving rise to the gain or loss of charged amino acid residues anywhere on the surface of the molecule would change the functional properties of the molecule by virtue of their effect on the functions (Lys - Arg - Glu - Asp - 8). Since charged amino acid residues have been found in about 60% of the sites in the molecule (33), this suggests that mutations giving rise to amino acid changes at a large proportion of the positions in the molecule could, under favourable circumstances, be selected for by virtue of their effect on the functional properties of the molecule.

References

1. Beetlestone, J.G., and Irvine, D.H. Proc. Roy. Soc., A, 277, 401 (1964).

2. Beetlestone, J.G., and Irvine, D.H. Proc. Roy. Soc., A, 277, 414 (1964).

3. Beetlestone, J.G., and Irvine, D.H. J. Chem. Soc., 5086 (1964).

4. Beetlestone, J.G., and Irvine, D.H. J. Chem. Soc., 5090, (1964).

5. Beetlestone, J.G., and Irvine, D.H. J. Chem. Soc., 3271, (1965).

6. Anusiem, A.C., Beetlestone, J.G., and Irvine, D.H. J. Chem. Soc. (A), 357 (1966).

7. Beetlestone, J.G., and Irvine, D.H. J. Chem. Soc. (A), 951 (1968).

8. Anuseim, A.C., Beetlestone, J.G., and Irvine, D.H. J. Chem Soc. (A), 960 (1968).

9. Anusiem, A.C., Beetlestone, J.G., and Irvine, D.H. J. Chem. Soc. (A), 1337 (1968).

10. Beetlestone, J.G., and Irvine, D.H. J. Chem. Soc. (A), 1340 (1968).

11. Beetlestone, J.G., Epega, A.A., and Irvine, D.H. J. Chem. Soc. (A), 1346 (1968).

12. Bailey, J.E., Beetlestone, J.G., and Irvine, D.H. J. Chem. Soc. (A), 2778 (1968).

13. Bailey, J.E., Beetlestone, J.G., and Irvine, D.H. J. Chem. Soc. (A), 2913 (1968).

14. Bailey, J.E., Beetlestone, J.G., and Irvine, D.H. J. Chem. Soc. (A), 241 (1969).

15. Beetlestone, J.B., and Irvine, D.H. J. Chem. Soc. (A), in press.

16. Watson, H.C., and Chance, B. In Hemes and Hemoproteins (R.W. Estabrook, . Yonetani and B. Chance, eds.), Academic Press, Inc., New York, 1966, p. 149.

17. Ogawa, S., and McConnell, N.M. Proc. Nat. Acad. Sci. U.S., 58, 19 (1967).

18. Muirhead, H., and Perutz, M.F. Nature, 199, 633 (1963).

19. Roughton, F.J.W., Otis, A.B., and Lyster, R.L.J. Proc. Roy. Soc. B, 144, 29 (1955).

20. Antonini, E., Wyman, J., Brunori, M., Taylor, J.F., Rossi-Fanelli, A., and Caputo, A. J. Biol. Chem., 239, 907 (1964).

21. Ives, D.J.C., and Marsden, P.D. J. Chem. Soc., 649 (1965).

22. Beetlestone, J., and George, P. Biochemistry, 3, 707 (1964).

23. Beetlestone, J.B., and Irvine, D.H. Unpublished observations.

24. Beetlestone, J.G., Irvine, D.H., and Adeosun, O.S. Unpublished observations.

25. Stryer, L., Kendrew, J.C., and Watson, H.C. J. Mol. Biol., 8, 96 (1964).

26. Beetlestone, J.G., Ige, R.O., and Irvine, D.H. Unpublished observations.

27. Hammes, G.C., and Schimmel, P.R. J. Amer. Chem. Soc., 87, 4665 (1965).

28. Perutz, M.F. J. Mol. Biol., 13, 646 (1965).

29. Bailey, J.E., Beetlestone, J.G., and Irvine, D.H. Unpublished observations.

30. Ogunmola, G.B. Ph.D. Thesis, Ibadan, 1968.

31. Beetlestone, J.G., Irvine, D.H., and Okonjo, K.O. Unpublished observations.

32. Dayhuff, M.U., and Eck, R.V. Atlas of Protein Structure, National Biomedial Research Foundation, Silver Spring, 1968.

33. Zuckerkandl, E. and Pauling, L. Evolving Hemes and Proteins (V. Bryson and H.J. Vogel, eds.), Academic Press, Inc., New York, 1965, p. 97.

DISCUSSION

Czerlinski: I would like to enquire about your complexing of azide to metmyoglobin. According to your scheme, the imidazole residue rotates around from the solution side to the interior. Have you performed pH difference titrations to demonstrate your additional pK_H upon azide binding? The data from the laboratory of Alberty (1) and our own (2,3) show no substantial shifts at any pH.

Beetlestone: There are two points here. The first is that the diagram I showed are completely schematic, and do not attempt to show the actual position of the distal imidazole ring.

The second point is more substantial. If I understand you correctly, you suggest that if the hydrogen atom moves to the inside position, you are left with a free nitrogen atom which can then pick up a proton. This is, indeed, observed. Measurements of proton uptake upon addition of azide ion below pH 6 suggest a pK in the region of 5.5. The experiment which should be done is the equivalent one with fluoride, where I postulate that to a large extent the proton does not move to the inside position. Unfortunately, this is technically difficult, as one must add large concentrations of fluoride ion and the consequent change in the liquid junction potential is probably comparable to the change in the EMF arising from the pH change.

References

1. Goldsalk, D.E., W.S. Eberlein, R.A. Alberty. J. Biol. Chem., 240, 4312 (1965); J. Biol. Chem., 241, 2653 (1966).

2. Czerlinski, G.H. In Hemes and Hemoproteins (B. Chance, R. Estabrook and T. Yonetani, eds.), Academic Press, New York, 1966, pp. 195, 207, and 209.

3. Duffey, D., B. Chance and G. Czerlinski. Biochemistry, 5, 3514 (1966).

SPECTROSCOPY OF HEMES AND HEMEPROTEINS BELOW 275 nm*

Arthur S. Brill

Department of Materials Science, University of Virginia

Howard E. Sandberg

Molecular Biophysics Laboratory
Washington State University

Patricia A. Turley

Department of Chemistry, Yale University

The shift of the Soret band to longer wavelengths with decrease in the magnetic moment of the ferric ion in heme proteins has been the subject of study for many years (1-4). A similar effect has now been observed for transitions of about twice the Soret energy: for all cases thus far examined, the formation of a complex or compound of lower spin is accompanied by the appearance of a positive band or bands between 250 and 210 nm in the difference spectrum which is referred to the high-spin free hemeprotein (5-8). In figure 1 are shown several ferrihemoglobin difference spectra to demonstrate the phenomenon.

The use of difference spectroscopy is necessitated by a rapidly increasing absorptivity, which arises from the contribution of many protein chromophores below 240 nm. The loss of the absolute spectrum is not particularly damaging, for one could not distinguish among the overlapping maze of transitions underlying the absorption spectrum in this region those which are affected when the spin state of the iron changes. In the hope of being able, ultimately, to distinguish in the hemeprotein difference spectra features primarily porphyrin in origin   from those of primarily protein origin,

*These studies were supported by U. S. Public Health Service Research Grants GM-09256 and GM-16504 from the Division of General Medical Sciences.

we took up an investigation of the ultraviolet spectral properties of heme and heme complexes. A monomeric (hemes separated as in hemeproteins) ferric porphyrin solution would be
suitable for such experiments. Unfortunately complexes of
ferriprotoporphyrin IX, the prosthetic group of the proteins
we most frequently study, are known to aggregate in aqueous
solution, the interactions involved having spectroscopic
effects.

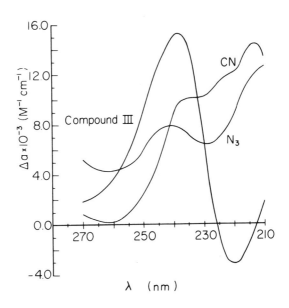

Figure 1. UV difference spectra of cyanide and azide complexes of ferrihemoglobin and of ferrihemoglobin compound III.
For ferrihemoglobin-CN, the heme concentration was 54 μM;
cyanide 30 mM; path length 0.05 cm; and pH 6.7. For ferri-
hemoglobin-N_3, the heme concentration was 60 μM; azide 3 mM;
path length 0.05 cm; and pH 6.7. The optimum formation of
ferrihemoglobin compound III (as determined by A_{424} nm)
occurred at pH 7.4. The maximum absorbance at 424 nm and the
maximum difference absorbance at 238 nm were observed 9 minutes after 10 ul of 4.3 mM H_2O_2 was mixed with 2.0 ml of
4.0 μM (heme) ferrihemoglobin in the absorption cell (1.0 cm
path length). The observed difference spectrum was corrected
for dilution attendant the H_2O_2 addition.

286

In alkaline aqueous solution (borate buffer, pH 9.5), heme (as hematin) is substantially depolymerized and spectral studies are possible. Hematin has a broad band with a maximum absorptivity at 254 nm of about 20,000 $M^{-1}cm^{-1}$, and a broad shoulder in the region 220 to 230 nm. The biscyanide complex is readily formed, and the difference spectrum versus hematin reveals an isosbestic point at 249 nm preceded by a trough with a minimum at 262 nm, the difference absorptivity being -2000 $M^{-1}cm^{-1}$. There is a single broad maximum located at 213 nm with a difference absorptivity of 6,400 $M^{-1}cm^{-1}$ (5-7). The azide complex does not form in sufficient concentration, even at saturation concentration of azide, for the difference spectrum to be observed.

We now have considerable spectroscopic and thermodynamic data on the association of hemin in alcohols (9,10). Hemin in methanol is an equilibrium mixture of monomer and dimer. Above 300 nm there are substantial spectroscopic differences between monomer and dimer, but below 300 nm the differences are small. There is a broad band with a maximum absorptivity at 260 nm of about 20,000 $M^{-1}cm^{-1}$ (similar to hematin), and another with a maximum absorptivity at 220 nm of about 30,000 $M^{-1}cm^{-1}$. Hemin in ethanol and n-butanol behaves much the same as in methanol but the dissociation constants for the dimer are considerably larger. In n-propanol the Soret absorptivities at both high and low concentration of hemin are smaller than in the other alcohols.

Our studies of spectral changes accompanying complex formation in alcohol solutions are just getting underway. In methanol the biscyano complex forms readily from hemin. The difference spectrum reveals an isosbestic point at 248 nm, there is a single broad maximum located at 215 nm with a difference absorptivity of 23,000 $M^{-1}cm^{-1}$, considerably greater than for ferrimyoglobin cyanide and ferrihemoglobin cyanide vs. the free hemeproteins (11,000-14,000), and very much greater than the 6,400 $M^{-1}cm^{-1}$ value given above for the corresponding measurement on heme in alkaline aqueous solution.

In n-propanol, cyanide is poorly soluble, and the cyanide complex cannot be formed. A complex with imidazole does form. The difference spectrum shows there to be an isosbestic point at 251 nm exhibits a well-defined trough with a minimum at 275 nm, and positive absorbance below 251 nm (our present spectra do not extend below 240 nm for this complex).

With regard to Soret changes accompanying the binding of strong field ligands (e.g. cyanide), the behavior of heme in alcohols is much the same as that of aqueous solutions of protein-bound heme. In the wavelength region below 250 nm, heme and hemeproteins behave similarly insofar as a decrease in spin is accompanied by an increase in absorptivity, but the maxima which occur at 235-248 nm in difference spectra from the hemeproteins has not yet been observed in unbound heme. While this favors our assignment of the latter absorption to a transition involving a histidine residue in the fifth coordination position of the iron, confirmation or denial of this interpretation awaits the results of additional experiments of several kinds.

References

1. Theorell, H. and A. Ehrenberg, Acta Chem. Scand. 5, 823 (1951).

2. Scheler, W., G. Schoffa and F. Jung, Biochem. Z. 329, 232 (1957).

3. George, P., J. Beetlestone and J. S. Griffith, in Symposium on Hematin Enzymes, (J. E. Falk, R. Lemberg and R. K. Morton, eds.) Pergamon Press, New York, 1961, p. 105.

4. Brill, A. S. and R. J. P. Williams, Biochem. J. 78, 246 (1961).

5. Brill, A. S. and H. E. Sandberg, Proc. Natl. Acad. Sci. (U. S. A.) 57, 136 (1967).

6. Sandberg, H. E., Ph.D. Thesis, Yale University, New Haven, 1967.

7. Brill, A. S. and H. E. Sandberg, Biophys. J. 8, 669 (1968).

8. Brill, A. S. and H. E. Sandberg, Biochemistry 7, 4254 (1968).

9. Brill, A. S. and P. A. Turley, Abstract Hc-4 Fifth Int. Cong. on Photobiology, Hanover, 1968.

10. Brill, A. S. and P. A. Turley, in preparation.

INFRARED EVIDENCE FOR TWO TYPES OF BOUND CO IN CARBONYL MYOGLOBINS*

Sue McCoy[+] and Winslow S. Caughey[‡]

Department of Chemistry, University of South Florida
Tampa, Florida 33620

The infrared difference spectrum for aqueous solutions carbonyl myoglobin vs. myoglobin exhibits two carbonyl stretch bands, one at 1944 cm^{-1} and a second weaker band at about 1931 cm^{-1} (Fig. 1). However, a similar spectra with normal hemoglobins have shown only one sharp carbonyl stretch band at 1951 cm^{-1} (Fig. 1). Consistent with only one type of bound CO at each of the heme binding sites (1-3). Since certain abnormal hemoglobin do exhibit two bands (3), we had earlier suggested that the multiple bands in the case of myoglobins might result from protein inhomogeneity (4). However, recent evidence suggests that this is not the case. Rather, carbon monoxide binds to myoglobins in two different forms.

The spectra were obtained with the apparatus used for azido metmyoglobins (5) and with solutions prepared as described earlier (3,4).

Spectra similar to the one in Fig. 1 were obtained for several different preparations of horse heart and sperm whale myoglobins as well as for an electrophoretically homogeneous (sperm whale) myoglobin from which several components had been separated by Professor E. Antonini and his colleagues, Drs. E. A. Magnusson and G. Amiconi. Furthermore, two bands were also noted for sperm whale myoglobins

---

*This work was supported by USPHS grant HE-11807.
[+]Present address: Department of Orthopedic Surgery, University of Virginia School of Medicine, Charlottesville, Virginia 22901.
[‡]Present address: Department of Chemistry, Arizona State University, Tempe, Arizona 85281.

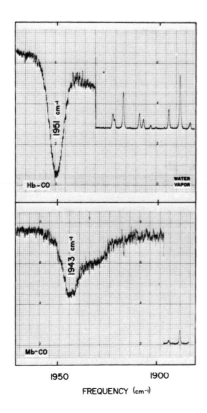

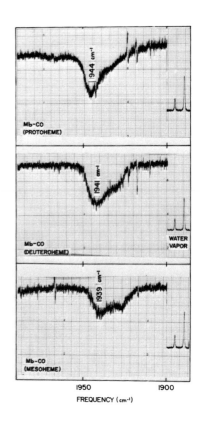

Figure 1. Top: Infrared difference spectrum of carbonyl hemoglobin (human, 5×10^{-3} M) vs. methemoglobin, in 0.1 M phosphate, pH 7.0

Bottom: Infrared difference spectrum of carbonyl myoglobin (sperm whale, 5×10^{-3} M) vs. metmyoglobin, in 0.1 M phosphate, pH 7.0.

Figure 2. The infrared difference spectra of the carbonyl derivatives of reconstituted myoglobins vs. myoglobin, in phosphate buffer, pH 6.4.

Top: with protoheme.

Middle: with deuteroheme

Bottom: with mesoheme.

reconstituted with protohemin, deuterohemin, and mesohemin by Professor Antonini and his colleagues (Fig. 2). In this case, the stronger band shifted to lower frequencies as the electron-withdrawing ability of the 2,4-substituent on the heme decreased. The shifts were of the same order of magnitude ($\sim$2 cm^{-1}) and in the same direction as were the shifts found for the pyridine carbonyl hemes (2). The reconstituted myoglobins also exhibited differences in the relative intensities of the two CO bands: the relative amount of lower frequency component increased in the order proto< deutero<meso. In the case of protoheme the area of the higher frequency band was nearly twice that of the other band; in the case of mesoheme, the bands are nearly equal in area. The shift in frequency was consistent with a lower CO band order with increased electron availability at the iron (2). With the assumption of comparable extinction coefficients for the two forms, increased electron availability at the iron also appeared to increase the relative amount of lower frequency component.

The relative amounts, but not the frequencies, of the two forms were affected somewhat by changes in pH. Spectra for solutions at pH 3, 5, 9, and 12 all showed a more intense band near 1944 cm^{-1} and a less intense band near 1931 cm^{-1}. The estimated relative areas of the bands varied from about 5:2 at pH 12 to 3:2 at pH 3.

If carbon monoxide binds to myoglobin in two forms, as does azide to metmyoglobin, then it is reasonable to expect that oxygen may be present in oxymyoglobin in two forms. This could provide an explanation for the greater oxidizability of oxymyoglobin compared with oxyhemoglobin which may exhibit only one type of oxygen binding.

References

1. Alben, J.O. and Caughey, W.S. In Hemes and Hemoproteins (B. Chance, R.W. Estabrook, and T. Yonetani, eds.), Academic Press, New York, 1966, p. 139.

2. Alben, J.O., and Caughey, W.S. Biochemistry, 7, 175 (1968).

3. Caughey, W.S., Alben, J.O., McCoy, S., Boyer, S.H., Charache, S., and Hathaway, P. Biochemistry, 8, 59 (1969).

4. Caughey, W.S., Eberspaecher, W.H., Fuchsman, S., McCoy, S., and Alben, J.O. Ann. N.Y. Acad. Sci., 153, 722 (1969).

5. McCoy, S., and Caughey, W.S., This volume, p. 295

DISCUSSION

Gibson: Dr. Caughey's very interesting contribution may perhaps explain a long-standing discrepancy in the kinetics of
myoglobin. In 1955, Prof. F.J.W. Roughton and I (1) pointed
out that if one studied the rate of replacement of oxygen by
carbon monoxide, using a range of oxygen concentrations, then
for a simple case such as that of myoglobin, if the reciprocal
of the rate of replacement is plotted against the oxygen concentration a straight line should be obtained. This procedure
worked out well for hemoglobin, where it was not necessary for
it to do so—but for myoglobin a curve rather than a straight
line was found, a result confirmed in a rather large number
of careful experiments by the Rome group and Dr. R.W. Noble
(2).

Quite a number of explanations were considered, mostly
involving conformation changes occurring within the same time
range as the ligand reactions, but in the experiments at Rome
none was able to satisfy tests designed to examine their predictions.

It was clear right along that heterogeneity would do
the trick, but there was not sufficient evidence arising from
observations other than those in the replacement reactions to
justify such an assumption. As a matter of fact, only a few
months ago Dr. L. Parkhurst and I (3) performed a series of
experiments on refractionated myoglobin using what we believed
to be single components according to Gurd's (4) findings in
the hope that this would provide an answer, but the curious
results persisted right along. Now Dr. Caughey is providing
evidence, if I follow him correctly, that there may be intrinsically more than one way of putting myoglobin together. It
is a fascinating prospect!

References

1. Gibson, Q.H. and F.J.W. Roughton. Proc. Roy. Soc. B.,
    143, 310.

2. Noble, R.W. Unpublished observations.

3. Gibson, Q.H. and L. Parkhurst. Unpublished observations.

4. Hapner, K.D., R.A. Bradshaw, C.R. Hartzell and F.R.N. Gurd. J. Biol. Chem., <u>243</u>, 683 (1968).

INFRARED BANDS FOR AZIDE IN THE HIGH AND LOW SPIN FORMS OF AZIDOMETHMYOGLOBIN*

Sue McCoy[+] and Windlow S. Caughey[+]

Department of Chemistry, University of South Florida
Tampa, Florida 33620

The infrared difference spectrum for aqueous solutions of azidometmyoglobin vs. metmyoglobin exhibits two absorption bands at 2045 and 2023 cm^{-1} due to asymmetric N-N stretching of bound azide (Fig. 1). The lower frequency band is of greater intensity and appears to represent the low-spin form whereas the higher frequency band represents the high-spin form.

The spectra were obtained by techniques similar to those used with carbonyl hemoglobins (1,2). Spectra were recorded with a Perkin Elmer Model 225 spectrometer equipped with Beckman-RIIC variable temperature cell holders. The use of calcium fluoride cells (pathlength 0.025 mm) permitted recording of both infrared and visible spectra of a given solution in the same cell. Metmyoglobin in a given medium was placed in the reference cell; a similar solution to which sodium azide had been added was placed in the sample cell. Temperatures were controlled within $\pm$5°C but variations in temperature over the range from 10 to 35° did not alter the results obtained.

Both the 2045 and 2023 cm^{-1} bands appear due to azido ligands bound to iron. The wave numbers are near those found for azido hemins (3). Excess azide in metmyoglobin solutions, azide ion in albumin solutions, and aqueous solutions of sodium azide give very broad absorptions, half-band

*This work was supported by USPHS Grant HE-11807.
[+]Present address: Department of Orthopedic Surgery, University of Virginia School of Medicine, Charlottesville, Virginia 22901.
[‡]Present address: Department of Chemistry, Arizona State University, Tempe, Arizona 85281.

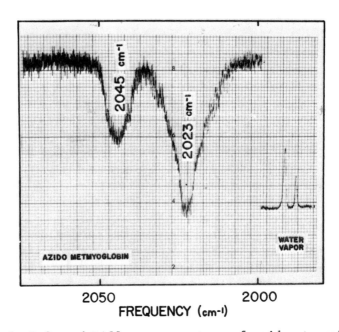

2050   2000
FREQUENCY (cm⁻¹)

Figure 1. Infrared Difference spectrum of azidometmyoglobin
$(10^{-2}$ M) in phosphate buffer (0.1 M, pH 7.0).

widths about 25 cm^{-1}, centered at 2047, 2043, and 2047 cm^{-1},
respectively, whereas the 2045 and 2023 cm^{-1} MbN$_3$ bands are
narrow bands, half-band widths ca. 7 cm^{-1}. The excess azide
from metmyoglobin solutions and the azide from albumin solutions are readily removed by dialysis, but the two MbN$_3$ bands
remain after dialysis against water for as long as 24 hours.

The two MbN$_3$ bands do not appear to result from protein
inhomogeneity. Several commercial preparations of sperm
whale and one of horse heart metmyoglobin exhibited similar
spectra, as did a sample of electrophoretically homogeneous
metmyoglobin from which several components had been separated
by Professor E. Antonini and his colleagues. Variations in
pH over the range from 3 to 11.6 did not significantly affect
the frequencies or relative intensities of the two bands.

Beetlestone and George (4) reported magnetic susceptibility evidence that azidometmyoglobin exists in two spin
states in a ratio of low to high spin forms of about 4 to 1
at the temperatures studied here. With the assumption of

296

comparable extinction coefficients for the two forms, comparison of the area under the MgN_3 bands shows the 2023 cm^{-1} band about four times as abundant as the 2045 cm^{-1} band suggesting their assignment to low and high spin forms respectively. Studies with simple hemin azides fully supported such an assignment.

Although it is difficult at present to assess any differences in sterochemistry expected between the two forms of bound azide, more than one location for azido ligand was not detected in X-ray studies of crystalline forms (5). Therefore, the structural differences may be very small and beyond the resolution of the X-ray methods. Also there may be differences in structure between dissolved and crystalline protein.

Preliminary studies with azidomethemoglobins have also revealed two azido bands which differ from the $MetMbN_3$ case, in that they are both at slightly higher wave numbers; also, the ratio of intensities of the lower frequency band to the higher frequency band is somewhat greater, consistent with the magnetic susceptibility findings of Beetlestone and George (4), which showed smaller proportions of high spin component in the hemoglobin than in the myoglobin derivative.

References

1. Alben, J.O. and W.S. Caughey. Biochemistry, 7, 175 (1968).

2. Caughey, W.S., J.O. Alben, S. McCoy, S. Charache, P. Hathaway, and S. Boyer. Biochemistry, 8, 59 (1969).

3. Sadasivan, N., H.I. Eberspaecher, W.H. Fuchsman, and W.S. Caughey. Biochemistry, 8, 534 (1969).

4. Beetlestone, J., and P. George. Biochemistry, 3, 707 (1964).

5. Stryer, L., J.C. Kendrew, and H.C. Watson. J. Mol. Biol., 8, 96 (1964).

KINETIC STUDIES ON THE ACID-ALKALINE TRANSITION IN METHEMOGLOBIN AND METMYOGLOBIN*

G. Ilgenfritz

Max Planck Institut für Physikalische Chemie
Göttingen, Germany

T. M. Schuster

Department of Biology, State University of New York
Buffalo, New York

Although the pH dependence of the optical spectra of metmyoglobin and methemoglobin are consistent with the assumption that the ionization of the water molecule bound to the sixth coordination position of Fe(III) is a simple protolytic reaction, there are equilibrium data (1,2) which suggest that this may not be the case.

The usual formulation of the reaction is:

$$Fe(III) \cdot H_2O \rightleftharpoons Fe(III) \cdot OH^- + H^+ \qquad (1)$$

$$or: Fe(III) \cdot H_2O + OH^- \rightleftharpoons Fe(III) \cdot OH^- + H_2O \qquad (2)$$

Since the kinetics of these reactions have never been studied, we felt that such an investigation might yield new information about the reaction mechanism. This paper presents kinetic evidence that in both hemeproteins the acid-alkaline transition is quite a complex reaction.

From these studies the following conclusions may be drawn:

1) Both reactions are very rapid, occurring in the time range of 1 to 10 μsec, in contrast to the much slower substitution reactions which occur with other hemeprotein Fe(III) ligands.

*This research was supported by USPHS HE-11425

2) In neither case can the kinetics be described by a simple one-step proton transfer process.

3) The kinetics of the ionization of the iron-bound water molecule in metmyoglobin and methemoglobin are similar in spite of opposite responses to the electric field.

4) There is no evidence for a rate limiting conformation change but there is a strong thermodynamic linkage of the water ionization with at least one other dissociable group in the protein.

Previous temperature-jump experiments with methemoglobin (MetHb) and metmyoglobin (MetMb) by Schuster and Brunori (unpublished results) revealed that there is no measurable relaxation in the time range covered by this method and that the entire change in optical absorbance was complete in the heating time of several microseconds.

Therefore in order to resolve the time course of these reactions we have used the more rapid electric field-jump relaxation method by which one may measure reactions as fast as 50 nanoseconds. (For a description of the method see G. Ilgenfritz, this volume.) In this method the external perturbing parameter is electric field strength which, ideally, should be a rectangular pulse. However since the instrument employed for these investigations was designed to give an exactly constant voltage for only 3 microseconds and these reactions are somewhat slower one must consider the decay of the electric field and the accompanying temperature increase which occurs in aqueous solutions. The electric field decays with a step function whereas the corresponding temperature increase is nearly linear and acts either in the same or opposite direction, depending on the relative signs of the change in electric moment and the enthalpy for the reaction. Examples are illustrated in Figure 1 where theoretical traces are shown for a rectangular field pulse followed by a linear temperature increase for chemical relaxations with positive, zero and negative reaction enthalpies. It is therefore difficult to analyze the chemical relaxation which occurs following the zero to high field perturbation. However, the field jump method offers the advantage that one obtains not only the relaxation from zero to high field but the back relaxation from high to zero electric field, as is illustrated in Figure 1.

If one observes a different optical density before and after the electric field pulse one can test whether this difference is due to heating by applying pulses of varying time

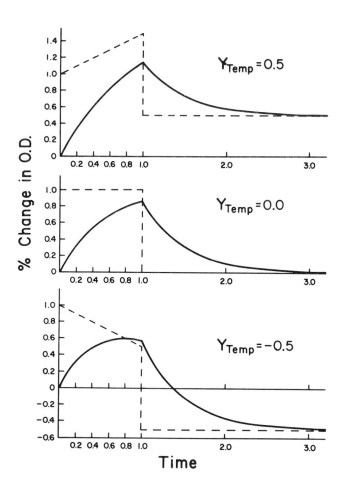

Figure 1. Calculated relaxation curves for a perturbation function containing a rectangular pulse of electric field strength and linear temperature increase. Pulse length, $t_1 = 1$; relaxation time, $\tau = 0.5$. $Y_{field} = 1$.

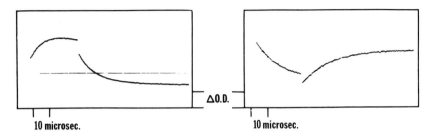

Figure 2. Relaxation curves for sheep methemoglobin at wave-lengths on either side of the equilibrium isosbestic point. Left-hand traces λ = 580 nm; ΔO.D. = 0.0077/cm; 9.5 x 10^{-5} heme, pH 8.7. Right hand traces λ = 405 nm; ΔO.D. = 0.0052/cm; 4.2 x 10^{-6} heme; pH 8.6. In both cases, E = 74 kV/cm; d = 10.7 mm; T = 20°C. No buffer.

duration. Figure 2 shows the observed optical density change in MetHb in an electric field-jump experiment.

The following results demonstrate that we are observing the acid-alkaline transition with this method.

The oscilloscope traces shown in Figure 2 demonstrate that the relative sign of the optical density change is in agreement with the changes observed in static spectophotometry i.e., opposite changes on either side of the isosbestic point of the acid-alkaline transition. Also, the measured time constants depend strongly on pH. There is an appreciable field effect for both MetMb and MetHb only in the pH range of the pK, as determined in equilibrium experiments (pK = 8.5 for sheep MetHb and pK = 9.0 for sperm whale MetMb), as one would expect for the corresponding protolytic reactions.

The relaxation curves for MetHb (see Fig. 2) and MetMb (see Fig. 6) cannot be described by a single exponential curve. Therefore the corresponding reaction must exhibit at least two relaxation times, from which we may conclude that the acid-alkaline transition in both proteins cannot be accounted for by a simple one step proton transfer reaction.

Because of the strong overlap of the relaxations we have been able to evaluate accurately only the slower relaxation, which has the dominant amplitude.

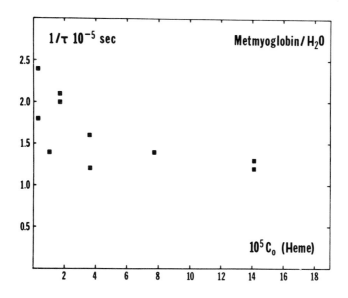

<u>Figure 3.</u> Spermwhale metmyoglobin dependence of $1/\tau$ on total protein concentration at pH 8.8, 15°C in 1.8 x 10^{-4} M borate. The same behavior is observed in the absence of buffer.

A bimolecular reaction, such as eq. 2, which is the correct formulation for this reaction in the pH range studied, should exhibit a single relaxation time which would obey the equation,

$$1/\tau = k_d + k_r [Fe(III)H_2O + (OH^-)]$$ (3)

and therefore the reciprocal relaxation time should increase with increasing protein concentration. However, the data in Figure 3 reveal that this is clearly not the case. Instead, there is a slight, but significant, downward trend of the data.

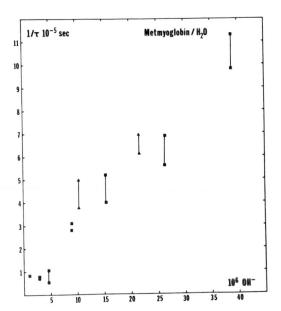

Figure 4. Sperm whale metmyoglobin. Dependence of $1/\tau$ on OH^- concentration. $\lambda = 580$ nm; $20°C$; no added buffer; 9.4×10^{-5} M heme.

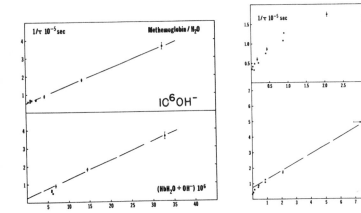

Figure 5. Sheep methemoglobin. Dependence of $1/\tau$ on OH^- concentration, $20°C$. No added buffer.

a) Total protein concentration = 7.3×10^{-6} M heme; $\lambda = 405$ nm upper curve, $1/\tau$ versus (OH^-); lower curve, $1/\tau$ versus $(HbH_2O + OH^-)$.

b) Total protein concentration = 8.5×10^{-5} M heme; $\lambda = 480$ nm.

In contrast to this result the relaxation behavior for both MetMb and MetHb at varying ligand concentration [OH⁻], appear to obey bimolecular kinetics, as is shown in Figures 4, 5a and 5b. In both cases $1/\tau$ varies linearly with [OH⁻]. These data are similar for both hemeproteins, with MetMb reacting about three times faster than MetHb.

There is, however, a significant difference in the response of these two proteins to the electric field, as is shown in Figure 6. Whereas the applied electric field shifts MetMb to the acid form, MetHb is shifted to the alkaline form at high field. However, the signs of the reaction enthalpies are the same and increasing temperature favors the acid form in both cases. In addition, MetHb exhibits a fast, nanosecond relaxation (see Figures 2 and 6) but MetMb does not (for a description of this fast effect, see G. Ilgenfritz and T. M. Schuster, this volume, p. 401)

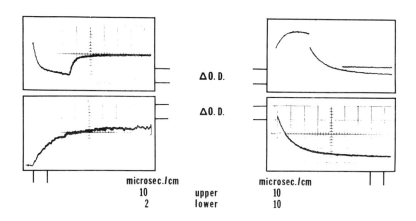

ΔO. D.

ΔO. D.

microsec./cm   microsec./cm
10   upper   10
2   lower   10

Figure 6. Relaxation curves of metmyoglobin sperm-whale (left) and sheep methemoglobin (right) showing opposite effects of electric field on the acid-alkaline transition. The left-hand traces were taken at pH 9.1 and show a decrease of O.D. with field, the acid form being favored: ΔO.D.=0.0027/cm; 9.5×10^{-5} M heme; E = 74 kV/cm; d = 10.7 mm. The right-hand traces were taken at pH 8.5 and show an increase of O.D. with field, the alkaline form being favored; ΔO.D.=0.0079/cm; 8.3 x 10^{-5} M heme; E = 70 kV/cm; d = 19.2 mm. The sweep speeds are indicated on the figure. The lower curves were obtained by triggering at the end of the field pulse.

305

It appears that the electric field jump relaxation method reveals significant differences in the properties of the two proteins. The reason for the different responses to the high field is not yet clear. It could be that different charged groups are involved in the reaction. For example, a reaction of the following type would be shifted toward the alkaline (higher charged) form by an applied electric field (as is the case in phenolphthalein):

$$A\ H^- + OH^- \longrightarrow A^{-2} + H_2O \qquad (4)$$

On the other hand, a reaction involving dissociation of an uncharged complex made up of oppositely charged ions would be shifted toward the acid form:

$$A\ H^+ + OH^- \longleftarrow A + H_2O \qquad (5)$$

One might expect a relaxation time to be independent of protein concentration if there were a rate limiting intramolecular reaction coupled to the ligand binding reaction. However, the following considerations rule out this possibility.

The general mechanism for an intramolecular reaction coupled to a ligand binding reaction is,

Acid Form $\quad Fe(III)H_2O + OH^- \overset{k_2}{\underset{k_1}{\rightleftharpoons}} [Fe(III)H_2O]^* + OH^-$

$$k_r \quad k_d \qquad\qquad k_r^* \quad k_d^*$$

Alkaline Form $\quad Fe(III)(OH^-) + H_2O \overset{k_4}{\underset{k_3}{\rightleftharpoons}} [Fe(III)(OH^-)]^* + H_2O$

equilibrium constants:

$$K = \frac{k_d}{k_r}$$

$$K^* = \frac{k_d^*}{k_r^*}$$

If the monomolecular steps are rate limiting, i.e., the binding reactions are fast compared to the conformation changes, the slow relaxation is given by:

$$1/\tau \;=\; \frac{k_2 + k_4 \dfrac{(OH^-)}{K} \;(B)}{1 + \dfrac{(OH^-)}{K}\,(B)} \;+\; \frac{k_1 + k_3 \dfrac{(OH^-)}{K^*} \;(B^*)}{1 + \dfrac{(OH^-)}{K^*}\,(B^*)} \tag{6}$$

where,

$$B \;=\; \frac{P + P^* + (OH^-) + K^*}{P\,(\dfrac{K^*}{K}) + P^* + (OH^-) + K^*}$$

$$B^* \;=\; \frac{P + P^* + (OH^-) + K}{P + P^* \,(\dfrac{K}{K^*}) + (OH^-) + K}$$

$$Fe(III)H_2O \;=\; P$$

$$[Fe(III)H_2O]^* \;=\; P^*$$

Equation 6 shows that in general the relaxation time will depend upon the concentration of protein $(P + P^*)$ due to the coupling of the concentration dependent binding reactions to the monomolecular steps. However, an apparent concentration independence may be achieved if the concentration terms entering the B and B* expressions are negligible, i.e., if $(OH^- + K^*)$ and $(OH^- + K)$ are each large compared to the protein concentration, such that $B = B^* \simeq 1$. This is clearly not the case in the range of protein concentrations and pH's at which our measurements were made. In addition, even if $B = B^* \simeq 1$, equation 6 does not account for the observed linear dependence of $1/\tau$ on (OH^-), particularly for the higher (OH^-) concentrations, where the relaxation time should become independent of (OH^-) concentration. Yet another consideration makes it rather unlikely that the observed protein concentration independence arises from a monomolecular reaction. We would have to postulate that the corresponding binding reactions are considerably faster than the monomolecular reaction. But if one calculates an apparent bimolecular recombination rate constant from the slopes of the curves

307

in Figures 4 and 5a, one obtains values of 0.5 to 1.0 x 10^{10} M^{-1} sec^{-1} for hemoglobin and 2 to 3 x 10^{10} M^{-1} sec^{-1} for myoglobin. To require that the true rate constants for binding be considerably faster brings us into conflict with the laws of diffusion which state that the upper limit (diffusion controlled reactions) for a bimolecular rate constant in aqueous solution is about 10^{10} M^{-1} sec^{-1}. Thus, even the observed apparent rate constants are unexpectedly high for a bimolecular reaction involving such large molecules.

It is possible, however, to account for the linear dependence of $1/\tau$ on (OH^-) and the near independence of $1/\tau$ on protein concentration if there is strong buffering of (OH^-). The fact that the maximum amplitude of the relaxation (as a function of pH) occurs at a pH lower than that expected for an unbuffered reaction agrees with this assumption (see Figure 7). Also, the calculated pH of the maximum amplitude

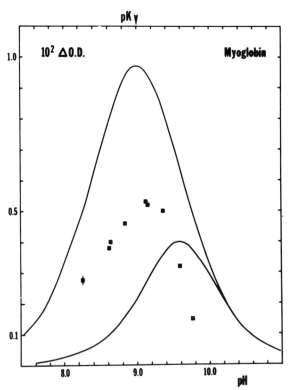

Figure 7. Sperm whale metmyoglobin. Observed and calculated amplitudes as a function of pH (pK = 9.0; Δln K = 0.08). Right hand curve, no buffering left hand curve, complete buffering. Total protein concentration, 9.4 x 10^{-5} M heme.

for a completely buffered reaction agrees fairly well with the observed pH as shown in Figure 7, although the magnitude of the amplitude (Δ0.D.) does not. But since these results were obtained without added buffer the source of the apparent buffering must be other than the solvent. (Addition of buffer does not change the relaxation behavior.) Buffering due to dissociable groups on the protein is unlikely since this would not agree with the observed kinetic behavior. The following mechanism represents such a case:

$$(H^+)RFe(III)H_2O \;+\; OH^- \underset{k_d}{\overset{k_r}{\rightleftharpoons}} (H^+)RFe(III)(OH^-) \;+\; H_2O$$

$$\Big\downarrow\Big\uparrow k_{buffer}$$

$$RFe(III)H_2O \;+\; H_2O$$

Assuming that the buffer reaction is very fast the expression for the relaxation time for the observable (slower) reaction is given by,

$$1/\tau \;=\; k_d + k_r\left[\frac{(OH^-) + P}{1 + \dfrac{(OH^-) + P}{K_{buffer}}}\right] \tag{7}$$

where $P = (H^+)\,Fe(III)H_2O$

This equation demonstrates that the above mechanism cannot be the correct one to account for the apparent buffering since equation 7 does not account for the observed protein and ligand concentration dependence of the relaxation time (figs. 3,4,5).

If, however, we assume a strong linkage of the ionization of the Fe(III) bound H_2O to the ionization of some other dissociable group, it is possible to explain the observed kinetic behavior. Such a thermodynamic linkage may be mediated by a very rapid conformation change. In terms of a mechanism this means that if a solvent hydroxyl ion takes a proton off one site, the other site releases a hydroxyl ion (or takes up a proton) in a consecutive step:

$$R^-Fe(III)H_2O + OH^- \rightleftharpoons R^-Fe(III)(OH^-) + H_2O$$
$$\rightleftharpoons H^+R^-Fe(III)(OH^-) + OH^- \tag{8}$$

309

In other words, there must be a large change in the binding constant of one site when the other site binds a ligand. The relaxation expression for the above mechanism, in the limiting case of complete buffering (i.e., negligible concentration of the intermediate species), is

$$1/\tau = (k_1 + k_2)OH^-$$

where k_1 and k_2 are the overall rate constants in the above two step mechanism.

However, this limiting case cannot be correct since it implies a pH independent spectrum. Therefore there must be a considerable concentration of the intermediate species in equation 8.

As yet we have not formulated a complete mechanism which explains adequately the equilibrium and kinetics results

From what is known about the structure of myoglobin it is reasonable to assume, as has been suggested previously (3), that the distal histidine, which on one side is hydrogen bonded to the Fe(III)-bound water and on the other side has a dissociable group in contact with bulk solvent, is involved in the acid-alkaline transition of these hemeproteins.

References

1. Theorell, H. and A. Ehrenberg, Acta Chem. Scand. 5, 823 (1951).

2. George, P. and G. Hanania, Biochem. J. 52, 517 (1952) and George, P. and G. Hanania, Biochem. J. 55, 236 (1953).

3. Goldsack, D. E.,W. S. Eberlein and R. A. Alberty, J. Biol. Chem. 241, 2653 (1966).

THE ACIDIC-ALKALINE TRANSITIONS IN METHEMOGLOBIN AND METMYOGLOBIN

T. L. Fabry, J. Kim,
S. H. Koenig, and W. E. Schillinger

IBM Watson Laboratory, Columbia University
New York, New York 10025

We report data on the pH dependence of the nuclear magnetic relaxation rate of solvent water protons (T_1^{-1}) in solutions of methemoglobin and metmyoglobin. The data, similar to but more extensive than work reported previously by Mildvan et al. (1) were taken as a function of temperature for the proteins in several different buffers.

Experimental

The hemoglobin was prepared from fresh human blood, oxidized with potassium ferricyanide, chromatographed on a Sephadex G-25 column, concentrated, and dialyzed against several changes of ice cold distilled water. Sperm whale myoglobin, obtained from the Gallard-Schlesinger Corp., was similarly treated.

The measurements of T_1^{-1} were all at a Larmor frequency of 50 MHz using procedures and apparatus described previously (2). The optical spectra were recorded on a Cary 14 spectrophotometer.

Results

Data were obtained on metmyoglobin and methemoglobin solutions in deionized water and in the following buffer systems: borate, phosphate, Tris, and Hepes (N-2-hydroxy-ethylpiperidine-N'-2-ethanesulfonic acid). The pH range covered was roughly pH 6 to pH 9, although not all the buffer systems could span this range.

Figure 1 shows the results for metmyoglobin at 25° C. The T_1^{-1} data are plotted as relaxivity, essentially the

311

relaxation rate per mole of protein with a correction for the
volume fraction of protein (2). The relaxivity of (diamagnetic) oxymyoglobin is indicated in the figure, which may be
considered as the diamagnetic contribution to the relaxivity
of the metmyoglobin. The pK for the acidic-alkaline metmyoglobin transition, as determined by optical measurements
(3), is also indicated. Figure 1 demonstrates that, for metmyoglobin, the functional form of the dependence of relaxivity on pH is very similar to a titration curve with a well-
defined pK, is essentially independent of protein concentration and buffering, and has the same pK as that obtained from
optical studies.

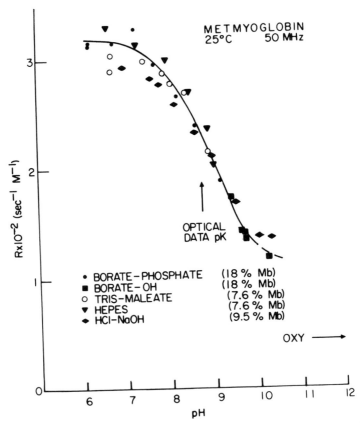

Figure 1. Relaxivity as a function of pH for metmyoglobin in
various buffer solutions at 25°. The buffer concentrations
are indicated on the figure. The relaxivity of oxymyoglobin
is indicated by an arrow.

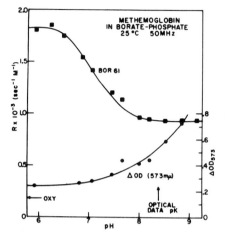

Figure 2. Relaxivity as a function of pH for methemoglobin in 0.5 M borate-phosphate buffers at 25° C. ■, variation of the optical density at 573 nm for the same series of solutions ●. Protein concentration, 11%. The relaxivity of oxyhemoglobin is indicated by an arrow.

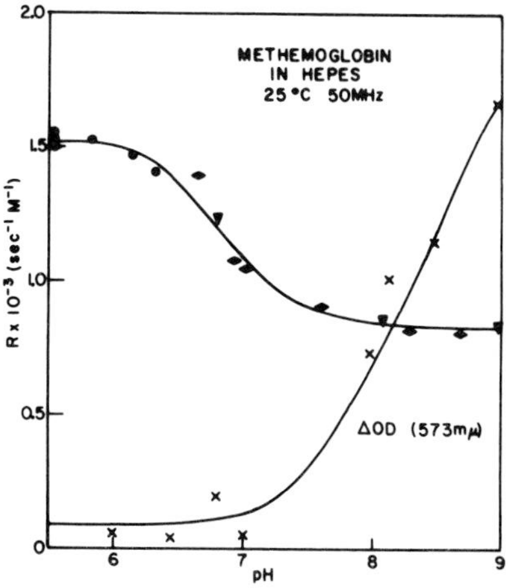

Figure 3. Relaxivity as a function of pH for methemoglobin in 0.1 M Hepes buffer. ●,◆,▼ represent different hemoglobin preparations. Variation of optical density of hemoglobin at 573 nm in the same buffer system (x). All data taken at 25°; protein concentrations, approximately 7%.

Figure 2 shows the variation with pH of the relaxivity of methemoglobin in borate buffer. The variation of optical density at 573 nm for this sample, and the optical pK previously reported (4) are shown. It is quite clear that, in contrast to metmyoglobin, the optical and magnetic resonance data are measuring different ionizations.

Figure 3 is similar to Figure 2, except that the buffer is Hepes. The relaxivity data in the two cases are similar, but quantitatively different. Again, the optical and magnetic resonance data are measuring different ionizations.

In the deionized water system, no measurements were made below pH 7 and in the phosphate system, the highest pH reached was 8.2. In these two systems, the high and low pH limits, respectively, were not reached. For the phosphate buffer, the dependence of relaxivity on buffer concentration is shown in Figure 4. For all the other data reported here, the buffer concentration was less than 0.5 M.

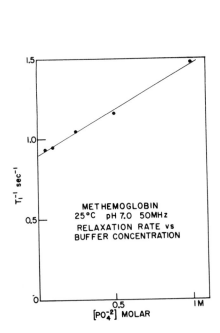

Figure 4. Relaxation rate as a function of buffer concentration at pH 7.0 at 25° C; protein concentration, approximately 4%

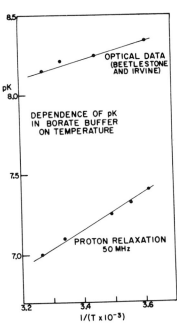

Figure 5. Temperature dependence of the apparent pK as determined by optical spectroscopy and NMR. Both measurements are in 0.5 M borate buffer.

Figure 5 shows the temperature dependence of the pK in methemoglobin, as determined from optical data (4) and T_1^{-1} data. It is quite clear from these results also that these pK's must refer to separate ionizations; the range of pK values do not begin to overlap.

Discussion

The main conclusions concerning the transition from the acidic to the alkaline form of methemoglobin and metmyoglobin are 1) the transitions are different for the two and 2) for hemoglobin, there are two ionizing groups and the buffer ion is directly involved in one of the ionizations. These distinctions resemble the Bohr effect in that the ligand-heme interaction is a function of pH for hemoglobin but not for myoglobin.

One may speculate on the mechanisms of the two ionizations in methemoglobin. Because the one at lower pH is buffer dependent, it could be due to the loss of a proton from the imidazole of the distal histidine, since this group hydrogen bonds to the buffer ion, as suggested by Stryer (5) and discussed by Mildvan (1). This exchanging proton could in principle contribute to the titration of the relaxivity, without effecting the optical data. The second ionization would be the ionization of a water bound in the sixth ligand position of the heme iron. The exchange rate of this proton to the solvent may be too slow to contribute to the relaxivity in methemoglobin, but not so slow in metmyoglobin.

References

1. Mildvan, A.S., N.S. Rumen, and B. Chance. Abs. Am. Chem. Soc., 156th Meeting, 32 (1968). This volume, p. 205

2. Koenig, S.H. and W. Schillinger. J. Biol. Chem., 244, 3283 (1969).

3. Hartell, C.R., R.A. Bradshaw, K.D. Hapner, and F.R.N. Gurd. J. Biol. Chem., 243, 690 (1968).

4. Beetlestone, J.G. and D.H. Irvine. Proc. Roy. Soc., 277A, 401 (1963).

5. Stryer, M., J. Kendrew, and H. Watson. J. Mol. Biol., 8, 96 (1964).

ANOMALOUS BINDING OF HYDROXIDE, CYANIDE, AZIDE, AND FLUORIDE IONS WITH A GLYCERA METHEMOGLOBIN

Bette Seamonds

Johnson Research Foundation, School of Medicine
University of Pennsylvania, Philadelphia, Penna.

The bloodworm, Glycera, appears to have two cellular hemoglobins of vastly different molecular weights. One of these hemoglobins, a monomer of molecular weight ∿17,000, exhibits myoglobin-like physiological properties (1). However, oxidation of the protein to methemoglobin produces certain anomalous characteristics.

The methemoglobin spectrum of Figure 1 shows marked shifts in absorption maxima compared with other hemoglobins and myoglobins. The Soret absorption band at ∿392 nm is shifted 13 to 17 nm to the blue, while the visible bands are

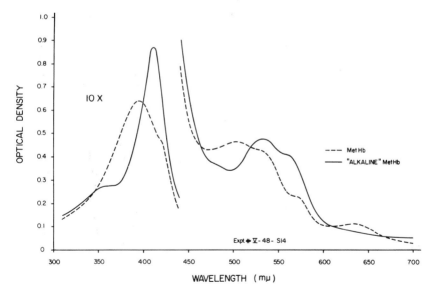

Figure 1. Spectral changes undergone during titration of monomeric methemoglobin with base (μ = 0.1).

317

shifted 5 to 7 nm to the red, in comparison with other hemoglobins and myoglobins (2). Inflections at 534 and 571 nm
are presumably due to the presence of low-spin hemoprotein.

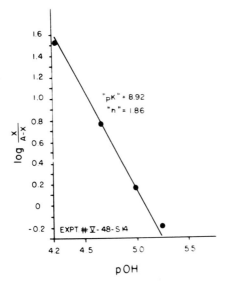

Figure 2. Titra-
tion of Glycera
methemoglobin with
base.

Titration of the protein with base (Figure 2) indicates
the formation of imidazole methemoglobin (3) instead of the
expected hydroxide complex. The pseudo pK for the reaction
(μ = 0.1) was 8.99 ± 0.06, with n = 1.94 ± 0.07, indicating
unusual stoichiometry. The true pK is possibly like that of
peroxidase (4), but could not be determined due to denaturation above pH 10.

Methemoglobin ligand binding studies with cyanide, azide
and fluoride are shown in Figure 3. The general spectral
features of the low spin cyanide and azide complexes were
not especially noteworthy. However, the fluoride spectrum
(Figure 4) indicates a 10 to 12 nm shift to the blue, to
593 nm, of the charge-transfer band present in other hemoglobins and myoglobins (2). This raises the question of
whether this is a true charge-transfer transition. In
addition, the peak which normally occurs at 480 nm is hardly
evident.

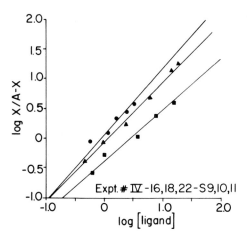

Figure 3. Determination of binding constants of monomeric methemoglobin with cyanide (▲), azide (■), and fluoride (●) at 20°. A and X represent the maximum and fractional changes in absorbance induced by the reagent in question at a specific wavelength. K_{eg} is computed in each case by evaluating log [free ligand] when log X/A-X = 0.

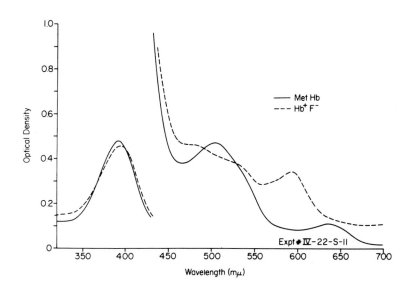

Figure 4. Spectral changes induced in monomeric methemoglobin by potassium fluoride.

Table I summarizes the binding constants for cyanide, azide and fluoride, together with the values for other hemoglobins and myoglobins under similar conditions (5-8). In the case of azide binding, no pH dependence between pH 6.45 and 7.3 was found. In addition, the relative affinities of Glycera methemoglobin for azide and cyanide were altered.

TABLE I

Binding Constants for the Combination of Hemoglobin and Myoglobin between pH 6.9 and 7.0 and the Glycera Monomer with Potassium Cyanide, Sodium Azide, and Potassium Fluoride

Ligand	Glycera monomer	Hemoglobin	Myoglobin
Azide	2.2×10^{-3} M	6.9×10^{-6} M(7)*	2.8×10^{-5} M(8)
Cyanide	1.2×10^{-3} M	3.6×10^{-8} M(5)	3.6×10^{-7} M(6)
Fluoride	9.4×10^{-1} M	4.7×10^{-3} M(5)	6.5×10^{-2} M(6)

*The numbers in parentheses indicate references

The data suggest that the affinity of monomeric Glycera methemoglobin for ligands is vastly different from that of other hemoglobins as well as myoglobins. Dr. Love, presenting his crystallographic data on Glycera cyanomethemoglobin, has indicated that this protein shows more non-planarity of the heme iron and nitrogens. These data support the postulate that Glycera monomeric methemoglobin possesses a peculiar heme geometry which is reflected in its reactivity towards ligands.

References

1. Rossi-Fanelli, A. and E. Antonini. Arch. Biochem. Biophys. 77, 478 (1958).

2. Brill A.S. and R.J.P. Williams. Biochem. J., 78, 246 (196

3. Beetlestone, J.G. and D.H. Irvine. J. Chem. Soc., A, 1340 (1968).

4. Theorell, H. Ark. Kemi. Min. Geol., 16A, 1 (1942).

5. Coryell, C.D., F. Stitt, and L. Pauling. J. Amer. Chem. Soc., 59, 633 (1937).

6. Theorell,H.and A.Ehrenberg. Acta Chem.Scand.,5,823(1951).

7. Krese, M. and H. Kaeske. Biochem. Z., 312, 212 (1942).

8. Duffey, D., B. Chance, G. Czerlinski. Biochem., 5, 3514 (1966).

IS THERE A STRUCTURE-SPIN STATE CORRELATION?
AN ANALYSIS OF TWO DIMENSIONAL FOURIER DIFFERENCE SYNTHESIS IN MYOGLOBIN DERIVATIVES*

B. Chance

Johnson Research Foundation, School of Medicine, University
of Pennsylvania, Philadelphia, Pennsylvania 19104

The several approaches described in the preceding papers,
afford overwhelming evidence in favor of structural changes
in myoglobin related to the liganding of the sixth coordination position and allow us to postulate that the extent of
such structural changes is related to changes in the spin
state of the iron atom. It is the general goal of studies
such as these to provide a structural basis for cooperativity
in the liganding of hemoglobin with oxygen by the transmission
of small structural changes from one subunit of the polymer
to another, thereby affording a foundation for our ultimate
understanding of cooperativity.

The pioneer work of Stryer, Kendrew, and Watson on the
Fourier difference synthesis for metmyoglobin and metmyoglobin azide (1) showed, at the error level set by them, no
structural changes due to liganding of myoglobin. Their
results suggested that, for some reason, structure-liganding
interactions did not occur in the azide derivative and possibly not in other derivatives.

The relationship of these studies of the hemoprotein
and those of the heme and its homologues (2,3) is of great
interest and evoked considerable discussion (4) as to whether
the high spin "out of plane" iron would become "in plane" in
high-to-low transition of the myoglobin derivatives. Along
the same lines, much less elegant studies than those of
Stryer, Kendrew and Watson gave an "indicative" correlation
between structure and spin-state changes (5). Fourier
difference synthesis on the cyanide and hydroxide derivatives
of myoglobin suggested that these low-spin compounds showed
significant structural differences from the high-spin aquo
and fluoride derivatives. More recently, Bretscher has identified structural changes of cyanide metmyoglobin in three

*Supported by GM 12202 by the USPHS.

dimensions (6) and Schoenborn has made a similar study of
the Xenon-hydroxide compound (7). Bretscher has considered
in detail the spin-state structure problem (6).

During my stay at the Laboratory of Molecular Biology
in Cambridge in 1966, I was privileged to be able to study
the original data on a variety of myoglobin derivatives investigated by Kendrew, Watson, Stryer, Schoenborn, Nobbs,
and others. Perhaps my unfamiliarity with the limitations
of x-ray crystallography made me bold enough to attempt some
comparisons of structural features of the Fourier difference
synthesis which might be regarded as trivial by profession-
al crystallographers. In all cases, electron density differences were measured in the h0ℓ projections at two pairs
of coordinates at which some regularity of the electron
density difference could be detected. The coordinates
selected for evaluation should meet four criteria. First,
there should be a consistent pattern of electron density
differences as one proceeds through a series of derivatives.
Second, the region selected should not coincide with the
region occupied by the ligand and, thus, the ligand itself
should not contribute to the electron density differences.
Third, the coordinates should pass through the edge of the
heme and indeed, close to the vinyl group region which
might be suspected to change in this series of compounds.
Fourth, the coordinates should be chosen to avoid regions
of superposed structural changes in the projection chosen,
wherever possible. Fifth, the coordinates should avoid the
regions in which heavy atom substitutions were made in myoglobin (8).

Two pairs of coordinates in the h0ℓ projections that
seem to meet these requirements are X = 21 and Z = 2, and
X = 25 and Z = 2. As shown in the two-dimensional Fourier
difference syntheses presented in <u>Hemes and Hemoproteins</u>,
the full scale for these diagrams is 48 in both the X and
Y directions and the scale factor is 3 units of X and Z =
2 Å. These regions showed distinctive positive and negative
peaks in the cyanide and hydroxide compounds that lie at the
top of the heme rectangle in Figure 1B and C of that paper
(5). Coordinate pairs Z = 11, X = 32, and Z = 13, X = 36
show similar positive and negative electron density peaks
at the bottom of the heme rectangle, which will not be
further described here.

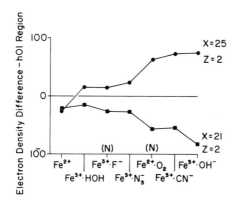

Figure 1. Electron density differences in h0ℓ projections
in the regions X = 25, Z = 2 and X = 21, Z = 2 for various
compounds of myoglobin as obtained from the data of Kendrew,
Watson, Stryer and Nobbs (1,8,9) and Watson and Chance (5).
The electron densities in these two regions represent the
sum of the contours from the maximum value to a level of 6
units, equivalent to 0.15 el/Å^3 or approximately 1/2 carbon
atom. 3Å = 2 units of X and Z.

Since the profiles of the peaks and troughs vary in the
different compounds because of different details of the
structural changes, I have sought here to evaluate the total
structural change by summing the electron density differences
in the region which includes the peak value; these range
down to twice the noise level (0.15 electrons/Å^3).

The substitution of the water molecule by the ligand was
suspected to be incomplete in the fluoride (5) and oxygen
(9) compounds. In the case of the fluoride, I had simply
underestimated the change of dissociation constant in the
crystalline suspension. The work of Watson and Nobbs (9)
clearly outlines the problem of the oxygen compound. In
both cases, I assumed the depth of the sulfate trough to
be independent of the nature of the ligand, and increased
the electron density differences computed at X = 21, Z = 2,
and X = 25, Z = 2, in proportion to the ratio of the sulfate

trough (X = 36, Z = 13) of a "fully occupied" compound such as azide (1) to that of the compound in question. These two compounds are identified in Figure 1 by the symbol N. In the case of the cyanide compound, the excessive depth of the sulfate trough is now recognized from Bretscher's data (6) to be the summation of structural changes, and thus no correction was made in this case.

Two compounds on which data were available were omitted --the imidazole (10) and the xenon-hydroxide (7). In these compounds, extensive alterations of the myoglobin structure were observed, and the superposition of electron density changes from other regions of the molecule would lead to a substantial overestimation of the localized structural change.

Figure 1 plots for the X = 21, Z = 2, and X = 25, Z = 2 coordinates the summed electron density differences against the name of the compound. The ordering of the compounds is arbitrary. Some points are worthy of discussion. First, considering the native metmyoglobin itself, the positive and negative changes represent the noise level of Figure 1 of ref. 5. It is clear that the fluoride compound is not significantly different. The azide compound also shows a very small structural alteration, in agreement with the previous conclusion (1); nevertheless a further investigation of the azide compound would be of interest. The compounds-- oxygen, cyanide and hydroxide--seem to show significant structural alterations with respect to the ferrous, ferric, aquo and ferric fluoride derivatives.

In order to better correlate these electron density differences with the properties of the myoglobin compound, Table I has been included which summarizes the data on composition, percentage occupancy of the spin state, computed number of unpaired electrons and the electron density differences reported in Figure 1. The fluoride compound is the ideal example of a high spin compound and the aquo and ferro forms have similar high spin states and show correspondingly small electron density differences with respect to one another in spite of the fact that the ferro myoglobin is five coordinated.

The low spin, one unpaired electron cyanide compound and the zero spin oxygen compound seem to fit well the idea that large structural changes occur in high to low spin transitions. However, the azide and the hydroxide compound seem to refute a general conclusion to this point. The azide compound has a low spin character, yet a small structural change. The hydroxide compound has a high spin character and a large structural change.* In the case of oxymyoglobin, two factors are involved; first are changes from five-coordinated to six-coordinated iron due to the oxygen ligand and secondly a change from high spin state to zero spin state occurs. Thus, it is not possible to attribute the structural change exclusively to a spin-state change.

In summary, it appears that the simplistic idea of a correlation of spin-state structure is not tenable and that the nature of the ligand and the coordination state of the iron must be considered. The results of the analysis are sufficiently indicative that I hope they will stimulate those gathered here, and others as well, to take advantage of the rich variety of compounds formed by myoglobin. Correlations between structure, as reported by the many techniques we have heard described, and the nature of the ligand or the spin state of the compound, by means of x-ray, nuclear magnetic resonance, and other methods offer an exciting challenge of the future.

*The possibility that the hydroxide compound is converted to the ammonia compound in the presence of the ammonium sulfate mother liquor of the crystals can be substantially refuted by unpublished EPR studies at low temperatures which indicate that a single crystal of myoglobin at alkaline pH shows hydroxide character (11). The temperature change itself would cause an increased tendency towards formation of an ammonium compound by analogy with catalase (13).

TABLE I.

Summary of Spin State and Structural Correlations.

Name of Compound	Composition	Spin States (S values) high	low	Percentage Occupancy (high spin)	Average Unpaired Electrons	Summed Electron Density Difference (hol) X=21, Z=2 (Fig.1) ($\delta = 0.15$ el/Å3)	Ref.
Myoglobin fluoride	MbF	5/2	--	100	5	+6	5
Myoglobin (aquo)	Mb+H_2O	5/2	--	98	5	+6	5
Ferromyoglobin	Mb	2	--	(100)	4	-10	8
Myoglobin azide	MbN_3	5/2	1/2	22	1.8	+12	1
Myoglobin hydroxide	MbOH	5/2	1/2	69	4.3	+75	5
Myoglobin-oxygen	MbO_2	--	0	--	0	+65	9
Myoglobin cyanide	MbCN	--	1/2	0	1	+72	5

Acknowledgements

The author gratefully acknowledges the steadfast aid of Dr. H.C. Watson and Dr. Benno Schoenborn and the considerable encouragement to this work afforded by Dr. Max Perutz and Dr. John Kendrew. In addition, the further work by P. Bretscher and B. Schoenborn has been an invaluable support to the conclusions that were drawn from the very preliminary study described here.

References

1. Stryer, L., J.C. Kendrew, and H.C. Watson, J. Mol. Biol., 8, 96 (1964).

2. Hoard, J.L. In Hemes and Hemoproteins (B. Chance, R. Estabrook, and T. Yonetani, eds.), Academic Press, 1966, p. 9.

3. Koenig, D., Acta Crystallogr., 18, 663 (1965).

4. General Discussion In Hemes and Hemoproteins (B. Chance, R. Estabrook, and T. Yonetani, eds.), Academic Press, 1966, p. 154.

5. Watson, H.C. and B. Chance, In Hemes and Hemoproteins (B. Chance, R. Estabrook, and T. Yonetani, eds.), Academic Press, 1966, p. 149

6. Bretscher, P. Ph.D. Thesis, University of Cambridge, 1969.

7. Schoenborn, B.P. This volume, p. 181.

8. Nobbs, C., H.D. Watson, and J.C. Kendrew, Nature, 209, 339 (1966)

9. Watson, H.C. and C. Nobbs, In Biochemistry of Oxygen (B. Hess, ed., 19 Mosbacher Coll., 1968), Springer-Verlag, 1969, p. 37

10. Nobbs, C.L. J. Mol. Biol., 13, 325 (1965).

11. Yonetani, T. Personal communication.

12. Ehrenberg, A., Arkiv. Kem., 19, 119 (1962).

13. Ehrenberg, A. and R.W. Estabrook, Acta Chem. Scand., 20 1667 (1966).

Note added in proof: Perutz (1) has now put forward full details of a number of tertiary structure changes of ferro-hemoglobin on liganding with oxygen. The striking result that the iron atom is 0.75 A out of plane in the high spin deoxyhemoglobin and 0.3 A out of plane in the high spin methemoglobin and metmyoglobin suggests that the nature of the ligand itself is of great importance in the structure changes.

Whether there are as well fundamental difficulties in this simple two dimensional Fourier difference synthesis method is not known, although Perutz (1) refers to "conflict-ing results" in the case of single chain hemoglobin.

(1) Perutz, M., Nature, 228, 726 (1970)

GENERAL DISCUSSION:
SPIN STATE AND STRUCTURE

Chance: Dr. Wuthrich, are the hyperfine shifts attributable
to the motion of the iron, or do you need to include the porphyrin and perhaps the F-8 histidine as well?

Wüthrich: The hyperfine shifts of the resonances of the heme
group in the proton NMR spectra of low spin ferric (Fe^{3+}
S = 1/2) and high spin ferrous (Fe^{2+}, S = 2) myoglobin might
be interpreted to indicate a motion of the iron relative to
the heme plane. We found that the size of the hyperfine
shifts is approximately the same in the two states. Since
the high-spin ferrous form has a larger total electronic spin,
the electronic wave functions of the iron and the porphyrin
ring thus seem to overlap less strongly in this state. This
might be expected if the iron were located essentially in
the heme plane in cyanoferrimyoglobin, and somewhat out of
the plane in deoxymyoglobin.

Chance: Perhaps I could also ask Dr. Love whether he can
yet attribute electron density differences to movement of the
iron atom, or whether movements of the porphyrin and histidine
may be involved.

Love: Our evidence with Glycera hemoglobin is that the presumed histidine and F helix do not move, but that the iron
and the heme can move relative to each other, i.e., that the
iron can move relative to the histidine. This is well exem-
plified in Figure 4 of our presentation. As to the question
of whether the motion is what is responsible for the things
that happen in these conformational changes, I do not know.
The data that we have are only to 3 Å resolution, are projections, and these are preliminary experiments. Work must
be carried out in three dimensions at high resolution, and
when you are finished and know what the situation is both
before and after, then maybe we can begin to ask questions
about how it happens.

Chance: I agree with you; one can imagine that if the histidine is tightly linked to the iron, they will move together. It seems to me that the qualitative distinction between motion of the iron atom and motion of the whole heme and portions of the protein should not be too difficult, providing the appropriate projections were available.

Dr. Wüthrich, would you comment on the changes induced at the #4 and #5 positions of the porphyrin ring induced by xenon and cyclopropane? These are interesting for further stud because while they do cause NMR changes, as yet no crystallographic changes are resolved.

Wüthrich: Recently Dr. Chance and Dr. Yonetani raised some questions about the mechanisms of heme-polypeptide interactions in heme proteins, and in particular about the role of the heme substituents. To comment on these questions I would like to compare the NMR spectra of a series of low spin ferric hemes and heme proteins which we have studied recently, and point out some empirical relations between heme environment and unpaired electron distribution in the heme group.

Figure 1 shows the positions relative to internal DSS of the hyperfine shifted ring methyl resonances in the NMR spectra of cyanoporphyrin iron (III) complexes (1,2), native a reconstituted cyanoferrimyoglobins (2,3), cyanoferrihemoglobin (4) and cyanoferricytochrome $\underline{c}$ (5) and ferricytochrome $\underline{c}$ (5). The size of the shifts is proportional to the spin densities o the neighboring ring carbon atoms. If the electronic wave functions had the D_{4h} symmetry of the porphyrin ring the four ring methyl resonances would be degenerate. The observed differences in the positions of the resonances come from perturbations of the electronic wave functions by the heme substituents, and by interactions with the polypeptide chains in the heme proteins.

There appear to be three quite different types of spectra for compounds with different axial ligands of the heme iron. (i) In cyanoporphyrin iron(III) complexes both axial ligands, CN^- and "x", might be cyanide ion, since the resonance positions are only little affected by change of the solvent. Only small perturbations of the symmetry of the spin distribution seem to come from variation of the 2,4 substituents, which are vinyl groups in Proto CN, protons in

Deut CN, and ethyl groups in Meso CN, and formation of the methyl or ethyl esters of the 6,7 propionates. (ii) The methyl resonances are quite similar in all the compounds with cyanide ion and histidyl in the axial positions, even though there are four different kinds of 2,4 substituents in MbCN, Deut MbCN, Meso MbCN, and Cytc CN, and the polypeptide chains are different in myoglobin, hemoglobin and cytochrome c. On tthe other hand the extent and symmetry of the spin delocalization in the heme groups of these proteins are markedly different from those in the cyanoporphyrin iron(III) complexes in various solvents. (iii) In cytochrome c, where the axial ligands are histidyl and methionyl (5), the four ring methyls are split into two groups of two resonances with greatly different hyperfine shifts.

It is quite apparent that in the heme proteins of Figure 1 at most a small part of the effects of heme polypeptide interactions on the spin distribution in the heme groups comes from interactions with the 2,4 substituents. On the other hand the axial ligands of the heme iron appear to be a dominant factor. However, as is seen from a comparison of MbCN and HbCN, different chemical nature of the axial ligands

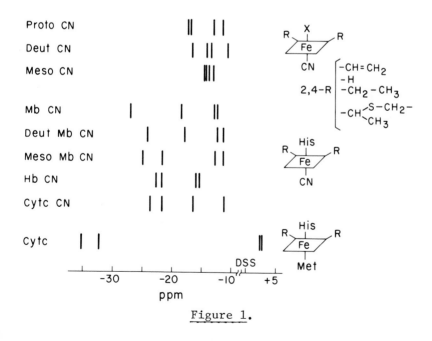

Figure 1.

331

cannot account for all the observed differences either. Different arrangements of the polypeptide chains in different heme proteins might cause the same axial ligands to bind at different angles. "Non-binding" interactions of the porphyrin ring with nearby amino acid residues, which could perturb the planarity of the heme group, or force it into a planar structure of lower symmetry might also be operative.

These observations might have been anticipated. One would expect from theoretical considerations that appreciable differences in the electronic wave functions of the heme groups would come from interactions with the axial ligands, and from possible variations of the configuration of the porphyrin ring. Modifications of the heme substituents might be important if they forced the heme group into a position which is different from that in the native heme protein. The axial ligands in the reconstituted compound might then bind at different angles or be entirely different, and the whole environment of the heme group would be different. This might happen in the experiments reported by Dr. Yonetani, where cytochrome c peroxidase reconstituted with modified heme was found to be greatly different from the native compound.

Brill: There were two effects and both of them played a role. Are there any other effects that could play roles?

Wüthrich: Since the atomic structure of protoporphyrin IX indicates that the 2,4-vinyls could be part of the conjugated π-bond system of the heme group one might have expected polypeptide-vinyl interactions to contribute a considerable part of the observed protein effect on the NMR spectrum of the heme groups. The present data imply quite clearly that in the compounds discussed above the vinyls are not "unique handles" for heme-polypeptide interactions.

It would certainly be nice if contributions from different mechanisms, for example coordination of the axial ligands and "non-binding" interactions with the porphyrin ring, could be further separated in future experiments, maybe with model systems.

Caughey: The striking solvent effects observed in proton NMR spectra of paramagnetic hemins suggest that we can expect rather marked changes in chemical shifts of heme moieties with changes in apoprotein. For example, with high-spin hemins in $CDCl_3$ meso-protons have been found at very high fields (e.g.,

35 to 60 ppm up-field from TMS) whereas the same compounds in DMSO-d_6 did not show these high field resonances, rather these protons now appeared at low fields. However, chemical shifts for protons at other locations on the porphyrin were not so markedly affected by the change in solvent. (These observations were in collaboration with Dr. LeRoy Johnson of Varian Associates.)

Shulman: My first question is whether the high spin to low spin transition occurs within a fixed crystal structure or whether it occurs as a result of the structural change. In Tasaki's data, the excited state always had a high statistical weight. In the past they have interpreted this in terms of a conformation change which involved many modes of freedom of the protein, giving the excited state a high statistical weight. That meant that conformation changes were an explanation of the spin changes. Is that still true, or are the large statistical weights now explained in terms of a temperature-dependent energy gap?

My second point is that one should distinguish the conformational changes caused by high spin to low spin transitions from the conformational changes which are necessary for the cooperativity during oxygenation. Because, in the latter case, Taylor showed that when you go from high spin ferro deoxy to high spin met, you still get cooperativity and almost the same amount of free energy of interaction as during oxygenation (6). Therefore, a high spin to high spin transition with ferrous going to ferric is perfectly capable of giving the conformational changes which, as we show below, are responsible for the cooperative oxygen binding in hemoglobin. The cooperativity seems to depend upon the 6 position going from empty to occupied, not upon the spin state.

References

1. Wüthrich, K., R.G. Shulman, B.J. Wylude and W.S. Caughey, Proc. Nat. Acad. Sci. U. S., 62, 636 (1969).

2. Shulman, R. G., K. Wüthrich, T. Yamane, E. Antonini and M. Brunori, Proc. Nat. Acad. Sci. U. S. 62 (1969).

3. Wüthrich, K., R. G. Shulman and J. Peisach, Proc. Nat. Acad. Sci. U. S. 60, 373 (1968).

4. Wüthrich, K., R. G. Shulman and T. Yamane, Proc. Nat. Acad. Sci. U. S. 61, 1199 (1968).

5. Wüthrich, K., Proc. Nat. Acad. Sci. U. S. 62 (1969).

6. Taylor, J. F. and A. B. Hastings, J. Biol. Chem. 131, 649 (1939).

LASER PHOTOLYSIS OF CRYSTALLINE CARBONMONOXYHEMOGLOBIN AT HIGH CARBON MONOXIDE PRESSURES*

T. Reed, J. Bunkenberg, and B. Chance

Johnson Research Foundation, University of Pennsylvania
Philadelphia, Pennsylvania

Increasing interest in the relation of conformation and enzyme activity has emphasized the importance of reactions of crystalline enzymes. In the case of carbonmonoxy hemoglobins, the use of crystals permits investigation of higher enzyme-substrate concentrations than are available with these substances in solutions. The photolysis of crystalline carbonmonoxy hemoglobin and the subsequent recombination was chosen for study using a laser to produce photolysis and high pressure techniques (1) to change the carbon monoxide concentration independently of the hemoglobin. The fact that this reaction had already been carefully examined by Parkhurst and Gibson (2) at atmospheric pressure using a xenon flash (white) source provided the additional advantage that overlap of observations would confirm the applicability of laser and pressure techniques.

The apparatus is shown in Figure 1. The first laser

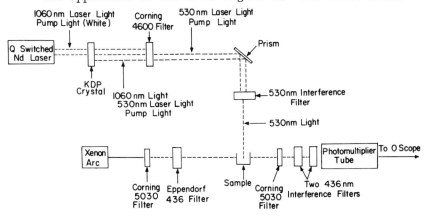

Figure 1. Optical details of the laser photolysis apparatus.

*This work was supported by NIH grants GM 12202 and 1237002.

photolysis of carbonmonoxy hemoproteins was conducted by Chance and Stark (3) with 347 nm light. The present system uses 530 nm light produced by frequency doubling the 1060 nm light from a Neodymium laser. Q-switching gives a flash duration of 30-50 nsec. The purity of the 530 nm light is ensured by fi ters which remove pump flash and 1060 nm laser light. Measuring light is 436 nm light. A blue glass guard filter and two 436 nm interference filters on the phototube virtually eliminate artifact due to laser light reflecting from the crystals.

Figure 2 shows a typical photolysis experiment. The suspension of crystalline sheep hemoglobin reduced with dithionite, but not containing CO was flashed with the laser giving a line corresponding to free hemoglobin and a small peak indicating the maximum artifact size. Note that the artifact goes downward which is the direction of light increase. The suspension was then equilibrated with CO at 1 atm and flashed again. The base line corresponded to unphotolized carbonmonoxy hemoglobin. At 436 nm, free hemoglobin has a greater optical denisty than carbonmonoxy hemoglobin. Hence the photolysis was indicated by movement in a direction of less light. The recombination appears to be biphasic.

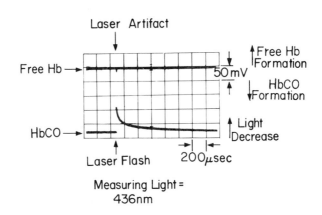

Figure 2. Overall view of photolysis experiment showing base lines corresponding to 100% photolysis (free Hb) and 0% photolysis (Hb CO) as well as maximum laser artifact and recombination of photolized Hb CO.

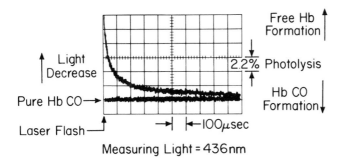

Measuring Light = 436nm

Figure 3. Recombination of free Hb and CO following photolysis. Increased gain used to show the biphasic nature of the recombination reaction more clearly.

The biphasicity is shown in more detail in Figure 3. The slow phase followed second order kinetics with a rate constant of 1.0×10^6 M^{-1} sec^{-1}. This rate constant was independent of degree of photolysis. The fast phase would not fit second order kinetics plots.

A plot of the log of the reciprocal of free hemoglobin concentration vs. times is shown in Figure 4. During the first

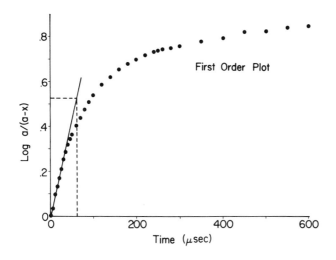

Figure 4. Plot of Log[a/(a-x)] vs. times for the rate curve shown in Figure 3.

337

twenty or thirty μsec after photolysis the recombination reaction follows a first order rate law, as is indicated by the linearity of this section of the curve. To get a clearer picture of what is going on in the fast phase and to confirm the nature of the slow phase, it would be useful to be able to change the concentration of carbon monoxide relative to that of free hemoglobin. A pressure of 40 atm CO was necessary to increase the fast reaction rate two-fold. Our pressure system, tested to 100 psi, precluded experiments at significantly higher pressures.

A high pressure photolysis experiment is indicated in Figure 5. In the first photograph a suspension of crystalline carbonmonoxy hemoglobin has been photolized at 1 atm CO. The total change corresponds to 15.4% photolysis. The reaction was biphasic and the rate constant of the fast phase was 2.6×10^4 sec^{-1}. Without disturbing the sample, the pressure in the reaction chamber was raised to 40 atm CO and equilibration established by stirring for an hour. The stirrer was stopped and the system flashed again, giving the trace shown in the lower photograph. It appeared that the slow phase was gone and the slope of the fast phase increased.

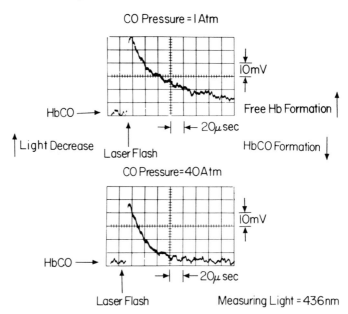

Figure 5. Recombination of hemoglobin and CO under atmospheric and increased CO pressures.

338

First order plots of the two reactions comprising the experiment are shown in Figure 6. In the recombination at one atmosphere, the reaction was a normal biphasic recombination as has already been noted. The rate constant of the fast phase was 2.6×10^4 sec^{-1}, that of the slow phase 1.0×10^6 M^{-1} sec^{-1}. At 40 atm, the first order plot appears virtually monophasic with an overall slope corresponding to a rate constant of 4.0×10^4 sec^{-1}. The slight deviations from monophasicity will be used in the present preliminary interpretation of the kinetic results. They reinforce the

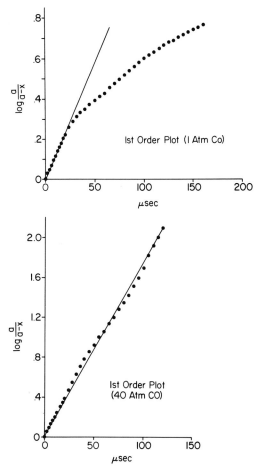

Figure 6. Plot of Log [a/(a−x)] vs. time for the rate curves shown in Figure 5.

view of localized concentration increases implicit in our explanation.

The rate constant data and concentration data are sumarized in Table I. Comparing the slow phase of the atmospheric reaction with what we may call the "slow phase" of the reaction at 40 atmospheres indicates that we have the usual conversion of a second order reaction to a pseudo first order reaction in the presence of a large excess of carbon monoxide. The concentration of CO in the intersticies of the crystals is assumed to be that of the surrounding saturated ammonium sulfate solution which is about 36 mM to which must be added .5 to 1mM CO from the as yet unrecombined photolytic products. This CO concentration readily accounts for the pseudo first order rate constant of the slow phase at high pressure.

TABLE I

Rate Constants and Concentrations During
The High Pressure Experiment

Fast Phase
For 15.3% photolysis at 1 atm; $k = 2.6 \times 10^4 \ sec^{-1}$

$$[CO] = 5.4 \ mM_{(phot)} + 0.9 \ mM_{(sol)} = 6.3 \ mM$$

Effective Concentration $= 4 \times (5.4 \ mM) + 0.9 \ mM = 23 \ mM$

For 10.6% photolysis at 40 atm; $k = 4.0 \times 10^4 \ sec^{-1}$

$$[CO] = 3.7 \ mM_{(phot)} + 36 \ mM_{(sol)} = 40 \ mM$$

Effective concentration $= 4 \times (3.7 \ mM) + 36 \ mM = 51 \ mM$

Slow Phase
For 15.3% photolysis at 1 atm; $k = 1.0 \times 10^6 \ M^{-1} \ sec^{-1}$

$$[CO] = 1.5 \ mM_{(phot)} + 0.9 \ mM_{(sol)} = 2.4 \ mM$$

For 10.6% photolysis at 40 atm; $k = 4.0 \times 10^4 \ sec^{-1}$

$$[CO] = 36 \ mM_{(sol)} + 1.0 \ mM_{(phot)} = 37 \ mM$$

In formulating a tentative interpretation for the fast recombination phase at atmospheric pressure we make use of ideas presented by Parkhurst and Gibson in their photolysis studies of horse hemoglobin crystals with xenon flash lamps (2). Although they did not resolve the fast phase of the recombination reaction they discussed possible concentration effects in crystals in great detail. Basing their calculations on Perutz's studies of crystalline horse hemoglobin they showed that if CO from photolysis were denied access to the main volume of the hemoglobin molecule, but only had access to the heme groups and the water of crystallization, its effective concentration would be about twice what one would calculate on the basis of the amount of CO produced by photolysis and unit cell volume. If the CO produced by photolysis was restricted to the bound water and denied access to the water of crystallization in the main body of the hemoglobin molecule, then its effective concentration would be four to five times that calculated from the amount of CO released and the volume of the unit cell.

If one considers the fast phase of the recombination at one atmosphere to be due to a localized, transient concentration effect, then a comparison with the pseudo first order rate of the slow phase of the high pressure reaction which is 4.0×10^4 sec^{-1} with the rate of fast phase of the one atmosphere reaction which is 2.6×10^4 sec^{-1} suggests that the fast phase of the 1 atmosphere reaction "sees" a transient concentration of CO equivalent to that at 26 atmospheres of CO pressure. This concentration would be 23 mM. Correcting for the fact that 0.9 mM of this CO is from equilibration with the suspending medium, the effective concentration resulting from photolysis is 22 mM. Since the concentration of CO calculated directly from photolysis is 5.6 mM, a factor of 4 would account for the apparent concentration increase.

Examination of the fast phase of the high pressure reaction tends to confirm this view. Using a four-fold correction factor on the CO produced by photolysis gives 15 mM, which added to the 36 mM concentration resulting from equilibrium with the suspending medium, yields an effective concentration of 51 mM of CO. Thus, the predicted rate constant for the fast section of the high pressure curve would be about 5×10^4 sec^{-1}. Although this is in fair agreement with the observed value, considering the uncertainties of CO concentration at high pressure, an even better agreement is suggested by examining the observed first order curve shown in detail in detail in Figure 6. Considering the fast phase of the high pressure curve separately from the rest of the curve, we find a rate of 4.8×10^4 sec^{-1}. This correspondence between the predicted and observed rate constants of the high pressure recombination of CO and hemoglobin suggests that

341

attributing the fast phase of the recombination of CO and crystalline sheep hemoglobin at atmospheric pressure to a localized, transient concentration effect is probably correct.

In summary, we have presented results based on the photolysis of crystalline carbonmonoxy hemoglobin at atmospheric and increased pressures using monochromatic light from a Q-switched laser. The results indicate a heretofore unobserved fast phase in the recombination kinetics. This is interpreted as resulting from localized, transient concentration increases. The size of this effect implicates some structure within the protein molecule itself, such as the bound water or the hydrophobic pocket near histidine E-7. We are continuing studies on this phenomenon.

References

1. Chance, B., D. Jamieson, and H. Coles. Nature, 206, 257 (1965).

2. Parkhurst, L. J. and Q. J. Gibson. J. Biol. Chem., 242, 5762 (1967).

3. Chance, B. and H. Stark. unpublished data.

4. Perutz, M. F. Trans. Faraday Soc., 428, 187 (1946).

PURIFICATION, PROPERTIES, AND LINKAGE SYSTEMS
OF SPERM WHALE MYOGLOBIN*

Melvin Keyes and Rufus Lumry

Laboratory for Biophysical Chemistry
Chemistry Department, University of Minnesota,
Minneapolis, Minnesota 55455

Up to the present time, it has not been possible in
this laboratory to prepare hemoglobin in a state suitable for
quantitative studies. A minimum requirement for a suitable
preparation is that it demonstrate quantitatively identical
behavior in oxygen binding by gasometric and spectrophotometric methods. Although only horse hemoglobin preparations
have been investigated in detail, considerable evidence has
been accumulated to suggest that isotherms obtained by gaso-
netric methods are consistent with Adair's formal equation
but that those obtained by spectrophotometry usually are not.
However, reliable experiments are extremely difficult to carry
out because the preparations have low reproducibility; ferric
iron forms, even in small concentrations, have a large effect
on isotherm shape, and the many important factors influencing
the behavior of the protein are by no means fully understood
and certainly not fully controlled.

Initially, myoglobin also proved to be unsatisfactory,
at least in its ferrous form. We have now learned how to remove the contaminants responsible for abnormal behavior and,
having done so, have collected a limited amount of information
about both contaminant-free and contaminated sperm whale muscle
myoglobin (1,2). It would be out of place to present the full
details of these studies here, but a summary of the results may
be of some interest. Complete details are available in Ref. 1
and will be published elsewhere (3).

*This is Paper #48 from this Laboratory. The research was
supported by N(onr)710(55) and AFOSR-1222-67.

Preparation and properties of "uncontaminated" and "contaminated" myoglobin. Sperm whale myoglobin was prepared directly as Fe(II) myoglobin by a modification of the procedure
of Yamazaki, Yakota, and Shikama for horse heart myoglobin (4).
The first eluted fraction from chromatography on DEAE Sephadex
has been reported by Hardman et al. (5) to be the homogeneous
major component which we shall call "contaminated" myoglobin.
"Contaminated" myoglobin was further purified by chromatography
on G-25 Sephadex to obtain a material from the leading edge of
the single main peak which will be called "uncontaminated" myoglobin.

The two preparations containing 1.0×10^{-4} M Tris buffer
demonstrate a spectral difference centered at 410 nm when KBr
is added to a concentration of 0.5 M (see Figure 1). A study
of the eluent from the G-25 Sephadex column shows that the contaminant (or contaminants) is a small molecule which moves with
a fraction having high electrical conductivity. Addition of
the fraction containing the contaminant restores the spectral
effect at 410 nm to uncontaminated myoglobin. By the same
method of recombination of eluent fractions, differences in
ligand binding behavior have been shown to be due to material
in this fraction, presumably the same contaminant. Partially
contaminated preparations give non-linear van't Hoff plots in
which the data points lying off the line are found to lie between the plots for fully contaminated and uncontaminated material. The thermodynamic changes in oxygen and carbon monoxid

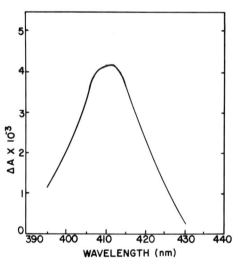

WAVELENGTH (nm)

Figure 1. The difference
in absorption spectrum
between contaminated and
uncontaminated Fe(II)
myoglobin in the pre-
sence of 0.5 M KBr.
Tris buffer, 1×10^{-3} M
at pH 8.5

TABLE I

Thermodynamic Changes in Ligand Binding to Fe(II) Myoglobin[1]

Myoglobin Type	Ligand	$\Delta H°$ (2) kcal/mole	$\Delta S°$ (2) e. u.	$\Delta H°$ (3) kcal/mole	$\Delta S°$ (3) e. u.
Uncontaminated (van't Hoff)	CO	-21.4 ±0.3	-65.7±0.8	-21.2	-38.5
Fully contaminated (van't Hoff)	CO	-25.9 ±0.6	-79 ±3	-25.7	-52
Uncontaminated (van't Hoff)	O_2	-18.1 ±0.4	-60 ±1	-17.9	-33
Uncontaminated (calorimetric)	O_2	-19 ±1			

---

[1] pH = 8.5; 0.001 M Tris buffer.

[2] standard state of one torr.

[3] The thermodynamic values of $\Delta H°$ and $\Delta S°$ obtained for a standard state of one torr have been corrected for large changes in entropy of solution of O_2 or CO in water by the following reaction (14):

$$x(gas) \rightleftharpoons x(org)$$

$$\Delta F° = -8.0 ±0.2 \text{ kcal/mole}$$

$$\Delta H° = -0.2 ±0.4 \text{ kcal/mole}$$

$$\Delta S° = -27.2 ±0.2 \text{ e.u.}$$

The (x) represents either O_2 or CO and the organic phase may be either CCl_4 or C_2H_5OH within the error limits given. The standard state for the two columns on the right of the Table is thus unit mole fraction of the ligand in the organic solvent.

binding obtained with uncontaminated myoglobin (Table I) agree with those reported by Theorell (6). The calorimetrically determined enthalpy change in oxygen binding obtained with this material agrees with the van't Hoff value (Table I). Data points obtained with contaminated myoglobin agree with the data reported by Rossi-Fanelli and Antonini (7) in those cases in which our conditions were similar to theirs. In these cases our preparations were partially contaminated.

The contaminant has not been identified, nor have its effects on Fe[III] myoglobin been determined. The high affinity of Fe(II) myoglobin for the contaminant and the fact that myoglobin prepared by the method of Yamazaki et al. (4) in our experience always contained some contaminant suggest that myoglobin in vivo may often be contaminated with "contaminant".

Thermodynamic changes in oxygen and carbon monoxide binding. In Table I are given the standard enthalpy and entropy changes for the binding of oxygen and carbon monoxide to myoglobin. The van't Hoff plots from which the enthalpy changes were computed were prepared from isotherm data which invariably were precisely fit by a mass law curve first-order in ligand concentration and in protein concentration. There was no detectable curvature in the van't Hoff plots. The error quantities are estimated ("least squares") standard deviations. The standard enthalpy of binding obtained calorimetrically is based on a limited series of experiments with the new LKB flow microcalorimeter and cannot yet be considered to be entirely reliable despite the excellent characteristics of the instrument. Although the agreement between calorimetric and van't Hoff enthalpy changes is good, the calorimetric values must be considered a lower bound, since we may not have prevented partial oxygenation of the deoxymyoglobin prior to mixing with the oxygen solution.

A standard state (see columns labeled $\Delta H°(3)$ and $\Delta S°$ (3) in Table I) has been chosen to minimize cratic entropy changes and unusual thermodynamic consequences of the interaction of dissolved oxygen and carbon monoxide with water. The entropy changes given in the last column of the Table are thus unitary entropy changes due primarily to the interaction of the ligand with the protein. The entropy changes obtained with uncontaminated myoglobin are nevertheless larger than might have been expected for combination of the ligands with a passive complex ion, even though X-ray diffraction results indicate that myoglobin in solution may have no sixth position ligand (8). In the presence of contaminant, both enthalpy and entropy changes are considerably larger than those found with uncontaminated material. In these experiments the protein was saturated with contaminant so that no deviations from linearity were found in the van't Hoff plot.

Linkage between zinc-ion binding site and oxygen binding site. In Table II are given the ratios of the O_2- and CO-affinity constants obtained in the presence of several divalent cations, divided by the value of the appropriate constant obtained in the absence of cations. A weak linkage between ligand and cation-binding site is demonstrated by these ratios. Zn(II) was the most effective among those ions tried. Cu(II) might have been more effective, but oxidized myoglobin. The relatively high single-site affinity constant which we find for Zn(II) binding to carboxymyoglobin and oxymyoglobin under experimental conditions similar to those employed in X-ray diffraction studies of Zn(II)·Fe(III) myoglobin suggests that the zinc ion was bound at the same site in our experiments as that found in the X-ray studies. This site is on the side of the protein molecule opposite to the heme-group position, thus suggesting a linkage path of considerable length. These experiments were carried out before the presence of contaminant was known and it is probable that the protein preparation was partially contaminated.

TABLE II

Effect of Metal Ions on Carbon Monoxide
Binding by Fe(II) Myoglobin

Metal Ion	Metal Ion Concentration (M)	K_{app}/ K_o*	Ligand
Ca^{2+}	1.68×10^{-3}	0.918	O_2
Ca^{2+}	1.43×10^{-3}	0.930	O_2
Co^{2+}	6.69×10^{-3}	0.995	CO
Hg^{2+}	complex	1.29	O_2
Zn^{2+}	2.14×10^{-3}	1.67	O_2
Zn^{2+}	2.07×10^{-3}	1.49	CO
Ni^{2+}	2.98×10^{-3}	1.28	CO

*$K_o(CO) = 50.4$; $K_o(O_2) = 2.09$

Tris buffer in total concentration from 0.002 to 0.047 M, pH 7 , at 20° was used. The equilibrium constants have been corrected for the reaction of the buffer anion with the metal ions and the interaction of the metal ions with water.

Linkage between xenon-binding and oxygen-binding sites. The binding of oxygen and carbon monoxide to uncontaminated myoglobin shows no dependence on the presence of xenon in solution. It has been shown by X-ray diffraction studies that there are at least two xenon binding sites in Fe(III) myoglobin (9). We have not yet been able to determine the affinity

constants for these sites or to establish whether or not xenon
is bound by uncontaminated Fe(II) myoglobin in solution. With
contaminated myoglobin the carbon monoxide affinity of the
protein measured by the affinity constant increases with increasing xenon concentration. Statistical analysis of our
data (1) suggests that there are three xenon sites linked to
the carbon monoxide binding site but the analysis is not inconsistent with the existence of only two sites. There is
no doubt that more than one site is linked to the heme group.
The calculation of the linkage free energy from our data requires a model. Assuming two xenon sites, the interaction
free energy between these sites is of the order of 1 kcal/
mole or less when the protein is in its CO-form.

Our results establish a weak linkage system in sperm-
whale myoglobin consisting of two or three xenon sites, the
oxygen site and the contaminant site. The zinc ion site is
linked to the oxygen site but its relationship to the other
sites has not been determined. Linkage between xenon sites
and sixth-ligand site is mediated by the contaminant site.
Xenon binding in the absence of carbon monoxide must be weak
and may be very weak in the absence of contaminants. As shown
by Schoenborn et al. (10) one xenon site is on the proximal
side of the heme group in contact with both the porphyrin plan
and the proximal imidazole group. This site might be expected
to interact with oxygen binding at heme iron. The absence of
an effect in uncontaminated myoglobin may be due to the absenc
of xenon binding in our experiments. Chance (personal communication) has found that the presence of xenon in Fe(III)
myoglobin solutions had no effect on azide-ion binding.

Discussion

In addition to the large entropy change in oxygen and
carbon monoxide binding which is of interest in suggesting
that changes in the protein-water system may occur, the large
value of the enthalpy of binding of these molecules appears
to demonstrate that their interaction with heme iron is not
a minor process. If so, in Fe(II) myoglobin the balance between high-spin and low-spin magnetic states may not be so
delicate as has been usually thought. Iizuka and Kotani (11)
have suggested that the enthalpy changes in ligand-binding
processes of ferric forms of hemoglobin and myoglobin are
due in major part to the primary-bond rearrangements between
ligands and heme iron and that these large negative enthalpie
are partially compensated by entropy changes in the protein

conformation so that the free energy change has the relatively small value. We believe that this proposal is attractive and it is consistent with earlier "rack" proposals for these proteins (12). However, it is necessary on the basis of recent findings to consider the possibility of contributions from water as well as from the protein conformation to both enthalpy and entropy changes in ligand binding (13). If for the Fe(II) myoglobin reaction with O_2, the adjusted entropy (Table I) is assumed to be due entirely to changes in the protein-water system and the ratio of enthalpy to entropy change has the usual value of $290^{\circ}K$ (13), of the total of -19 kcal of enthalpy change, about -5 kcal can be attributed as a lower limit to the ligand-iron interaction and -14 kcal to the protein-water rearrangements. It is interesting to note that when the difference in enthalpies of CO binding to contaminated and uncontaminated myoglobin is divided by the difference in the corresponding entropies of binding, the ratio is $338^{\circ}K$ which is not far from the range of compensation temperatures observed in some other protein reactions as is described elsewhere in this volume (14).

A rapid and effective procedure for preparing uncontaminated myoglobin in quantity has been developed recently. This procedure, which will be published shortly (3), takes advantage of the fact that the contaminant is only weakly bound when reducing agents such as ascorbic acid are present, presumably because it has itself become reduced. All our experience to date indicates that the material thus prepared is highly reproducible, free of contaminants and suitable for re-examination and future study of the reactions of myoglobin.

Recent evidence, particularly that of Caughey and his associates (15) and Wüthrich and Shulman (16), seems to demonstrate that linkage among oxygen-binding sites of hemoglobin ('heme-heme interaction') is not transmitted through the proximal ligands and porphyrin planes of the heme group, in which case attention must be fixed on rearrangements in the regions of the protein distal to the heme groups and on the internally sequestered water molecules as well as the water molecules near the exposed surfaces of the protein (14). If these rearrangements are usually effected through sub-unit reorganization, linkage will be found most commonly in multiple sub-unit protein complexes such as hemoglobin. However, though it is probably true that delicacy in feedback control and in other manifestations of linkage can be improved by increasing the number of linked sites, it is not obvious a priori that multiple sub-unit proteins are necessary. In this connection myoglobin provides an interesting example of a complex linkage system in a single small globular protein and is thus an excellent experimental subject for separating aspects of linkage due to internal conformational rearrangements from those due to external conformational rearrangements involving water.

Acknowledgments

The authors with to express their thanks to Mrs. Meredith Falley for her invaluable technical assistance in the preparation of protein materials.

Addendum*

Dr. Arnold Wishnia (17) has established the existence of at least two binding sites for pentane in myoglobin. Evidence based in part on direct binding studies and in part on competition between xenon and pentane shows that there is also more than one site for xenon binding. In addition, Wishnia has sho that there is linkage between oxygen-binding sites and pentane binding sites in myoglobin.

References

1. Keyes, M., Ph.D. Dissertation, University of Minnesota, 1968.

2. Keyes, M. and R. Lumry, Fed. Proc. 27, 895 (1968).

3. Keyes, M. and R. Lumry, submitted to J. Am. Chem. Soc.

4. Yamazaki, I., K. Yokota and K. Shikama, J. Biol. Chem. 239, 4151 (1964).

5. Hardman, K., D. Eylar, D. Ray, L. Banaszak and F. Gurd, J. Biol. Chem. 241, 432 (1966).

6. Theorell, H., Biochem. Z. 268, 73 (1934).

7. Rossi-Fanelli, A. and E. Antonini, Arch. Biochem. Biophys 77, 478 (1958).

8. Nobbs, C., H. Watson and J. Kendrew, Nature 209, 339 (1966); Nobbs, C. in Hemes and Hemoproteins (B. Chance, R. Estabrook and T. Yonetani, eds.) Academic Press, New York, 1966, p. 143.

*Note added in proof: August 23, 1969.

9. Schoenborn, R., J. Mol. Biol., in press.

10. Schoenborn, B., H. Watson and J. Kendrew, Nature 207, 28 (1965).

11. Iizuka, T. and M. Kotani, Biochim. Biophys. Acta, 194, 351 (1969)

12. Lumry, R. (ed.), P. Boyer, H. Lardy and K. Myrback, Enzymes 1, 157 (1959); Biophysics (Japan) 1, 1 (1961); Review: Lumry, R., in Treatise on Electron and Coupled Energy Transfer in Biological Systems (T. King and M. Klingenberg, eds.) Dekker, New York, in press. Lumry, R. and R. Biltonen, in Structure and Stability of Biological Macromolecules (S. Timasheff and G. Fasman, eds.) Dekker, New York, 1969

13. Lumry, R. and S. Rajender, Macromolecular Reviews 4, (1969) in press.

14. Lumry, R., this volume, p. 353

15. Caughey, W., J. Alben, S. McCoy, S. Boyer, S. Carache and P. Hathaway, Biochemistry 8, 59 (1969)

16. Shulman, R., K Wüthrich, and J. Peisach, this volume, p. 195.

17. Wishnia, A. Personal communication.

DISCUSSION

Chance: How strong is the interaction between the Xenon-binding sites?

Keyes: Our data show that there must be two Xenon-binding sites in carbon-monoxide myoglobin. Moreover, analysis by the use of a statistical F-test indicates that there is a high probability (~90%) that there are three sites. Since the values of the interaction free energies depend upon whether there are two or three sites, no accurate values can be given at this time.

Schoenborn: What type of myoglobin was it?

Keyes: Sperm whale myoglobin.

Peisach: What is the impurity you mentioned?

Keyes: We don't know. We do know some of the properties of the contaminant, however. First, since it is separated by chromatography on G-25 Sephadex it must have a molecular weight of less than 5000. Furthermore, the eluent which contains the contaminant has a high conductance and we therefore suspect that it is charged. Finally, preliminary experiments indicate that the contaminant has both an oxidized and a reduced state, and that it is the oxidized state which binds tightly to the protein.

ENTHALPY-ENTROPY COMPENSATION IN RELATION TO PROTEIN BEHAVIOR*

Rufus Lumry

Laboratory for Biophysical Chemistry
Department of Chemistry, University of Minnesota
Minneapolis, Minnesota 55455

Heme-heme interaction in the oxygenation of hemoglobin is the classic example of linkage among the binding sites of a protein or protein complex. Contaminated myoglobin demonstrates a linkage system of complexity nearly as great as that of hemoglobin (1). Linkage is a consequence of free-energy exchange among binding sites (2,3) and for many years has been attributed to mechanical or electro-mechanical pathways of conformation change (3-6). There is a variety of evidence which can be interpreted as indicating changes in polypeptide conformation during oxygenation but there is no proof that such changes as do occur in the system are limited to changes in polypeptide or indeed that the polypeptide changes are significant in terms of the free-energy criteria with which their importance must be assessed. There is at least one un-understood and perhaps generally important factor which has not heretofore received much attention in considerations of heme-protein behavior. The existence of this factor suggests an oversimplification in our considerations of protein mechanism. It also makes the translation of quantitative information about protein mechanisms into chemical descriptions a hazardous undertaking and it appears not improbable that some information previously interpreted as due to protein conformation changes is rather due to changes in water about and, perhaps in proteins. This factor is the subject of the present communication.

As is reviewed elsewhere (7), there are many processes involving small solutes in water for which it is found that

---

This is Paper #49 from this laboratory. The research was supported by grants from the National Science Foundation (GB-7896) and the National Institutes of Health (AM-05853).

the enthalpy and entropy changes partially or completely compensate each other so as to leave the net free energy of the system unchanged or nearly unchanged. In some instances the relationship is linear, in others it is not. Linear compensation between measured enthalpy and entropy changes is described by Eq. 1 which is not in any way an implicit consequence of thermodynamic laws.

$$\Delta H = \alpha' + T_c \Delta S \tag{1}$$

Pairs of enthalpy and entropy changes are obtained in small-solute systems by varying the chemical description of the solutes to form homologous series or by adding small amounts of co-solvents of which the small alcohols have been most popular. When Eq. 1 applies to thermodynamic or rate processes, the slope, T_c, of the "compensation plot" of ΔH versus ΔS is usually found to lie in the range from about 270°K to 290°K. Compensation behavior has been attributed generally and probably correctly to solvation part-processes of the total processes (8) and in some cases, such as the dissociation of weak acids, it is necessary to subtract contributions from electronic rearrangements - chemical part-processes - to show the characteristic compensation behavior (9).

The variability of the examples clearly suggests that the source of this extra-thermodynamic behavior lies in water itself so that it might have been anticipated that some protein reactions would demonstrate compensation behavior with T_c values in the range from 279-290°K. Some years ago Likhtenshtein and Sukhorukov (11) collected data from the literature which showed just this behavior in a large number of protein systems and their collection included enzymic rate and equilibrium processes as well as thermal denaturation. There are non-trivial sources of error in measurements of enthalpy and entropy change which tend to demonstrate the enthalpy-entropy compensation pattern when it is not in fact present (7) and there is little doubt that a number of the examples treated by these authors give false indications of compensation. However, recently collected data of high precision for a few protein systems have been shown by appropriate statistical tests to reveal true linear enthalpy-entropy compensation. Not only are the deviations from a straight-line fit of ΔH^- versus ΔS^- values remarkably small but the enthalpy and entropy contributions from the compensation part-process are found to be as high as 30 kcal and 100 e.u.

in some specific binding processes and even higher in some denaturation reactions.

Beetlestone and Irvine and their coworkers (12-16) have demonstrated compensation in the binding of anions to FeIII hemoglobin and FeIII myoglobin on varying pH. The ($\Delta H°$, $\Delta S°$) points obtained in this way conform to Eq. 1 thus demonstrating linear compensation. Representative compensation plots of their data are given in Figure 1. As pH is increased, the ($\Delta H°$, $\Delta S°$) points move to the right along the compensation line. Near the isoelectric point, there is a "turn-around" point and the ($\Delta H°$, $\Delta S°$) points move back along the line to the left or more correctly back along a line of slightly larger T_c value as shown. Figure 2 is a compensation plot of all their data for the binding of N_3^-, CN^-, and SCN^- to FeIII hemoglobin from man, dog, pigeon, and guinea pig, to FeIII

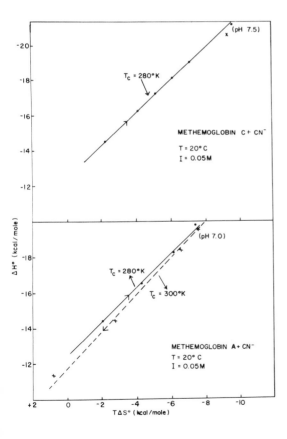

Figure 1. A representative compensation plot of the form ΔH vs $T\Delta S$ for CN^- binding by FeIII hemoglobin. Data from Anusiem et al.(12) The coordinates of the points are the standard enthalpy and entropy of binding x 293° K at different pH values. Points obtained with increasing pH as indicated by the arrow. There is a "turn-around" pH value at about pH 7.5 (Fe (III) Hb C) or 7.0 (Fe (III) Hb A). The points on the dashed line were obtained at pH values above the turn-around value. The solutions contained 0.05 M neutral salt.

whale myoglobin, and to an α-chain preparation from human
FeIII hemoglobin*. The straight compensation lines of Figure
2 are fitted only to the data obtained at pH values up to the

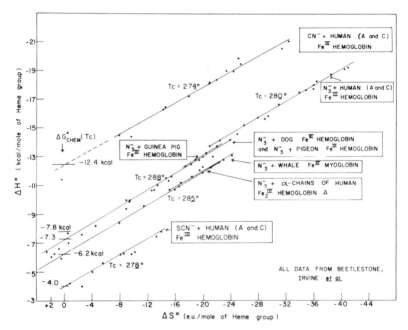

Figure 2. Compensation plots for the binding of anionic ligands by proteins of the hemoglobin-myoglobin family. All
data from Beetlestone, Irvine, and their collaborators. The
($\Delta H°$, $\Delta S°$) points each represent one pH value. The lines are
fitted only to the points for pH values on the acid side of the
"turn-around" points. The points for pH values at and above
the turn-around values are included, but would be fitted to a
different set of compensation lines (cf Figure 1). References
Human A and C FeIII Hb plus CN⁻ and SCN⁻, (12); human A and C
FeIII Hb plus N_3^-, (13,14); dog, pigeon, and guinea pig FeIII
Hb plus N_3^-, (14); α-chain preparation from human A FeIII Hb
plus N_3^-,(15); FeIII myoglobin (sperm whale) plus N_3^-, (16).

---

*The data for F⁻ binding to FeIII human hemoglobin also demon-
strate this behavior, but the changes are too small relative
to experimental errors to make this example useful. Although
OH⁻ may not show the characteristic pattern of binding to FeII
human hemoglobin, it does manifest approximately the same behavior with FeIII myoglobin (16).

356

turn-around points of Figure 1. There is more variability in the data obtained at higher pH values, but a similar set of curves having higher slopes can be demonstrated by fitting to these points (Fig. 1). Although the variations in T_c values for the different anions are probably significant, we have no more explanation for this variation than for the fact that they lie in the same rather narrow range of values. Within error, the N_3- binding family of curves form straight lines of the same slope but different intercepts.

The T_c values which can be calculated from the collected data of Likhtenshtein and Sukhoroukov scatter considerably but tend to lie in the interval 270-290° K (11). Belleau and Lavoie (17) report that by varying the chemical composition in a series of 30 inhibitors of the tetramethylammonium type for acetylcholinerase, a compensation line of ($\Delta H°$, $\Delta S°$) points with a slope of about 285° K is obtained. There is an unavoidable but relatively large scatter in their thermodynamic data points, and

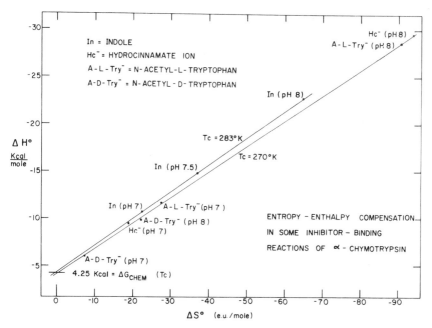

Figure 3. Compensation plots for inhibitor binding to α-chymotrypsin. Data from Reference 19.

it is probable that their data do not conform to the high
standards for such a fit proposed by Exner (18,7). However,
in Figure 3, compensation plots for binding of several inhibitors to α-chymotrypsin are given (19,7). The precision of
the points in this figure is high-estimated standard deviation
of 1%. The data of Figure 3 were obtained by the van't Hoff
method using the rate of the direct proton-transfer reaction
between the imidazole group of HIS 57 and an indicator freely
dissolved in solution. Calorimetric data for N-acetyl-L-
tryptophan (20) and hydrocinnamate ion (21) are found to be
in excellent agreement with van't Hoff results.* Thes studies
of relative binding of hydrocinnamate ion and hydrocinnamoyl
alcohol to α-chymotrypsin and the effects of these inhibitors
on the A-substates of this protein (22), show that there is
little quantitative difference in their effects. Hence we may
conclude that the presence or absence of a charge is of secondar
importance in its effect, perhaps that indicated by the difference between indonl compensation line and the compensation line
for the anionic inhibitors in Figure 3. Rajender, Han and
Lumry (23) have found that the points formed by the standard
enthalpy and standard entropy of formation of the three major
metastable intermediates in the chymotryptic hydrolysis of
N-acetyl-L-tryptophan ethyl ester are distributed along a
line which, within error, is identical with the indole compensation line of Figure 3. This is true at all pH values
from 6 to 9.5 (24). The turn-around behavior observed with
increasing pH values by Beetlestone et al. for ligand binding to the FeIII heme-proteins may be present in the binding of inhibitors to α-chymotrypsin. Rajender has found a
turn-around point with N-acetyl-L-tryptophan which lies not

* Recent calorimetric studies of the binding of Yapel's inhibitors to α-chymotrypsin carried out by D. F. Shiao and J.
Sturtevant (27) do not yield $\Delta H°$ and $\Delta S°$ values quite so large
at a given pH as those obtained in this laboratory. These
differences may be due in part to the corrections made for
dimerization of the protein by these workers. Yapel and
Rajender worked at protein concentrations in which dimerization does not occur under the conditions of the experiments.
In any event, the ($\Delta H°$, $\Delta S°$) points found by Shiao and
Sturtevant for Yapel's inhibitor lie slightly above the
compensation line of Figure 3.

far below the isoelectric point of α-chymotrypsin in her experiments (20). However, there may be quite another explanation for the latter observation since it is known that the binding of inhibitors and substrates to this protein shows an apparently simple pH dependence with an effective pK_a of about 9 (25).

Anusiem et al. (12) and Belleau and Lavoie (17) have inferred from the studies of Ives and Marsden on weak-acid dissociation (10) that their compensation patterns are manifestations of some general characteristic of water. The former authors attribute the influence of the binding reaction on nearby external water to charge rearrangements in the protein surface. The latter authors speak of displacements of water from the binding site by inhibitor or alternatively, of an increase in amount of water sequestered in the binding site or improved order in such sequestered water resulting from inhibitor binding. Bernhard and Rossi have emphasized the displacement of water by inhibitor or substrate in the case of chymotryptic catalysis though they have not taken into account the compensation characteristic of the experimentally observed process (26). Yapel's findings (19) have been interpreted (7) in terms of changes in bulk solvent on the basis of the hypothesis that compensation phenomena in water with T_c values in the range from 270-290° are all manifestations of the same property of water whether the solute system contains a protein or only small molecules. Furthermore the large enthalpy and entropy changes attributable to this property in Yapel's work as well as the high degree of linearity of Yapel's plots (Figure 3) suggest that a number of water molecules in the bulk or surface solvent phase is altered by inhibitor binding. Likhtenshtein (11) attributes compensation behavior to a shell of solid or semi-solid water about the protein and proposes that this shell act as an energy-storage reservoir in enzymic catalysis.

The removal of non-polar groups from water on binding the inhibitors of Yapel to chymotrypsin can be expected to make a positive contribution of as much as 15 e.u./mole to the entropy of binding.* This small quantity demonstrates that the compensation behavior cannot be attributed to direct interaction between inhibitor and water.

*The thermodynamic changes in the transfer of a non-polar side chain of a protein from water into the protein can be estimated in terms of side-chain transfer expressions provided by Brandts (28).

359

Discussion

It is apparent that much interesting information can be overlooked when studies of proteins are carried out at one temperature. This may be particularly true when that temperature is 25°C or less since the enthalpy and entropy contributions to the free energy from compensation part-processes may nearly cancel each other at these temperatures. It is also of some importance that compensation contributions considerably complicate any attempt to obtain information about chemical changes in protein reactions manifesting compensation behavior. This can be seen in a simplified scheme in which we recognize chemical part processes, which give information about the reaction of the ligand and the binding site, and solvation or conformation part-processes, which measure the extent of the compensation process whether it involves water alone or water and protein conformation. The details of the resulting analysis are given in Scheme I in which $L_1P(1)$ and $L_2P(2)$ are the normal forms of the binding compounds formed between ligand L_1 or L_2, and the conformational system which we define to include solvent and conformation. $L_2P(1)$ is the compound in which ligand 2 is bound

OVERALL REACTION: $\quad L_2 + L_1P(1) \rightleftharpoons L_2P(2) + L_1$

PARTIAL REACTIONS:

$$L_2 + L_1P(1) \rightleftharpoons L_2P(1) + L_1 \qquad \text{"CHEMICAL"}$$
$$L_2P(1) \rightleftharpoons L_2P(2) \qquad\qquad \text{"CONFORMATIONAL"}$$

$$\Delta G^\circ = \Delta G^\circ_{CHEM} + \Delta G^\circ_{CON}$$
$$\Delta H^\circ = \Delta H^\circ_{CHEM} + \Delta H^\circ_{CON}$$

$$\overline{\Delta H^\circ_{CON} = \alpha + T_C \Delta S^\circ_{CON} \qquad \text{LINEAR COMPENSATION}}$$

$$\Delta H^\circ = (\alpha + \Delta H^\circ_{CHEM} - T_C \Delta S^\circ_{CHEM}) + T_C \Delta S^\circ$$
$$\Delta H^\circ = \alpha + \Delta G_{CHEM}(T_C) + T_C \Delta S^\circ$$

Scheme I. Simplified formal breakdown of a process manifesting linear compensation behavior into chemical and conformational part-processes. See text.

protein in the conformational state of $L_1P(1)$. Our part process mechanism is an oversimplification but is sufficiently detailed for present purposes since it shows that it is only the experimental value of the enthalpy change when the experimental entropy change is zero, which can give uncomplicated information about the chemical part-process. This enthalpy change consists of the term α from the conformational part process and the free-energy of the chemical part process evaluated at T_c. Unless α can be shown to be very small or can be estimated, even this intercept enthalpy-change value is complicated. In any event, it is not possible to obtain information about the enthalpy and entropy contributions from the chemical part-process from this type of experiment. Only the free energy is obtainable in favorable cases. Furthermore, since the total enthalpy and entropy changes cannot be divided into contributions from the two kinds of part processes, absolute information about these quantities for the conformational part processes are also not obtainable from this type of experiment. In the examples discussed, the compensating enthalpy and entropy changes are usually negative. The significance of the sign depends on the absolute enthalpy and entropy contributions from the chemical part-processes which have not yet been determinable. It is thus not clear whether the contributions from the conformational part-processes indicate a maximum change in these processes at maximum negative values of the contributions or a minimum change.

The free-energy changes for the chemical part-processes in ligand binding to heme-proteins are shown as horizontal bars at the left of Figure 2. The ordering on this scale is the same as that on the unresolved free-energy scale but the experimental total enthalpy and entropy changes are seen to be unreliable as a basis for such ordering. It is interesting that ΔG°_{chem} for N_3^- binding to the α-chains of human hemoglobin is somewhat smaller than that for N_3^- binding to the total molecule, indicating either a difference between the α and β chains in N_3^--binding or that there is a subunit interaction in forming the total molecule which increases the average (negative) ΔG°_{chem} value.

It would be most hazardous to attempt an explanation for the special compensation process which appears to be a rather general consequence of the properties of bulk water. Elsewhere we have attempted in a very preliminary fashion to relate it to volume changes in the solute systems (7). Belleau

and Lavoie have dealt with the role of the volume of their
inhibitors in the acetylcholinesterase example (17). We can
offer as a possible reconciliation of the compensation process and the phenomenological observations on water a tentative proposal based on the well-known two-state analysis
of water behavior (29) now most strongly championed by Walrafen (30). On the assumption that water manifests the simple behavior of a two-state system, an assumption strongly
supported by his data and that of some other investigators
(29), Walrafen through studies of temperature and pressure
dependence of the Raman spectra arrives at a set of "phenomenological" entropy, enthalpy and volume changes in this
process. These numbers with our interpretation of them are
shown in Figure 4. Walrafen's description of the molecular
event involved as a hydrogen-bond breaking process is not a
unique phenomenological consequence of his data. However,
we interpret Walrafen's numbers to mean that in order to make
available 7 ml of volume from the free volume of ligand water
as is required to provide space for an expanding solute, it
is necessary to increase the enthalpy of the system by 2.5
kcal. In this process the entropy increase in the system is
8.5 e.u. The average value of T_c shown in Figure 4 is 300°K
and thus in as good agreement as might be expected with experimental T_c values. These matters together with additional
theorems, examples and supporting material are discussed elsewhere (7,31). In view of the complexity of the problem our
interpretation should be considered only as an illustrative
possibility. It does, however, suggest that there may be

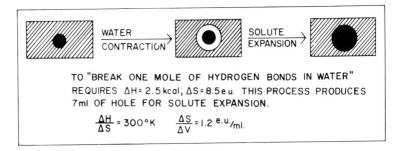

TO "BREAK ONE MOLE OF HYDROGEN BONDS IN WATER"
REQUIRES $\Delta H = 2.5$ kcal, $\Delta S = 8.5$ e.u. THIS PROCESS PRODUCES
7 ml OF HOLE FOR SOLUTE EXPANSION.

$$\frac{\Delta H}{\Delta S} = 300°K \qquad \frac{\Delta S}{\Delta V} = 1.2 \, e.u./ml$$

Figure 4. A possible phenomenological rationalization of compensation behavior on the basis of Walrafen's enthalpy, entropy
and volume changes for a two-state "hydrogen bond breaking"
process in pure water. See text.

significant changes in volume, protein-water interfacial area and the polarity of this interface during some reactions of proteins. It can be shown (7) by an application of the Gibbs-Duhem equation that such changes are a thermodynamic requirement, but it has not yet been possible to determine whether they are large or small, and whether or not they are closely related to the enthalpy-entropy compensation phenomenon.

In conclusion, we suggest that the compensation process may be important in ferrous as well as ferric forms of hemoglobin and myoglobin. Keyes and Lumry (1) find that the xenon-CO-contaminant linkage system of this protein manifests compensation through an apparent T_C value of about 330° K. Dr. John Beetlestone has pointed out to us that the thermodynamic data of Roughton, Otis, and Lyster for the first three steps of oxygen binding to sheep hemoglobin at pH 9.1 (32) show compensation. Figure 5 is a plot of the standard-enthalpy vs. standard-entropy changes in the four steps of oxygenation. Considering the unavoidably large errors which resulted in this work from the need to evaluate the four equilibrium constants of the Adair equation at each temperature, the apparent T_c value of 260° K is quite encouraging. Clearly, Figure 5 suggests that the first three steps of oxygen binding are

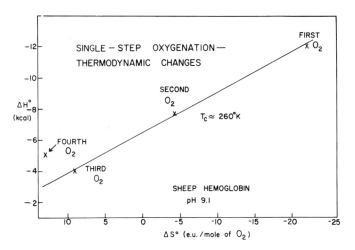

Figure 5. A compensation plot of the standard thermodynamic changes in the single steps of oxygenation of sheep hemoglobin. Data from Ref. 32, but with 1 M standard states and corrected for statistical factors by George and Lyster (33).

363

related by the compensation process we have discussed. In view of the NMR data of Shulman and Wüthrich (34) which appear to exclude a mechanism for heme-heme linkage which depends on changes in the fifth-position ligands and the heme planes, this suggestion deserves considerable attention. Rearrangements of water sequestered within the tetramer, of water around the protein surface and changes in the amount and character of protein surface which occur when sub-units rearrange during the oxygenation steps provide attractive possibilities for explaining linkage among sixth-position ligands via the regions of the protein distal to the heme groups. Beetlestone et al. (12,16) have already proposed pictures of conformation and solvent rearrangement in the distal region in explaining the manifestations of the compensation process in the ferric forms of these proteins.

References

1. Keyes, M. and R. Lumry. This volume, p. 344

2. Wyman, J. Bull. Soc. Chimie biologique, 46, 1577 (1964).

3. Lumry, R. J. Biophys. (Japan), 1, 1 (1961).

4. Lumry, R. and H. Eyring. J. Phys. Chem., 58, 110 (1954).

5. Lumry, R. Enzymes, 2nd ed., 1, 167 (1959).

6. Lumry, R. In A Treatise on Electron and Coupled Energy Transfer in Biological Systems (T. King and M. Klingenberg, eds.), Dekker, in press.

7. Lumry, R. and S. Rajender. Macromolecular Revs., 4, (1969), in press.

8. Arnett, E.M. In Physico-Chemical Processes in Mixed Aqueous Solvents (F. Franks, ed.), Heineman Educational Books, London, 1967, p. 105.

9. Leffler, J. and E. Grunwald. Rates and Equilibria of Organic Reactions, John Wiley and Sons, New York, 1963.

10. Ives, D. and P.J. Marsden. J. Chem. Soc., 649 (1965). See also Helpler, L. J. Am. Chem. Soc., 85, 3089 (1963); Helpler, L. and W.J. O;Hara. J. Phys. Chem., 65, 811, 2107 (1961).

11. Likhtenshtein, G., Biofizika 11, 23 (1966); Likhtenshtein, G. and B. Sukhorukov, Zhur. Fiz. Khim. 38, 747 (1963); Biofizika 10, 925 (1965).

12. Anusiem, A., J. Beetlestone and D. Irvine, J. Chem. Soc. A, 960 (1968).

13. Beetlestone, J. and D. Irvine, J. Chem. Soc. A, 951 (1968).

14. Anusiem, A., J. Beetlestone and D. Irvine, J. Chem. Soc. A, 1337 (1968).

15. Bailey, J., J. Beetlestone and D. Irvine, J. Chem. Soc. A, 2778 (1968).

16. Bailey, J., J. Beetlestone and D. Irvine, J. Chem. Soc. A, 241 (1969).

17. Belleau, B. and J. Lavoie, Can. J. Biochem. 46, 1397 (1968).

18. Exner, O., Coll., Czech. Chem. Com. 29, 1094 (1964).

19. Yapel, A., Dissertation: A Kinetic Study of the Imidazole Groups of Chymotrypsin, University of Minnesota, 1967.

20. Rajender, R., unpublished observations from this laboratory.

21. Canady, W. and K. Laidler, Canadian J. Chem. 36, 1289 (1958).

22. Kim, Y. D., Dissertation: The Effect of pH, Virtual Substrates and Inhibitors on the Conformational States of -chymotrypsin, University of Minnesota, 1968.

23. Rajender, R., M. H. Han and R. Lumry, submitted to J. Am. Chem. Soc.

24. Rajender, R., M. H. Han and R. Lumry, to be submitted.

25. Oppenheimer, H., B. Labouesse and G. P. Hess, J. Biol. Chem. 241, 2720 (1965).

26. Bernhard, S. and G. Rossi, in Structural Chemistry and Molecular Biology (A. Rich and N. Davidson, eds.) Freeman, San Fransisco, 1968.

27. Shiao, D. and J. Sturtevant, to be published.

28. Brandts, J., J. Am. Chem. Soc. 86, 4291, 4302 (1964); in Structure and Stability of Biological Macromolecules (S. Timasheff and G. Fasman, eds.) Dekker, New York, 1969, p. 213.

29. Ives, D. and T. Lemon, Roy. Inst. Chem. Revs. 1, 62 (1968).

30. Walrafen, G., in Equilibria and Reaction in Hydrogen-Bonded Solvent Systems (Covington and Jones, eds.) Taylor and Frances, London, 1968, p. 9.

31. Lumry, R. and R. Biltonen, in Structure and Stability of Biological Macromolecules (S. Timasheff and G. Fasman, eds.) Dekker, New York, 1969, p. 65.

32. Roughton, F., A. Otis and R. Lyster, Proc. Roy. Soc. B 144, 29 (1955).

33. George, P. and R. Lyster, in Conference on Hemoglobin, Nat. Acad. Sci. U. S. A. and Nat. Res. Council publication 557, Washington, D.C., 1958, p. 39.

34. Shulman, R., K. Wüthrich, and J. Peisach, this volume, p. 195.

KINETICS OF OXYGEN BINDING TO HEMOGLOBIN: TEMPERATURE JUMP RELAXATION STUDIES*

T. M. Schuster

Biological Sciences Group
University of Connecticut, Storrs, Connecticut 06268

G. Ilgenfritz

Max Planck Institut für Physikalische Chemie
Göttingen, Germany

Summary of Conclusions

Efforts to explain the cooperativity of ligand binding to hemoglobin have attempted to include various structural properties of hemoglobin. Two principal kinds of protein-protein interaction have been proposed previously to describe the cooperative interactions. One is a model involving cooperativity mainly in the dimer (1-4) and the other involving the entire tetramer as the minimum cooperative unit in hemoglobin (5,6).

With the use of chemical relaxation methods, it has been possible to characterize three kinetic phases of the oxygen reaction (7,8). The slowest of these is the dissociation of tetrameric hemoglobin to dimers, from which we have obtained rate constants for the dissociation of tetramers and association of dimers. For sheep hemoglobin at pH 9.1, in 0.2 M borate at 19°C, the values obtained by fitting the data to the Guidotti model are: for deoxyhemoglobin, k_D = 4 sec^{-1}, k_R = 4 x 10^6 M^{-1} sec^{-1}; for oxyhemoglobin, k_D = 27 sec^{-1}, k_R = 5 x 10^5 M^{-1} sec^{-1}. The two faster relaxations involve ligand binding. The fastest process corresponds to a bimolecular oxygen binding rate constant of about 5 x 10^7 M^{-1} sec^{-1}. Kinetic data obtained under subunit dissociating conditions strongly suggest that there is cooperativity within the free dimer and that the cooperativity is significantly less than that corresponding to a Hill coefficient of 2,

*This research was supported by USPHS HE 11425.

367

since there is a considerable concentration of intermediate species with one molecule of bound oxygen. In addition, it is concluded that there is a large amount of free dimers even at 0.2 M ionic strength (for sheep hemoglobin at pH 9.1). These conclusions, taken together with other data, suggest that there is also cooperativity in the dimer within tetrameric hemoglobin.

Any general mechanism which attempts to explain the kinetics of ligand binding in the concentration range of 10^{-6} to 10^{-4} M (heme equivalents) must take into account, explicity the tetramer-dimer dissociation-association reaction. However, the magnitude of the tetramer-dimer equilibrium constant (4) precludes this reaction as contributing to cooperative ligand binding at physiological concentrations.

The conformation change (or changes) which accompany cooperative ligand binding in hemoglobin and, presumably, give rise to dimer dissociation, are much faster than oxygen binding (50 x 10^{-6} sec) in both tetramer and dimer in the usual range of protein concentrations, 10^{-6} to 10^{-4} M.

Further investigation of a previously discovered very rapid change in absorbancy ($<10^{-6}$ sec) using the electric field jump method revealed that this process has a relaxation time of 60-80 nsec, but is not directly related to the conformation changes accompanying ligand binding. It is instead dependent on the charge state of the protein in a unique manner that may reflect the linkage between the heme groups and the Bohr groups (cf Ilgenfritz and Schuster, this volume, p. 40).

Results

The principal barrier to the elucidation of the mechanism of ligand binding in hemoglobin has been the difficulty in determining the rate constants of each reaction step in the overall reaction.

From the theory of chemical relaxation, it can be shown that all elementary steps in a multistep reaction are manifested as a linear superposition of exponential reactions:

$$\Delta(\text{O. D.}) = \sum A_i \, e^{-t/\tau_i} \qquad (1)$$

where the τ_i's are the relaxation times and the A_i's are the corresponding amplitudes. In general, an individual relaxation time, τ_i, cannot be identified with a particular relaxation step. However, the fastest relaxation can, to a very good approximation, be correlated with the fastest reaction step. In order to identify the slower relaxation times, it is necessary to take into account the coupling to all faster steps.

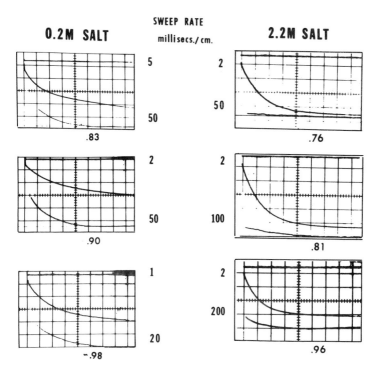

Figure 1. Temperature jump relaxation spectra of sheep hemo-globin at varying degrees of oxygen saturation, as indicated. Dual sweep oscilloscope traces obtained at low and high ionic strengths under otherwise similar conditions: 3.3×10^{-5} M heme, pH 9.2; 3.3 mm optical path; $\lambda = 436$ nm; 10°C; $\Delta T = 4$°C.

Figure 1 shows relaxation spectra of hemoglobin par-tially saturated with oxygen. These traces show three dis-tinct relaxations. Previous measurements at lower detection sensitivity were adequately described by two relaxation times, since the two faster processes could not be resolved (7,8). (The higher resolution results from improvements in the tem-perature-jump method which have been introduced by C. R. Rabl.) This complex behavior is in contrast to the simple relaxation kinetics seen in non-cooperative hemoproteins such as myoglobin and the isolated α and β subunits of hemoglobin (7).

On the other hand, since hemoglobin has four oxygen binding sites one might expect at least four relaxation times. Such an apparent degeneracy may arise either from a non-detectable concentration of one or more species or from coincidentally identical relaxation times. The former case has been found in the helix-coil transition of short chain oligonucleotides, where the reaction is an all-or-none process (11). True degeracy may arise in a system containing identical binding sites, as in a protein composed of identical subunits. Such a case of true degeneracy has been found in the binding of NAD to each of the two conformations of glyceraldehyde-3-phosphate dehydrogenase (9).

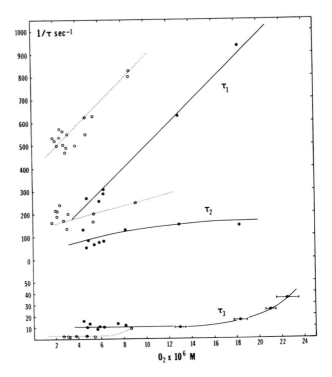

Figure 2. Dependence of $1/\tau_1$, $1/\tau_2$ and $1/\tau_3$ on concentration of free oxygen at low (solid curves, 0.2 M borate) and high (dotted curves, 0.2 M borate plus 2.0 M NaNO$_3$) ionic strengths. Sheep hemoglobin, 3.3 x 10^{-5}M heme, at pH 9.1, 10° C; 3.3 mm optical path;λ = 436 nm; ΔT = 4° C.

Before considering a detailed model for the mechanism of ligand binding to hemoglobin, it is first necessary to identify qualitatively the nature of the reaction(s) associated with each of the observed relaxation times. This was done by studying the protein and ligand concentration dependence of each of the relaxation times and corresponding amplitudes.

Analysis of such data, shown in Figures 2 - 4, reveals that τ_1 and τ_2 must be associated with the binding of oxygen to hemoglobin. However, the slowest relaxation, τ_3 which is in the range 5 to 50 sec-1, is clearly not involved in the oxygen binding reactions. Instead, τ_3 arises from a protein-protein interaction which itself does not exhibit a spectral change but is indicated by the faster oxygen binding steps to which τ_3 is coupled.

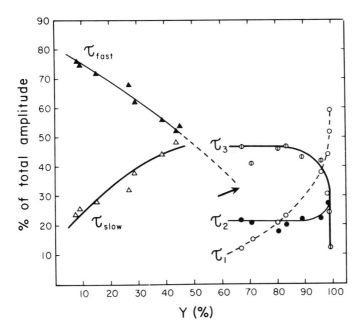

Figure 3. Dependence of amplitudes on Y at low ionic strength. Data on right hand side of Figure are for relaxation data of Figure 2. Data at left-hand side are for relaxation data obtained under similar conditions but at 19°C and lower detection sensitivity, in which τ_1 and τ_2 appeared as τ(fast) (cf. ref. 8).

371

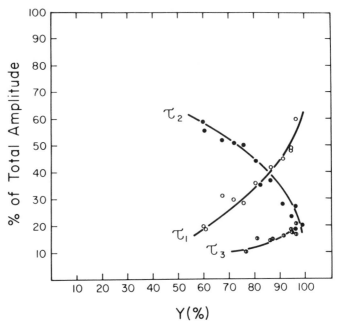

Figure 4. Dependence of amplitudes on Y for relaxation data at high ionic strength shown in Figure 2.

This conclusion is drawn on the basis of the analysis of the amplitude behavior of τ_3, which is outlined below. Consideration of amplitude data can provide information concerning the concentration of the species involved.

Analysis of A_3, the amplitude corresponding to the relaxation time, τ_3. It can be seen in Figure 3 that at low saturation, the fast phase of the relaxation spectrum (composed of τ_1 and τ_2) has a considerably larger amplitude than does the slow phase, τ_3. This is a significant result, since one would expect just the opposite if τ_3 corresponded to an oxygen binding step. This can be understood by the following reasoning. The slower steps in a multistep binding reaction are coupled to all faster steps in the overall reaction. This is clearly seen by considering the amplitude behavior at low saturation. It can be shown that τ_1 probably corresponds to the last oxygen binding step in the overall reaction (see Discussion of Ligand Binding, below). This means that those

372

intermediate species involved in τ_1 must be present in low concentration. Therefore, the relative amplitude corresponding to τ_1 must be very small, especially at low saturation, which means that the relative amplitude of the slower relaxations will dominate. This conclusion can be understood by considering the simple case of a two-step autocatalytic binding reaction:

$$Hb_2 + 2O_2 \rightleftharpoons Hb^*_2O_2 + O_2 \rightleftharpoons Hb^*_2O_4$$

The amplitude of the slowest relaxation in the lower saturation range is given to a very good approximation by:

$$A_{(slow)} = \frac{[O_2]}{[O_2] + [Hb^*_2O_2]} \tag{2}$$

Therefore, $A_{(slow)}$ is considerably less than unity only when the concentration of $Hb_2^*O_2$ is considerable. For a total hemoglobin concentration of 4×10^{-5} M, $[O_2] > [Hb_2^*O_2]$ at low saturation, and $A_3 > 0.5$ (see Figures 6 and 10).

The results of exact calculations for both a two-step and a four-step binding model, using kinetic constants which describe the overall reaction (6) verify this conclusion. Figures 5 and 6 show the results of the exact calculation of the relative amplitudes and relaxation times for the two- and four-step models. In view of the fact that the amplitude of the slowest relaxation, A_3, contradicts these theoretical predictions for a binding reaction, we must exclude the latter as being associated with τ_3.

Yet another feature of τ_3 which is not understandable in terms of oxygen binding reactions is the finding that at low ionic strength (0.2 M), A_3, the amplitude of τ_3 in the upper saturation range increases with decreasing protein concentration (from 30 to 7×10^{-6} M heme) whereas at high ionic strength (2.2 M), A_3 decreases with decreasing protein concentration. Furthermore, A_3 vanishes entirely at 4×10^{-4} M heme in 0.2 M ionic strength, whereas at 2.2 M ionic strength A_3 disappears at about 7×10^{-6} M heme.

These considerations show that τ_3 cannot be associated directly with oxygen binding steps and must be explained on the basis of some other kind of protein reaction.

At higher concentrations of free oxygen (about 10^{-5} M) where a monomolecular reaction would be independent of protein concentration, τ_3 increases from 30 sec^{-1} at 10^{-5} M to 100 sec^{-1}

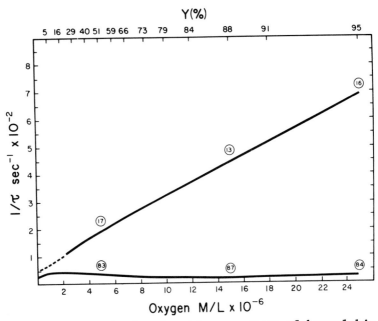

Figure 5. Theoretical relaxation spectrum of hemoglobin, dimer model. 4×10^{-5} M heme, 19° C, Hill coefficient, n = 1.85. Solid lines indicate detectable relaxations; dashed lines, relaxations with amplitudes 10%. Numbers in circles, relative amplitudes.

at 10^{-4} M protein concentration. On the basis of having excluded oxygen binding steps as well as intramolecular changes, we are forced to consider that τ_3 is an intermolecular protein reaction, namely, an association-dissociation reaction.

It is well known that hemoglobin dissociates into $\alpha\beta$ dimers at low protein concentrations or high ionic strength (4). The corresponding reaction must appear in the relaxation spectrum if the process is either thermodynamically linked to oxygen binding and/or has a finite reaction enthalpy. The same evidence which argues against τ_3 being an oxygen binding reaction is in agreement with the interpretation that it is instead a tetramer-dimer dissociation-association reaction.

If the dissociation to dimers is ligand-linked, as has been demonstrated (4) one would expect A_3 to go through a maximum as a function of Y. Figure 3 shows that this is true at 0.2 M ionic strength. In addition, lowering the protein

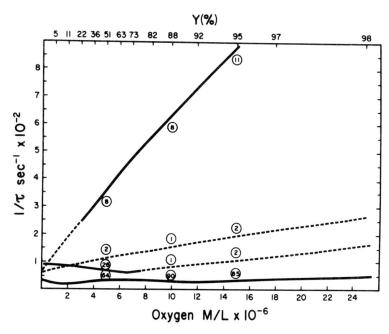

<u>Figure 6.</u> Theoretical relaxation spectrum of hemoglobin, tetramer model.

concentration under strong dissociation conditions (2.2 M ionic strength) should cause A_3 to decrease, since the tetramer-dimer equilibrium is greatly displaced toward the dimer. Similarly, at low ionic strength (0.2 M), increasing the protein concentration displaces the equilibrium completely toward tetramer, thereby again reducing A_3. It is therefore possible to understand the dependence of A_3 on protein concentration, oxygen concentration, and ionic strength on the basis of a single equilibrium reaction.

Comparison of the low and high ionic strength relaxation data of Figures 2 - 4 strongly suggests that at 0.2 M there is a considerable concentration of dimer at these protein concentrations. Figure 7 shows a comparison of the relaxation spectrum at low and high ionic strength, under otherwise similar conditions. Here it can be seen that the principal features of the relaxation spectrum are quite similar. In both cases, three relaxations can be seen. Each of the two fast relaxations, τ_1 and τ_2, are in similar time ranges and increase similarly with oxygen concentration, at low and high ionic strength. The <u>relative</u> amplitudes of τ_1 and τ_2 shown in Figure 7 are also similar. These similarities suggest that basically the same kinds of reactions are involved at low and high ionic strength.

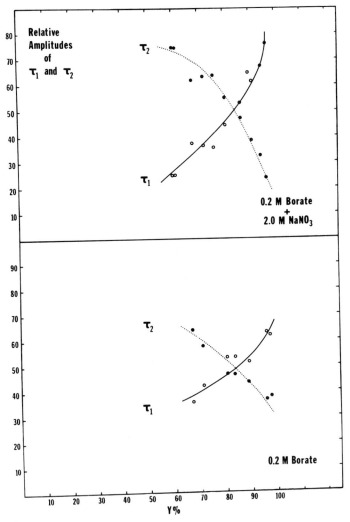

Figure 7. Relative amplitudes of τ_1 and τ_2 at low and high ionic strength, at 10°C. Data from Figures 3 and 5.

The one striking difference between the relaxation behavior at low and high ionic strength is the small amplitude of τ_3 at high ionic strength. As pointed out above this cannot be understood if τ_3 were a binding step but is easily understood in terms of the displacement of the tetramer-dimer equilibrium since the amplitude of the corresponding relaxation is determined by the lowest concentration in the equilibrium.

These conclusions are not in gross disagreement with the thermodynamic results as can be seen by considering the fit of O_2 binding data to a model which explicitly includes the tetramer-dimer equilibrium (4) as is shown in Figure 8.

Discussion of Dissociation. It is apparent from the above observations that a pure binding multistep model is inadequate. Therefore any model which seeks to explain the kinetics of ligand binding in the concentration range of 10^{-6} to 10^{-4}M heme must take into account explicitly the tetramer-dimer dissociation-association at each intermediate degree of ligand binding.

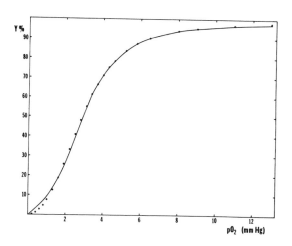

Figure 8. Saturation isotherm of sheep hemoglobin, 19°C, in 0.2 M borate, pH 9.2, 0.1 to 0.4% hemoglobin. Solid line from experimental data of Roughton (10). Points, calculated best fit of Guidotti model (4). See text for details.

Recently a model has been proposed which does take into account dissociation (4). In his model Guidotti considered the dissociation to dimers of the deoxy, half-oxy, and full-oxy species. We have calculated the expected relaxation kinetics behavior of the tetramer-dimer dissociation-association reaction of this model:

$$
\begin{array}{ccc}
& \mathrm{Hb}_4 \;\underset{k_R^o}{\overset{k_D^o}{\rightleftharpoons}}\; \mathrm{Hb}_2 + \mathrm{Hb}_2 & \\
K_1 \;\Big\updownarrow & \Big\updownarrow & \Big\updownarrow\; K \\
& \mathrm{Hb}_4 O_4 \;\underset{k_R^2}{\overset{k_D^2}{\rightleftharpoons}}\; \mathrm{Hb}_2 O_4 + \mathrm{Hb}_2 & \\
K_2 \;\Big\updownarrow & & \Big\updownarrow\; K \\
& \mathrm{Hb}_4 O_8 \;\underset{k_R^4}{\overset{k_D^4}{\rightleftharpoons}}\; \mathrm{Hb}_2 O_4 + \mathrm{Hb}_2 O_4 &
\end{array}
$$

The exact equation used in this calculation is too cumbersome for visual inspection. In order to gain some insight into the form of the dependence of the relaxation time on ligand concentration one may assume quasi-buffering of ligand, which does not change the principal features of the result shown below.

$$
\frac{1}{\tau_3} = \frac{k_D^o + k_D^2 \dfrac{2F^2}{K_1} + k_D^4 \dfrac{F^4}{K_1 K_2}}{1 + \dfrac{2F^2}{K_1} + \dfrac{F^4}{K_1 K_2}} + 4P_o \frac{k_R^o + k_R^2 \dfrac{F^2}{K} + k_R^4 \dfrac{F^4}{K^2}}{1 + \dfrac{F^2}{K}} \tag{3}
$$

where P_o = hemoglobin concentration and
      F = oxygen concentration.

Inspection of this equation shows that in general $1/\tau_3$ will vary strongly with ligand concentration. However, the form of the dependence can be increasing or decreasing with increasing O_2, depending on the rate constants. Had we included the dissociation of the one- and three-liganded tetramer states (not considered by Guidotti) the form of the ligand dependence would not change significantly.

Figure 9 gives examples of the ligand dependence of $1/\tau_3$. It can be seen that with a proper choice of rate constants one may observe a ligand-independent relaxation time for tetramer dissociation.

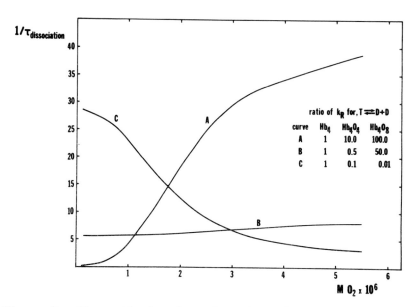

<u>Figure 9</u>. Theoretical relaxation times for tetramer-dimer equilibrium in Guidotti model (4) for various ratios of rate constants.

On the basis of thermodynamic constants obtained by fitting the Guidotti model to the O_2 binding data of Roughton, shown in Figure 8, we can fit our data for τ_3 with the rate constants in Table I.

TABLE I

Estimated rate constants for sheep hemoglobin; pH 9.1; 0.2M borate; 19°C (cf. ref. 8).

$\dfrac{k_D}{k_R}$	$k_D(sec^{-1})$	$k_R(M^{-1}sec^{-1})$	$K(M^{-1})$
$Hb_4 \rightleftharpoons Hb_2 + Hb_2$	4	4×10^6	10^{-6}
$Hb_4O_4 \rightleftharpoons Hb_2O_4 + Hb_2$	24	2×10^5	12×10^{-6}
$Hb_4O_8 \rightleftharpoons Hb_2O_4 + Hb_2O_4$	27	5×10^5	54×10^{-6}

379

The equilibrium constants in Table I were obtained by first fitting the Guidotti model to the high protein concentration (3%) data of Roughton (10) where presumably there is negligible dissociation. This provided independently the binding constants to the tetrameric form of hemoglobin. Using these constants we then sought the best fit to the low protein concentration (0.1 to 0.4%) binding data of Roughton which resulted in a set of equilibrium constants given in Table I.

The obviously inadequate fit to the binding data obtained in this manner (cf. Fig. 8) results no doubt from the simplifying assumptions in the Guidotti model. Nevertheless, the kinetic constants obtained by fitting our relaxation data to these equilibrium constants are probably within the correct order of magnitude.

Discussion of Ligand Binding. The fast phase seen in the relaxation spectrum contains two relaxation times which must be attributed to oxygen binding steps. This is seen clearly for $1/\tau_1$ which increases linearly with free O_2 (cf. Fig. 2.), which implies that τ_1 is a binding step and, in addition, that the concentration of hemoglobin associated with that step is much smaller than the concentration of free oxygen.

Rapid flow kinetic measurements have shown that the overall rate of ligand binding to hemoglobin increases with the extent of reaction and that this is probably due mainly to an increase in the recombination rate constants (6). Provided this latter point is correct it is possible to identify τ_1 with a later binding step (perhaps the last) in the overall reaction. In addition, this conclusion follows independently from the linearity of $1/\tau_1$ with O_2 since if τ_1 corresponded to the first binding step the concentration of free hemoglobin would not be small compared to the concentration of free oxygen and the linear increase of $1/\tau_1$ with O_2 would not be observed.

Since the fastest step can be considered, to a very good approximation, to be uncoupled from the slower reactions it must obey the relation

$$\frac{1}{\tau_1} = k_D + k_R \left([Hb] + [O_2]\right) \tag{4}$$

where [Hb] is the concentration of that unliganded hemoglobin species involved in this particular binding step. Attribution of τ_1 to the fastest binding step in a multistep binding reaction requires that there be at least one relaxation time with appreciable amplitude which represents the equilibrium of all slower steps, together with the re-equilibration of the faster steps. Moreover, Equation 2 and Figures 6 and 7 show that this slower relaxation should have the dominant amplitude. Therefore, τ_2 must also represent O_2 binding.

Since τ_1 corresponds to the fastest binding step, we can estimate a recombination rate constant from Figure 2 for this step. The slope of the curve $1/\tau_1$ versus O_2 yields a recombination rate constant of $5 \times 10^7 M^{-1} sec^{-1}$, which is in good agreement with the value reported by Gibson (6) for the last binding step. For the relaxation time τ_2, a similar approximation cannot easily be made since τ_2 does not correspond to a single step. Nevertheless, the rate constants involved in τ_2 must be much smaller than $5 \times 10^{-7} M^{-1} sec^{-1}$.

Inasmuch as the intramolecular changes in hemoglobin appear to be faster than the binding steps ($<50 \times 10^{-6}$ sec) in the usual range of concentrations used for kinetic studies, and the cooperativity of binding is quite high, it is not yet possible to distinguish between concerted and sequential conformational changes by studies of ligand binding alone, provided there is no spectral change associated with the conformation change.

Discussion of Possible Mechanisms. Although it has not yet been possible to describe theoretically all aspects of the relaxation kinetics data there are two central points concerning the mechanism about which some conclusions may be drawn.

The results at high ionic strength suggest that the free dimer exhibits cooperative ligand binding and that the degree of cooperativity is significantly less than that corresponding to a Hill coefficient of $n = 2(2-4)$.

Throughout most of the range of saturation two O_2 binding relaxations (τ_1 and τ_2) are observed with rather large amplitudes. If the dimer were either non-cooperative ($n = 1$) or extremely cooperative ($n = 2$) only one relaxation would be observed. The fact that two relaxations are seen tells us immediately that there is a detectable concentration of the intermediate species with one O_2 molecule bound. Furthermore, if there were no intermediate present ($n = 2$) there would be one relaxation time for the overall reaction of free dimer with two O_2 molecules. Since the concentration of free dimer is rather large one would observe strongly non-linear behavior of $1/\tau_1$ with O_2. The latter argument also excludes a non-cooperative dimer in which each subunit binds with different rate constants.

Model dimer calculations, summarized in Figure 10, show that the ratio of the amplitudes of the slow to the fast relaxations decreases with decreasing values of the Hill coefficient, n. The results allow us to exclude the Guidotti model, in which it was proposed that there is extreme cooperativity within the free dimer since the presence of a considerable concentration of intermediate means a Hill coefficient significantly less than 2.

Based upon the very strong similarity of the relaxation spectrum at low and high ionic strength one might speculate that the mechanism of binding in the hemoglobin tetramer is very similar to that in the free dimer. Although we base to some extent this suggestion on physical reasoning it is difficult to imagine that each dimer pair in the tetramer does not possess some cooperativity. This would mean that the increase in the change of free energy accompanying ligand binding does not occur in only one step of the overall reaction in the tetramer.

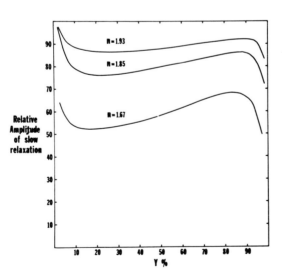

Figure 10. Relative amplitude of slow relaxation time for cooperative dimer model, for various values of Hill coefficient (cf. Ref. 9).

References

1. Schuster, T. M. and M. Brunori, in International Symposium on Comparative Hemoglobin Structure and Function, (A. Christomanos, ed.) Thessaloniki, Greece, 1966, p. 129.

2. Wyman, J., J. Am. Chem. Soc. 89, 2202 (1967).

3. Antonini, E., Science 158, 1417 (1967).

4. Guidotti, G., J. Biol. Chem. 242, 3685 (1967).

5. Perutz, M., et al., Nature 219, 131 (1968).

6. Gibson, Q. H., Prog. Biophys. Biophys. Chem. 9, 1 (1959).

7. Brunori, M. and T. M. Schuster, "Kinetic Studies of Ligand Binding to Hemoglobin and its Isolated Subunits by the Temperature-Jump Method". J. Biol. Chem., 244, 4063 (1969).

8. Schuster, T. M. and G. Ilgenfritz, in Nobel Symposium 11: Symmetry and Function of Biological Systems at the Macro-molecular Level, (A. Engstrom and B. Strandberg, eds.) Interscience, 1969, p. 181.

9. Kirschner, K., in Regulation of Enzyme Activity and Allo-steric Interactions, Proc. of the Fourth Meeting of the Federation of European Biochemical Societies, Oslo, 1967, (E. Kvamme and A. Pihl, eds.) Academic Press, 1968, p. 39.

10. Roughton, F. J. W., Clinical Chemistry 9, 682 (1963) and J. Gen. Physiol. 49, 105 (1965).

11. Pörschke, D., Diplom-Thesis, Georg-August University, Göttingen, Germany, 1967.

DISCUSSION

Eigen: For glyceraldehyde phosphate dehydrogenase, the con-formation change between the R- and T-form is slow compared to the binding reaction. For hemoglobin, the conversion from one conformation to the other is probably fast, since it comes along with the uptake of oxygen. Now Dr. Schuster reports the dimer-tetramer equilibrium to be the slowest process, but I would put "slow" in quotation marks because if this process involves any conformation change it has to be fast indeed.

Actually, the rate constant of the dimer-tetramer association suggests that it would occur within about 10^{-5} sec, i.e., fast compared to the oxygen release.

Gibson: I thought there would be two numbers from each relaxation component, a concentration-dependent rate and a concentration-independent rate, and that this was the point of making the $1/\tau$ vs. concentration plots, but we don't have quite enough numbers.

Schuster: These are the experimental relaxation times we find. In order to derive rate constants from these data, one must have a mechanism which tells you how to transform the observed relaxation time to the coupling of the appropriate rate process involved. Now, since the fastest relaxation is the fastest relaxation step, and since it is linear with the oxygen concentration, this is a bimolecular step and is, effectively and to a first approximation, uncoupled from the remaining steps. It is possible to estimate a bimolecular recombination rate constant from the slope of $1/\tau_1$ vs. free oxygen. When we do this, we obtain 5×10^7 M^{-1} sec^{-1} for the fastest oxygen binding step. The corresponding dissociation rate constant cannot be obtained as yet, since this is given by the intercept of the same plot, i.e., zero oxygen concentration, and it is not yet possible to make this extrapolation reliably because the concentration of non-liganded hemoglobin sites cannot be neglected at low oxygen concentration, in contrast to high oxygen concentration.

A similar estimate of rate constants for τ_2 cannot be made from this plot of the data, since this step is coupled to all faster processes.

Berger: Brunori and I studied the oxygen uptake reaction in the rapid flow apparatus in the same concentration range as Dr. Schuster, i.e., about 10^{-5} M human hemoglobin, and about 0.2 atm of oxygen before mixing. When you go to 0.5 atm of oxygen the slope steepens appreciably. Thus it appears that the bimolecular rate constant varies with oxygen saturation, just as it does with carbon monoxide. I hope that Dr. Schuster can somehow or other account for this; I do not know how to do so.

CONFORMATIONAL STUDIES ON HEMOGLOBINS MODIFIED WITH MONO- AND BIFUNCTIONAL MALEIMIDES

Sanford R. Simon

State University of New York at Stony Brook
Stony Brook, New York 11790

To facilitate understanding of the relationship between cooperative interactions, changes in the values of linked functions, and the conformational rearrangement associated with binding and release of oxygen to hemoglobin, Dr. W.H. Konigsberg and I have modified the native hemoglobin molecule with a bifunctional reagent which freezes horse hemoglobin into a conformation identical to that of the normal oxy-protein, even when deoxygenated. This bifunctional reagent, bis(N-maleimidomethyl)ether (BME), forms internal bridges within individual beta chains, linking cysteine $\beta93$ to histidine $\beta97$. We have already heard from Drs. Gibson and Edelstein that dissociation of the normal hemoglobin molecule into subunits is a "linked function," occuring less readily in deoxy- than in oxyhemoglobin. Dr. Schuster has presented relaxation data which suggest that this ligand-linked dissociation, as well as the conformational transformation associated with ligand binding, can be detected by their effects on kinetics of oxygenation and deoxygenation. In Dr. Eigen's labs, Dr. Schuster and I have examined the modified BME-hemoglobin in the T-jump apparatus. Only a single relaxation is observable. The dependence of the relaxation time constant upon concentration of hemoglobin and oxygen with which it is partially saturated, gives values for rates of binding and release of oxygen to the protein. These are comparable to the rates obtained for myoglobin and estimated for hemoglobin in an "oxy-" conformation. There is no kinetic evidence for any conformational transformation. BME-hemoglobin can dissociate into subunits, but, as shown by osmometric measurements, the oxy- and deoxy-proteins both dissociate to the same extent. Since subunit dissociation is no longer linked to ligand binding, it is not reflected in an increased number of relaxation times.

385

In collaboration with Drs. J.K. Moffat and M.F. Perutz, we have obtained crystallographic evidence for the identity of conformations of oxy- and deoxy-BME-hemoglobin. A 2.8 Å difference Fourier analysis of horse oxy-BME-hemoglobin shows the BME molecule, attached to cysteine β93 and histidine β97. The conformation of the modified protein is virtually identical to that of normal horse oxyhemoglobin, except for a displacement of tyrosine β145, penultimate to the C-terminus, and a distortion of the FG corner of the beta chains. The C-terminus of the beta chains takes up a new position, closer to the center of the hemoglobin molecule, and away from contact with the surrounding medium. This is consistent with the abnormally slow rate of digestion of horse BME-hemoglobin by carboxypeptidase A. Recent crystallographic studies by Perutz and Moffat on another conformationally constrained hemoglobin, reacted with N-α-bromoacetoxymethyl maleimide, reveal similar displacement of tyr β145 from its normal position, and concomitant distortion in the FG corner. Monofunctional maleimides, on the other hand, even with bulky N-substituents, such as N-phenyl, N-isopropyl, N-naphthyl, and N-tosyl maleimides, do not abolish cooperativity in oxygen binding when reacted with hemoglobin, but only reduce interactions partially. The difference Fourier of one such modified hemoglobin, reacted with N-ethylmaleimide, shows no displacement of tyr β145, no distortion of the beta chain FG corner; the conformations of the crystalline oxy- and deoxy-derivative are different in this case. We can tentatively conclude that the FG corner of the beta chains, apparently stabilized by tyrosine β145 in the native hemoglobin molecule, plays a critical role in the mechanism of cooperative ligand binding. Distortion of the FG corner in chemically modified hemoglobins which have lost all cooperative interactions results in failure to undergo the normal ligand-linked conformational rearrangement.

The difference Fourier of BME-hemoglobin shows some slight shifts in other regions of the molecule, but these are less than 1/2 A in magnitude, and could be expected to have little, if any effect, on the properties of the molecule. The way in which these shifts are transmitted through helical segments of the polypeptide chains, however, may prove to be of more general significance for understanding allosteric interactions in other oligomeric proteins. The F, G, and H helices of the beta chains, which attach directly to the displaced residues in the FG corner and the C-terminal segment, all seem to move a rigid rods, as evidenced by symmetrical regions of positive and negative density, in the difference map, framing the individual turns of helix seen in the map of native horse oxyhemoglobin. Similar movements along the screw axis of helical regions in other proteins might provide a pathway for transmitting the effects of conformational perturbation in one region of the molecule to other more distantly removed regions.

Finally, in relation to Dr. Shulman's results on the nature of the "heme-heme" reactions in native hemoglobin, Dr. C.R. Cantor and I have compared circular dichroism spectra and normal and BME-modified hemoglobins at different degrees of oxygen saturation to look for evidence for interactions in the conformational transformation. In the region around 285 mμ, the CD change going from deoxy- to oxy-protein is sharply attenuated in BME-hemoglobin, indicating that this area of the spectrum is sensitive, at least in part, to the ligand-linked conformation change in normal hemoglobin. The CD change in this region is strictly linear with fractional saturation with oxygen, for both native and modified proteins, suggesting that the conformational change in normal hemoglobin, which is reflected in at least 40% of the CD change at 285 mμ, is also linear with oxygenation. Theories of cooperative interactions like that of Monod, Wyman, and Changeaux, in which there is a concerted conformational change within the entire oligomeric molecule, would predict a non-linear relationship between conformational change and ligand binding. Models like those of Koshland, Nemethy, and Filmer, on the other hand, in which conformational change within each individual subunit necessarily accompanies binding of ligand to the subunit, are more fully consistent with our data. These results, combined with those of Dr. Ogawa on the ESR patterns of spin-labeled hemoglobin, all favor a picture of conformation change within the native hemoglobin molecule which is largely confined to subunits which themselves have bound ligand, and which may not be transmitted to adjoining unliganded neighbors.

References

1. Simon, S.R. and W.H. Konigsberg. Proc. Nat. Acad. Sci. U.S., 56, 749 (1966).

2. Simon, S.R., W.H. Konigsberg, W. Bolton and M.F. Perutz. J. Mol. Biol., 28, 451 (1967).

3. Simon, S.R. and C.R. Cantor. Proc. Nat. Acad. Sci. U.S., 63, 205 (1969).

4. Arndt, D.J., S.R. Simon, T. Maita and W.H. Konigsberg. J. Biol. Chem., in press.

5. Moffat, J.K., S.R. Simon and W.H. Konigsberg. J. Mol. Biol., in press.

OXYGEN CHANGES HEMOGLOBIN'S BREATHING

S.W. Englander

Department of Biochemistry, School of Medicine
University of Pennsylvania, Philadelphia, Pennsylvania 19104

Mrs. Catharine Mauel and I have been looking at hemoglobin with our hydrogen-tritium exchange techniques. We find a rather novel kind of conformational change, a change in the dynamic structure, that is, in the "breathing" behavior of the protein.

Figure 1 shows some exchange-out data for hemoglobin that has been fully labelled with tritium, _i.e._, exposed to a long time equilibration in tritiated water. Some hydro-

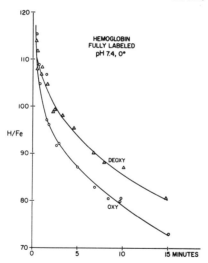

Figure 1. Hydrogen-tritium exchange-out data for hemoglobin labelled by equilibration in tritiated water for 22 hours at 37°. Some hydrogen exchange faster in the oxy than in the deoxy form.

*Supported by USPHS AM11295 from the Institute of Arthritis and Metabolic Diseases.

gens exchange faster in the oxy than in the deoxy form. We would like to look more specifically at the hydrogens which are altered on liganding. In order to do so, we perform an exchange-in/exchange-out experiment, in which the hemoglobin is exposed to tritiated water for only a limited time so that fast exchanging hydrogens are preferentially labelled, and the slower hydrogens are not labelled at all. Results are shown in Figure 2.

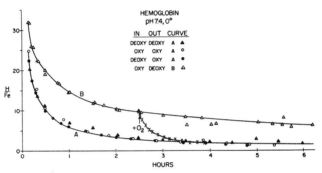

Figure 2. Exchange-in/exchange-out data. Faster exchanging hydrogens were preferentially labelled by limiting exchange-in to 1 hour at 0°. About 10 H, fast in oxyhemogloin, become slow upon removal of oxygen, and again fast on reoxygenation.

If hemoglobin is tritiated for one hour in the oxy form, then switched over to deoxy and exchanged out, we get the top curve. Any other combination of exchange-in/exchange-out conditions gives us the bottom curve. Thus, there are about 10 hydrogens fast in the oxy form that are locked into a slowly exchanging form when oxygen is removed. When oxygen is given back to the protein, these hydrogens again become fast.

Figure 3 makes two points. The hydrogens that we are looking at here represent, as we know from myoglobin studies, hydrogen-bonded peptide hydrogens. We believe that the exchange rate of hydrogen-bonded hydrogens is controlled by the kind of opening-closing equilibrium shown, so that hydrogens can only exchange when hydrogen bonds are open. We can then write down these equilibrium constants for this opening-closing reaction in both oxy and deoxyhemoglobin, because we know the kc rate from model studies, and we can measure the overall rate of exchange from the data shown in Figure 2.

THE PEPTIDE GATE

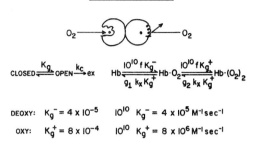

$$\text{CLOSED} \underset{K_g}{\overset{K_g}{\rightleftharpoons}} \text{OPEN} \xrightarrow{k_c} \text{ex} \qquad \text{Hb} \underset{g_1 \, k_x \, K_g^+}{\overset{10^{10} \, f \, K_g^-}{\rightleftharpoons}} \text{Hb·O}_2 \underset{g_2 \, k_x \, K_g^+}{\overset{10^{10} \, f \, K_g^+}{\rightleftharpoons}} \text{Hb·(O}_2)_2$$

DEOXY: $K_g^- = 4 \times 10^{-5}$ $10^{10} \, K_g^- = 4 \times 10^5 \, M^{-1} sec^{-1}$

OXY: $K_g^+ = 8 \times 10^{-4}$ $10^{10} \, K_g^+ = 8 \times 10^6 \, M^{-1} sec^{-1}$

Figure 3. Possible modulation of ligand binding by a peptide gate mechanism. The left column shows how measured hydrogen exchange rates relate to an equilibrium constant (K_g) for H-bond "opening," i.e., measured rate constant = $K_g k_c$, and k_c is known. The right column suggests how ligand binding would respond to a gate. With kg known, maximum binding rate constants can be predicted, as indicated. Rate ratios and equilibrium ratios may also be obtainable.

My last point is illustrated by the oversimplified dimer model for hemoglobin shown in Figure 3. The X-ray diffraction data show that the heme binding site is out of contact with solvent; that means that some structure, some gate, must open for oxygen or any other ligand to get in. Is it possible that the hydrogen exchange data are showing us the opening of this gate, and is it further possible that the opening of such a gate might modulate ligand binding? If so, then we could write down the rate of ligand binding as 10^{10}, the collision frequency, times a factor, f, equal to 1 or less, times the fraction of time the gate is open; Kg^- for deoxy-Kg^+ for oxyhemoglobin. If f is taken equal to one, we are led to the maximum possible binding rate constants. This model predicts the maximum binding rate constants written down here. These predicted values turn out to be close to correct. The model makes other predictions about binding equilibria, which we do not have time to discuss here.

With respect to the discussion we have had earlier, it should be pointed out that here we are looking at a kind of protein structural change, which could determine binding parameters. It is not a heme electronic change, not a kind of structural change that you would seen by NMR. It only exists 10^{-3} or less of the time, so that you would not see it by X-ray diffraction either.

I would like to refer to this model as the "peptide gate" model, and suggest that it looks interesting enough to consider further.

ON THE COOPERATIVE OXYGENATION
OF HEMOGLOBIN

R. G. Shulman

Bell Telephone Laboratories, Incorporated
Murray Hill, New Jersey

My colleagues and I have reported elsewhere (1) experimental studies bearing upon the cooperative oxygen binding to hemoglobin. Wyman (2) has shown that to saturate hemoglobin with oxygen, the free energy of interaction, ΔF_I, is of the order ot 3 Kcal per mole of oxygen which is about 30% of the total free energy of oxygenation (3). Various models, of course, have been proposed (4-7) to explain the oxygen binding.

Leaving models aside for the moment to consider mechanisms, we may break up the free energy of interaction into a heme part plus a protein part, i.e., $\Delta F_I = \Delta F_{IH} + \Delta F_{IP}$. The questions we asked were: Does this interaction energy come about from the ΔF_{IH} term because one heme which is oxygenated somehow reaches over, presumably by a structural change, and affects the second heme, so that the second heme now has a different affinity for oxygen? Or must we look for the free energy of interaction in the protein part of the molecule, i.e., in ΔF_{IP}? Presumably a strong contribution to ΔF_{IP} would come from the different subunit interfaces which Perutz sees in the oxy and deoxy forms (8). The experiments which we have done show that the ΔF_{IH} term is negligible, and therefore the only significant contribution comes from the ΔF_{IP} term.

Figure 1 shows the low field resonances of cyanoferric hemoglobin. They are temperature-dependent, indicating that they are shifted by hyperfine interactions (9). They describe the electronic distribution very accurately and the spectral resolution is good, although not as good as in myoglobin. However, one interesting difference between hemoglobin and myoglobin, which bears directly upon the

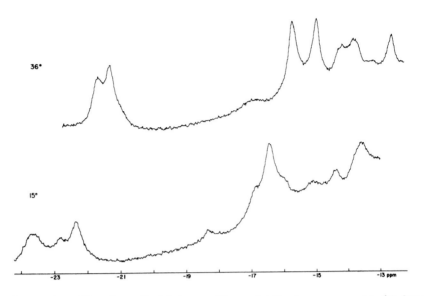

Figure 1. The low field hyperfine shifted resonances in human cyanomethemoglobin measured at 220 MHz at 36° and 15°. From ref. (9).

present study, is that there are measurable differences in the spin densities between hemoglobin from different species even for the liganded derivatives (10). Thus, even though cyano-ferric hemoglobins from human, sheep, bovine, rabbit, porcu-pine, and dog give similar NMR spectra, there are differences observed. In myoglobin, on the other hand, there is no change between species in the spin densities at the heme (11). Hence in hemoglobin the NMR measurements are extremely sensitive to amino acid changes even some distance from the heme. It is also known that NMR measurements are very sensitive to small structural changes such as those introduced by xenon and cyclopropane near the heme in myoglobin (12), but it is clear that changing the amino acids with species does not af-fect the heme environment.

We have approached the question of cooperativity by using mixed state hemoglobin in which like chains have like ligands, but the two kinds of chains have different ligands. We have used different combinations of mixed state, such as α-ferric with water or cyanide as ligands, and β-ferrous. We have prepared these mixed state samples in two different ways: first, by reassembling them from the chains and secondly, by using Huishman's (13) chromatographic separation of aged

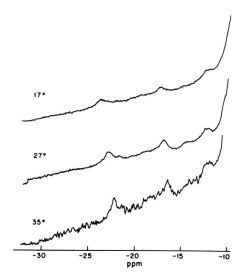

Figure 2. Temperature dependent low field shifted resonances of deoxy hemoglobin, pD 8.0, at 17°, 27°, and 35°.

hemolysate. The important point is that this kind of mixed state hemoglobin -- β-deoxy, α-ligated -- has a high affinity for oxygen by comparison with the low affinity of the deoxy tetramer, (14), whose NMR spectrum is shown in Figure 2. Therefore, by comparing the NMR spectra of the deoxy hemes in their high and low affinity forms, we can look for differences at the heme associated with the differences in affinities. We did these experiments, expecting that there would be large differences.

Surprisingly, however, we saw no differences between the NMR spectra of the high affinity deoxy and the low affinity deoxy. Figure 3 shows the NMR spectrum of the mixed state; in this case, the α-chain is deoxy and the β-chain is in the met form. It is compared with the spectrum of the deoxy tetramer and with that of the met tetramer, and it may be seen that the mixed state spectrum is merely the sum of the spectra of the two constituents.

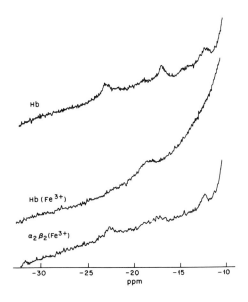

Figure 3. Comparison of the 220 MHz NMR low field spectrum of $[\alpha^{III}(H_2O)\beta^{II}]_2$ with $[\alpha^{III}(H_2O)\beta^{III}(H_2O)]_2$ and $[\alpha^{II}\beta^{II}]_2$ at pD 7.0 and 25°, showing that the mixed state represents a sum of the other two.

Nor are there any differences in any of the other states, i.e., there is no difference between the met form in this mixed state and the met form in the tetramer nor of the cyanide form in the mixed state and the cyanide in the tetramer. When we oxygenate the β-chains, it does not change the hemes of the α-chains.[*] Nothing changes at the heme and yet there is high

[*] Note added in proof: Recent experiments (18) have shown this to be true at pH > 7.3. However, at pH < 7.3 or in the presence of DPG, small changes are observed at the α-chain heme group in $(\alpha\ CN\ \beta O_2)_2$ upon deoxygenation which, while they do not disagree with the present report, since they do not correspond to significant energy changes, have been interpreted as reflecting quaternary structural changes.

affinity. Therefore, the high affinity form has the same wave
functions, spin densities and structure in the immediate environment of the heme as does the low affinity form, within
the experimental accuracy. The accuracy is much higher than
the 30% change in the energy during ligation that we are concerned with explaining.

In summary, there are no changes at one heme when the
other heme is oxygenated or deoxygenated. On the other hand,
we see many changes at the heme when we go from the chains to
the tetramers. This is well-known optically for the deoxy
Soret band (15), and we see many more of these changes by magnetic resonance. Chain to tetramer, there are heme changes;
oxy to deoxy, there are no heme changes at the neighbor.

Ogawa and McConnell's (16) spin label experiment shows
that there are changes at β-93 upon oxygenation, which are
not transmitted from the next subunit when it is oxygenated.
By NMR we have shown specific structural changes in myoglobin
upon oxygenation (17), and there are similar indications of
structural changes within the subunit of hemoglobin upon
oxygenation (10). It can be shown from these results that
the ligand induced conformational changes, expected within a
subunity in Koshland's (5) induced fit model, do occur. Furthermore these tertiary structural changes must be the source
of the cooperativity since there are no changes at the heme
large enough to explain the observed changes in affinity.
Hence we can also conclude that the highly restrictive symmetry
hypothesis of Monod et al. (4), i.e. the subunit is the same
regardless of ligation is not supported by these experiments.
Recently the importance of a modified Monod model, incorporating tertiary and quaternary structural changes has been demonstrated (18).

References

1. These experiments are described in more detail in
    R.G. Shulman, S. Ogawa, K. Wüthrich, T. Yamane, J.
    Peisach and W.E. Blumberg. Science, 165, 251 (1969).

2. Wyman, J. Adv. Protein Chem., 19, 223 (1964).

3. Roughton, F.J.W., A.B. Otis and R.L.J. Lister. Proc.
    Roy. Soc. London, B144, 29 (1955).

4. Monod, J., J. Wyman and J.P. Changeux. J. Mol. Biol., 12, 88 (1965).

5. Koshland, D.E., G. Némethy and D. Filmer. Biochemistry, 5, 365 (1966).

6. Noble, R. J. Mol. Biol, 39, 479 (1969).

7. Guidotti, G. J. Biol. Chem., 242, 3704 (1967).

8. See Perutz, M.F., H. Muirhead, J.M. Cox and L.C.G. Goaman. Nature, 219, 139 (1968) for earlier references.

9. Wüthrich, K., R.G. Shulman and T. Yamane. Proc. Nat. Acad. Sci. U.S., 61, 1199 (1968).

10. Yamane, T., K. Wüthrich, R.G. Shulman and S. Ogawa. J. Mol. Biol., 49, 197 (1970).

11. Wüthrich, K., R.G. Shulman, T. Yamane, B.J. Wyluda, T. Hugli and F.R.N. Gurd. J. Biol. Chem., 245, 1947 (1970).

12. Shulman, R.G., J. Peisach and B.J. Wyluda. J. Mol. Biol., 48, 517 (1969).

13. Huishman, T.H.J., A.M. Dozy, B.F. Horton and C.M. Nechtman. Jour. Lab. Clin. Med., 67, 355 (1966).

14. Enoki, Y. and S. Tomita. J. Mol. Biol., 32, 121 (1968); Brunori, M., G. Amiconi, E. Antonini, J. Wyman and K. Winterhalter. J. Mol. Biol, 49, 461 (1970).

15. Benesch, R., Q.H. Gibson and R.E. Benesch. J. Biol. Chem., 239, PC1668 (1964); Brunori, M., E. Antonini, J. Wyman and S.R. Anderson. J. Mol. Biol, 34, 357 (1968).

16. Ogawa, S. and H.M. McConnell. Proc. Nat. Acad. Sci. U.S., 58, 19 (1967): Ogawa, S., H.M. McConnell and A. Horwitz. Proc. Nat Acad. Sci. U.S., 61, 401 (1968).

17. Shulman, R.G., K. Wüthrich, T. Yamane, D.J. Patel and W.E. Blumberg. J. Mol. Biol., 53, 143 (1970).

18. Ogawa, S. and R. G. Shulman. Biochem. Biophys. Res. Com.

STUDIES ON A NEW ELECTRIC FIELD INDUCED TRANSITION IN HEMOGLOBIN

G. Ilgenfritz

Max Planck Institut für Physikalische Chemie
Göttingen, Germany

T. M. Schuster
Department of Biology, State University of New York,
Buffalo, New York

During the course of an investigation of hemoglobin-oxygen binding kinetics by the temperature-jump relaxation method (1,2), it was noted that there was a very fast ($<10^{-6}$ sec) phase of the relaxation spectrum, τ_0, which remained unresolved by that method. Since the results of the temperature-jump experiments demonstrated that any structural changes associated with cooperative ligand binding must be very fast, we studied this effect further with the electric field-jump method (3) which has a dead time of about 30 nsec and employs a spectrophotometric detection system.

The results of this study reveal that τ_0 is a very rapid (60 to 80 x 10^{-9} sec) intramolecular process which is not directly related to the intramolecular change(s) accompanying ligand binding, but may be related to the thermodynamic linkage between the Bohr groups and the heme groups.

Figure 1 shows the resolution of τ_0. The relaxation time, 60 to 80 x 10^{-9} sec, for sheep oxyhemoglobin at pH 9, is independent of protein concentration in the range investigated, 7 to 70 x 10^{-6} M heme equivalents, and therefore τ_0 is a monomolecular process. The relaxation time is also independent of the degree of oxygen saturation. Measurements of the temperature dependence of τ_0 yielded an apparent acti-

*This research was supported in part by PHS HE 11425.

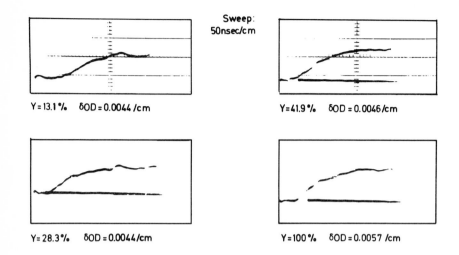

Sweep: 50nsec/cm

Y = 13.1 %   δOD = 0.0044 /cm

Y = 41.9 %   δOD = 0.0046/cm

Y = 28.3 %   δOD = 0.0044/cm

Y = 100 %   δOD = 0.0057 /cm

Figure 1. Resolution of the fast relaxation, τ_0, in sheep oxy-hemoglobin using the electric field jump method. Time constant as a function of saturation with oxygen. 4.4×10^{-5} M heme; 4×10^{-3} M borate, pH 9.0 at 10° C. d = 19.2 mm; λ = 580 nm; E = 69 kV/cm.

vation energy of 3 ± 1.5 Kcal/mole for oxyhemoglobin. The observed linear dependence of the optical density change with electric field (2) excludes the possibility that τ_0 is a molecular orientation effect.

This relaxation appears to be unique to hemoglobin, since no such effect is seen in myoglobin, either liganded or unliganded, or in metmyoglobin or uroporphyrin III. It is, however, seen in deoxy and oxyhemoglobin, and in methemoglobin. This "high electric field form" of hemoglobin exhibits a characteristic difference absorption spectrum, which is shown in Figure 2.

The high electric field spectrum of oxyhemoglobin is unlike that of deoxy or methemoglobin, which suggests that this is a new form of hemoglobin.

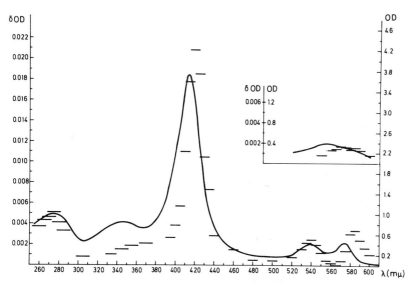

Figure 2. Difference spectrum of sheep hemoglobin at zero and high electric field. Solid curve refers to the right-hand ordinate and is the spectrum of sheep oxyhemoglobin at zero electric field. The data bars refer to the left-hand ordinate and are the high field minus zero field difference spectrum. The inset graph shows a similar comparison for deoxyhemoglobin in a limited wavelength range (same abscissa as for hemoglobin). 3.0×10^{-5} heme, 5×10^{-3} borate, pH 9.25 at 10° C. d = 10.0 mm, E = 79 kV/cm.

Although we cannot detect a change in the relaxation time at different oxygen concentrations, there is a significant change in the amplitude of τ_O. This is shown in Figure 3, which reports the change in amplitude at pH values of 6.3, 7.1, and 9.0 as a function of oxygen saturation. The linear amplitude behavior seen in Figure 3 indicates that the intramolecular change(s) associated with τ_O is not a conformation change that is directly coupled to oxygen binding. If τ_O were such a conformation change, one would expect a non-linear dependence of A_O on Y. The observed dependence can be explained as a transition which occurs within each subunit independently. The mechanism shown on the next page can account for these results if ligand binding is slower than the monomolecular process.

401

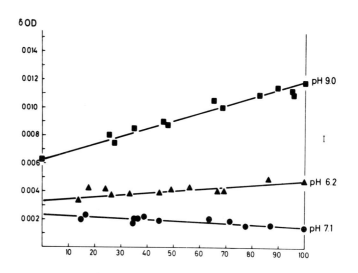

Figure 3. **Dependence** of the amplitude, A_o, of the fast transition on the degree of oxygen saturation at three pH values. Top curve, pH 9.0 in 4 x 10^{-3} M borate; middle curve, pH 6.2 in 2 x 10^{-3} M phosphate, 1 x 10^{-3} M borate; bottom curve, pH 7.1 in 1.75 x 10^{-3} M phosphate, 1 x 10^{-3} M borate. 4.4 x 10^{-5} M heme; λ = 580 nm; d = 19.2 mm, E = 72 kV/cm; 10° C.

Scheme 1:

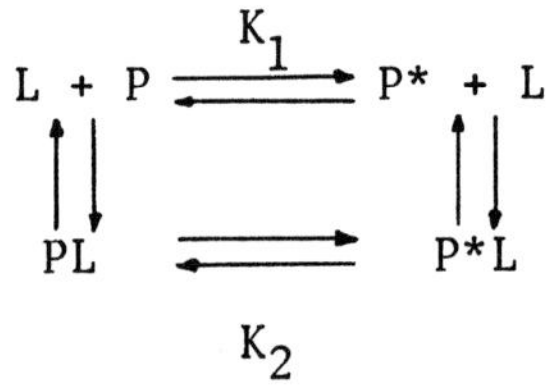

$$L + P \underset{K_1}{\overset{}{\rightleftharpoons}} P^* + L$$

$$PL \rightleftharpoons P^*L$$

$$K_2$$

where L indicates ligand and P indicates protein.

For such a mechanism the observed amplitude is given by

$$\Delta A = C_o [a + bx]$$

where

$$a = \frac{K_1}{(1 + K_1)^2}(e_P^* - e_P)\delta\ln K_1$$

$$b = \frac{K_2}{(1 + K_2)^2}(e_{PL}^* - e_{PL})\delta\ln K_2 - \frac{K_1}{(1 + K_1)^2}(e_P^* - e_P)\,\delta\ln K_1$$

A being the absorbance and e, e^* the molar extinction of the corresponding species.

If the ligand reactions are much slower than the horizontal re-equilibrations, one would expect two relaxations for the horizontal equilibria:

$$1/\tau_{o1} = k_{P,P*} + k_{P*,P} \quad ; \quad 1/\tau_{o2} = k_{PL,P*L} + k_{P*L,PL}$$

Since the observed $1/\tau_o$ is independent of Y, $1/\tau_{o1} \simeq 1/\tau_{o2}$

However important such a high electric field form may or may not be in shedding light on the mechanism of binding and properties of hemoglobin, it is at least somehow linked to oxygen binding and also to the charge state of the protein. The unusual pH dependence of the amplitude of τ_o in oxyhemoglobin is shown in Figure 4.

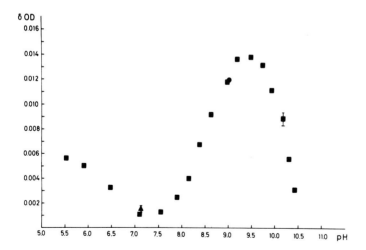

Figure 4. Dependence of the amplitude, A_o, of the fast transition upon pH. 7.9×10^{-5} M heme in 3.5×10^{-3} M borate, 6×10^{-4} M phosphate at 10°. $\lambda = 580$ nm; d = 10.7 mm; E = 79 kV/cm.

It should be pointed out that τ_0 is observed optically and that we are monitoring changes in the heme spectrum.

One well-known thermodynamic linkage of heme properties to pH is the Bohr effect. However, we may exclude that we are looking directly at the dissociation of protons, which certainly occurs in the strong electric field, since proton dissociation would occur at a much slower rate. In addition, if field-induced dissociation of protons were causing the observed spectral changes, one would expect the spectrum of oxyhemoglobin to vary markedly with pH, which is not the case.

A heme spectral change may arise from extremely small alterations in the position of the iron atom or the electronic environment. We would expect this change to be dependent upon the local charge distribution. And, if the Bohr groups contribute to this charge distribution (possibly in salt-bridge linkages) we would expect the spectral changes to somehow parallel the Bohr effect since the electric field acts by displacing charged groups.

If the magnitude of A_0 is related to and truly linear with the Bohr effect we would expect it to be linear with Y as well (see Figure 3) since the Bohr effect is itself linear with Y (4).

The solid line in Figure 5B shows the Bohr effect for human hemoglobin, i. e., the change in proton binding ($\Delta \bar{H}^+ = \bar{H}^+_{Hb} - \bar{H}^+_{HbO_2}$) as a function of pH. The dashed lines represent the pK's of two theoretical ionizable groups which describe very well both the alkaline and acid Bohr effects (4). Analogous data for sheep hemoglobin are not available. The absolute values of $\Delta \bar{H}^+$ for each of the two dashed curves in Figure 5B have been summed (assuming equal contributions to the change in heme extinction coefficient per change in unit charge) in order to obtain the dashed curve of Figure 5A. Two such curves have been computed, using published data for horse and human hemoglobin (4).

Figure 5A shows an empirical correlation between the pH dependence of A_0 and the Bohr effect (2). Further investigation is required to test our suggestion that this fast transition is related to the Bohr effect.

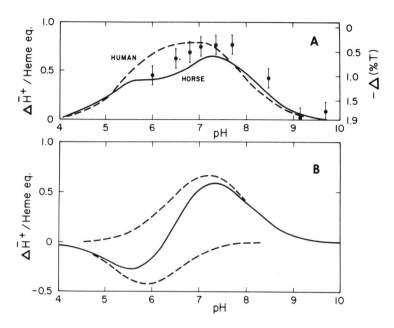

Figure 5. Dependence of the amplitude, A_0, of the fast transition on pH compared with the Bohr effect. **A**, points are experimental values, curves are calculated from **B** (see text). **B**, Bohr effect for human hemoglobin, from Ref. 4. Solid line represents experimental results. Dashed lines are theoretical curves for two ionizable groups, the pK's of which shift upon oxygen binding to produce the change in proton binding shown.

References

1. Brunori, M. and T.M. Schuster. J. Biol. Chem., 244, 4046 (1969).

2. Schuster, T.M. and G. Ilgenfritz. In Symmetry and Function of Biological Systems at the Macromolecular Level (Nobel Symp. 11, A. Engstrom and B. Strandberg, eds.), Interscience, 1969, p. 181.

3. Ilgenfritz, G. Doctoral dissertation, Georg-August University, Göttingen, Germany.

4. Antonini, E., et al. J. Biol. Chem., 240, 1096 (1965); Antonini, E., et al. J. Biol. Chem., 242, 4360 (1967).

Discussion

Chance: Are you also getting a pH jump with the field effect?

Ilgenfritz: Surely I get a pH jump but this pH jump can never be so fast. It can only be in the time range smaller than 1 microsecond. It is absolutely impossible that the observed effect involves a dissociation of protons. Furthermore, the spectrum of hemoglobin is independent of pH, so there would be no effect anyway, due to a change of pH.

KINETIC MEASUREMENTS ON THE OXYGEN-DEOXYHEMOGLOBIN REACTION*

Quentin H. Gibson

Section of Biochemistry and Molecular Biology
Cornell University, Ithaca, New York 14850

The experiments on the hemoglobin-oxygen reaction with which I am concerned today really started off almost 15 years ago when F.J.W. Roughton and I had completed a laborious study of the time course of the reaction of deoxyhemoglobin with carbon monoxide (1). It was natural, when that problem had been taken as far as our methods permitted, to turn to the reaction of oxygen with deoxyhemoglobin. Technically, this is a much more severe problem, both because the rates are about 10-20 times greater than those for the carbon monoxide reaction, and because reduction of the data presents a formidable obstacle, since the ligand dissociation velocities, which can be so conveniently neglected in the case of carbon monoxide, are anything but negligible for the oxygen reaction. In the hope of acquiring sufficient information to restrict to some degree the choice of rates and models we obtained, with the help of the Agricultural Research Council, sheep with hemoglobin of known electrophoretic type which could be bled when required and performed kinetic and equilibrium experiments simultaneously in our two laboratories. Attempts at analysis of the early results were discouraging. The apparent rates of combination at low oxygen concentrations were too low to fit in with the classical experiments of Hartridge and Roughton (2), and the initial reaction was followed by a much slower reaction still, with a $t_{1/2}$ of about a second, which we referred to as "the drift." At the time, the stopped flow apparatus had a deadtime of about 5 msec so the results were rationalized by imaginative reconstructions of what might be happening during this time. Perhaps wisely, we did not publish much about these findings (3).

*Supported by United States Public Health Service Grant GM 14276-05

When, in 1960, a small analog computer became available it was possible to try fitting the experiments on a more extensive scale, changing rate constants by potentiometer setting, and using X-Y recorder. The results continued to be much the same and no satisfactory fits were ever obtained to a family of experimental curves such as that shown in Figure 1. It was only possible to fit oxygen dissociation and any one oxygen combination curve, with the constraints imposed by the equilibrium findings (4). In 1966 a much faster analog computer was obtained with sufficient logical control to permit systematic trials of various combinations of parameters, and a number of schemes were tried, again with uniformly slight success. By then, the four step consecutive reaction scheme of Adair (5) was moving into eclipse, but dimer schemes were even less successful.

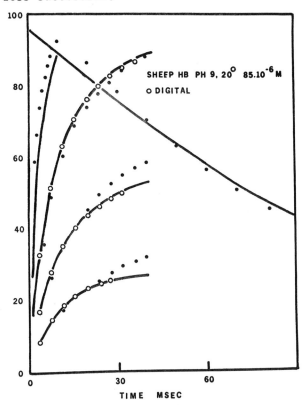

Figure 1. Comparison of results given by hybrid computation (TR 48 and PDP 8/I) (continuous lines) and digital computation (o). Experimental points (●).

Quite recently two new factors have combined to encourage a further attack on the problem. One was the upgrading of the performance of the stopped-flow apparatus and its connection to on-line to a small computer. This reduced the deadtime to 0.7 msec and made the collection and processing of replicate runs both more precise and much faster. The other was the completion of interfacing for the digital computer which allowed it to control the analog, and so permitted the use of efficient minimum seeking programs in comparing observation and computation.

The shortening of the deadtime permitted the use of more concentrated solutions of hemoglobin and a wider range of concentrations of oxygen. One of the first results was to confirm earlier suspicions that "the drift" concerned the dissociation of hemoglobin subunits on ligand binding, and so was only indirectly linked to the problem under study; the phenomenon is thus closely related to the appearance of "rapid" hemoglobin following binding of carbon monoxide to deoxyhemoglobin (6), and, as shown in another paper in this volume (7), is probably due to dissociation of tetrameric ligand bound hemoglobin into dimers.

The tetramer-dimer relation, although not immediately relevant to the kinetics of ligand binding by hemoglobin solutions stronger than about 1 μM, is of importance in choosing equilibrium data to correlate with kinetic findings. Roughton's (4) beautiful gasometric work on dilute solutions of sheep hemoglobin has resulted in an equilibrium curve at pH 7 which is rather distantly related to that of Roughton, Otis and Lyster (8) obtained with strong hemoglobin solutions. The extent of the difference is illustrated in Figure 2. If it is accepted that this effect is due to the dissociation of ligand-bound tetramer to ligand-bound dimer (which results in a displacement of the hemoglobin-oxygen equilibrium because of the increase in high-affinity species), then it is appropriate to correlate the kinetic results, even when obtained with dilute solutions of hemoglobin, with the equilibrium results for strong hemoglobin solutions. Thus, ironically, the equilibrium determinations which were made with such care on exactly the same solutions used for the kinetic work, turn out to be unsuited for purposes of comparison.

The pH to be used also seemed to deserve reconsideration. In the earlier work, most attention was given to pH 9, partly because borate is a preservative as well as a buffer, partly because it appeared advantageous to work at a pH far removed

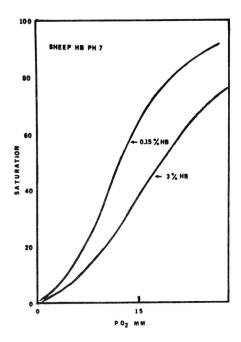

SHEEP HB PH 7

← 0.15 % HB

← 3 % HB

SATURATION

P O₂ MM

Figure 2. Equilib-
rium curves deter-
mined by Roughton
for sheep hemoglo-
bin at pH 7 for 3%
solutions and 0.15%
solutions. The
curves were drawn
using the best-fit-
ting values of the
coefficients in the
Adair equation.

from the isoelectric point, and partly because the use of
dithionite to measure the rate of dissociation from oxyhemoglobin seemed less open to objection at the higher pH (9).
In the recent work, more attention has been given to experiments at pH 7 because of the desirability of minimizing possible effects due to diphosphoglycerate. As shown by Benesch
et al. (10) DPG has little effect in the presence of 0.1 M
phosphate, and so most attention has been given to experiments
at pH 7 in phosphate buffer.

The next question considered was what firm data, if any,
could be incorporated into the curve fitting operations. It
seems that two rate constants are known, one for the dissociation of the first molecule of oxygen from oxyhemoglobin,
the other for the binding of the last ligand molecule by hemoglobin with one vacant gas binding site. The dissociation
rate has been measured in experiments which do not involve
the use of reagents other than oxygen and carbon monoxide (9),
and the combination rate has been measured directly by flash
photolysis, and confirmed by comparisons of kinetic and equilibrium measurements at the top of the saturation curve (11).
The appropriate numerical values were therefore incorporated

in all comparisons between reaction schemes and experiment.

The final point to be settled was the kinds of model to be used, and the order in which they should be examined. It is obvious that schemes which consider the dimer as the unit of structure and function (12) cannot give a fully satisfactory representation of the equilibria and kinetics if only because n in Hill's equation is observed to be greater than 2, and a scheme in which the highest state of aggregation of subunits is 2 can only give n = 2 as a limiting case. As it was intended to included comparisons of equilibria, the basic scheme tried was a 4 step consecutive reaction model, the kinetic equivalent of the equilibrium equations developed by Adair (5). If the rate constants are allowed to vary freely, such a scheme which is itself primarily formal, has the advantage of including several possible physical models if conformation changes are considered to occur quickly as compared with the rate of ligand binding. Among those included is the basic Monod model (13); multi-path models are not adequately represented, e.g. schemes in which a differentiation between α- and β-chains is admitted.

With these preliminaries out of the way the experiments with hemoglobin were performed, as well as control experiments with myoglobin, and mixing experiments with unbuffered solutions of phenol red at different pH values within the range of the indicator. The control experiments showed that at 20°, mixing was more than 99% complete with a deadtime of 0.7 msec, and that the second order rate constant for the combination of myoglobin with CO was the same over a measured range of rates from 33 to 340/sec.

The results were examined by hybrid computation using the method of steepest descents with many sets of starting points, obtained by prefixing the program with a random number generator. Using 50 data points, a single cycle of solution of the set of differential equations and associated generation of a sum of squared residuals occupied about 150 msec when 4 oxygen concentrations were examined, and location of a minimum required from 30 sec to 3 min. The correctness of the hybrid solutions was checked by comparison with the results of digital computation, and good agreement obtained--a point of great importance in view of the abundant opportunities for error offered by the computing procedures (Figure 1).

The results at pH 9 tended to be regrettable, as Figure 1 shows, exceedingly poor fits being obtained even with six

411

freely disposable parameters. Experiments at pH 7 could be better represented so far as kinetics were concerned (Figure 3A), but when the equilibrium curves corresponding to the kinetic constants were calculated (Figure 3B) the results were scarcely satisfactory, the kinetically derived curve showing a much higher value of n, and departing widely from the results of Roughton et al. (4). Thus the results of the present series of experiments agree with those of earlier work in suggesting that an Adair scheme cannot adequately represent the reaction of sheep hemoglobin with oxygen.

Attention is currently being transferred to multi-path schemes, which, however, offer formidable difficulties in practical application. The situation is, however, somewhat less hopeless than might appear on first sight because analogy with the carbon monoxide reactions suggests that the combination rate constants for the various pathways must be rather similar, otherwise kinetic heterogeneity would be seen in the

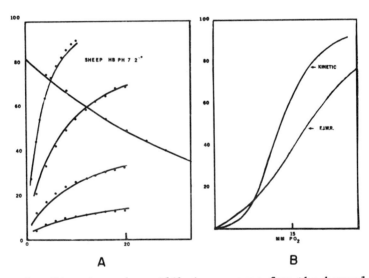

A   B

Figure 3. Kinetic and equilibrium curves for the hemoglobin-oxygen reaction at pH 7. A. Fitting kinetic oxygen curves by hybrid computation. The rising curves represent, in order of increasing steepness, the combination with oxygen at 16, 31, 62, and 124 μM; the falling curve, the oxygen dissociation reaction. B. Comparison of the equilibrium curve obtained by Roughton, Otis, and Lyster (8) with that calculated from kinetic curve fitting experiments.

deoxyhemoglobin carbon monoxide reaction. As a result, quite a modest number of independent rate constants may be sufficient to permit trials of simple alternate pathway schemes, which should help in moderating the extreme cooperativity shown in Figure 3B. The results described here are preliminary only, and further work on weighting functions will be required in order to screen model schemes rapidly and effectively.

The work described so far has all been somewhat negative and, for this reason, positive findings are especially welcome. In the curve fitting operations it was noticed that quite large (of the order of 1000/sec) dissociation velocity constants were regularly selected during kinetic curve fitting. An attempt has been made to visualize these large rate constants in direct experiments as illustrated in Figure 4. They are ordinarily unobservable because when dithionite is mixed with O_2Hb, the rate of the reaction is governed by the initial relatively slow step from $Hb_4(O_2)_4$ to $Hb_4(O_2)_3$. The intermediates of interest have been generated in combined flow-flash experiments in which a mixture of O_2Hb with $COHb$ was mixed with dithionite and then, before the deoxygenation

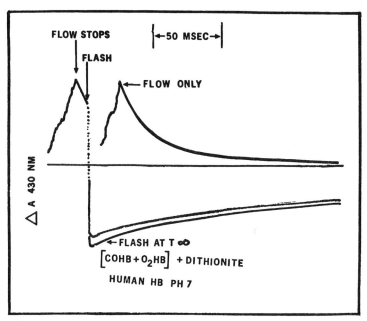

Figure 4. Flow-flash experiments with human hemoglobin at pH 7: approximately 50% CoHB-O_2Hb mixture flash fired at arrow.

413

reaction had proceeded far (left trace of Figure 4), a photo-lysis flash was fired. This removed CO photochemically, gen-erating a mixture of oxygen containing intermediates. These intermediates, however, were highly unstable and released oxygen rapidly, giving deoxyhemoglobin corresponding to more than 90% reduction of the total pigment present within 1 msec after photolysis. This was checked by firing a second photo-lysis flash after all oxygen had been removed by dithionite, displacing all CO from combination (lowest trace of Figure 4). This result suggests that the high rate constants re-quired in the computational work may well correspond to a physical reality, though some uncertainty must remain because of the use of dithionite in the experiments.

Two further points perhaps deserve brief mention. In the last few years the temperature jump method has been applied to the study of the reaction using blood from the original sheep mentioned earlier (14). So far, little correlation between mixing and relaxation results has been achieved be-cause of the failure to interpret either set of experimental results in an acceptable way. It does seem that some of the reasons for this difficulty may now be growing clearer. As discussed in another contribution in this book (7) the dimer is a unit with functional properties quite different from those of the tetramer, and dimers, moreover, are much more readily formed from ligand-bound than from deoxyhemoglobin. As the mixing experiments necessarily start from deoxyhemo-globin and oxyhemoglobin in varying proportions, the effects of dimerization will be quite different in the two cases and, as Figure 2 suggests, may be far from negligible. In spite of these troubles, however, one of the temperature jump find-ings is of great significance for interpretation of the mix-ing experiments; only a small number of relaxation times can be observed and so, although only tetramer models can be con-sidered realistic, they will have to be constructed out of rather simple kinetic building blocks. This gives some hope for the future.

The final point is that by sufficiently careful spectro-photometric measurements Dr. Parkhurst and I (15) have been able to distinguish the contributions of the α- and β-chains in native methemoglobin and to follow their reactions with ligands separately. The obvious extension to gas-binding reactions is now being attempted, and should it be success-ful, the uncertainties which have for so long plagued the interpretation of hemoglobin kinetic work may be materially reduced.

References

1. Gibson, Q.H., and Roughton, F.J.W. Proc. Roy. Soc. B., 146, 206 (1957).

2. Hartridge, H., and Roughton, F.J.W. Proc. Roy. Soc. A, 104, 395 (1923).

3. Gibson, Q.H. Adv. in Biophysics and Biophysical Chem., 9, 1 (1959).

4. Roughton, F.J.W. Clin. Chem., 9, 682 (1963).

5. Adair, G.S. J. Biol. Chem., 63, 529 (1925).

6. Gibson, Q.H., and Antonini, E. J. Biol. Chem., 242, 4678 (1967).

7. Edelstein, S.J., and Gibson, Q.H. This volume, p.417

8. Roughton, F.J.W., Otis, A.B., and Lyster, R.L.J. Proc. Roy. Soc. B, 144, 29 (1955).

9. Gibson, Q.H., and Roughton, F.J.W. Proc. Roy. Soc. B., 143, 310 (1955).

10. Benesch, R., Benesch, R., and Yu, C.I. Proc. Nat. Acad. Sci. (U.S.A.), 59, 526 (1968).

11. Roughton, F.J.W. J. Physiol., 126, 359 (1954).

12. Antonini, E. Science, 158, 1417 (1967).

13. Monod, J., Wyman, J., and Changeux, T.P. J. Mol. Biol., 12, 88 (1965).

14. Schuster, T. This volume, p.367

15. Gibson, Q.H., and Parkhurst, L.J. J. Biol. Chem., in press.

DISCUSSION

Chance: Can you explain what happened to the oxygen?

Gibson: On firing the flash, since only CO is photosensitive, you get a mixture of oxygen-containing intermediates: $Hb_4(O_2)_4$, $Hb_4(O_2)_3$, $Hb_4(O_2)_2$, and $Hb_4(O_2)$. The proportions are determined by the binomial distribution and by the proportion of CO initially present.

Then, if you assume that there is a rapid conformation change by the molecules which have vacant ligand binding sites, and that the dissociation rate from the deoxy state is high, the observations are explained. Of course, this is an ad hoc explanation; having gotten in so deep, it might not be out of place to go a little further and attempt to identify the conformation change with the subunit rotations observed by Perutz (1).

Perham: When you spoke of sheep hemoglobin, I take it you meant hemoglobin A or hemoglobin B, and not a mixture.

Gibson: All the experiments were performed using hemoglobin of the electrophoretically fast type.

Perham: I assume it would complicate matters if you did use a mixture.

Gibson: Yes.

Drabkin: I am pleased that Dr. Gibson has resurrected the relatively neglected possible consequences of "bound" water (defined as water inoperative in dissolving salt) in hemoglobin. It is of interest that in the distant past the relevancy of bound water to cellular phenomena was questioned by A. V. Hill (2).

Some twenty years ago, I published a detailed paper (3) on the hydration of macrosized crystals of human hemoglobin, crystallized by my method (4) in the presence of 2.8 M phosphate buffer, pH 6.8. By direct analytical measurements upon crystals (with long axes up to 1.3 cm) the total hydration was found to be 0.61 gm per gm protein, and 55.6% of this water, or 0.339 gm per gm was "bound" (inoperative in the above sense). I furthermore deduced that this value for bound water was operative in hemoglobin within the red blood cell.

References

1. Muirhead, H., J.M. Cox, L. Mazzarella and M. Perutz. J. Mol. Biol., 28, 117 (1967).

2. Hill, A.V. Proc. Roy. Soc. London B., 106, 477 (1930).

3. Drabkin, D.L. J. Biol. Chem., 185, 231 (1950).

4. Drabkin, D.L. J. Biol. Chem., 164, 703 (1946); Arch. Biochem., 21, 224 (1949).

WEIGHTS AND RATES IN HEMOGLOBIN: ABSENCE OF COOPERATIVITY IN THE DEOXY-DIMER*

Stuart J. Edelstein and Quentin H. Gibson

Section of Biochemistry and Molecular Biology
Cornell University, Ithaca, New York 14850

I. INTRODUCTION

In spite of the considerable amount of attention devoted towards hemoglobin in recent years, the prime result of which is the structural model developed by Perutz and his associates, it is remarkable that a satisfactory understanding of cooperative ligand binding by hemoglobin has not emerged. Nevertheless, on at least one point all students of hemoglobin would agree: solving the role of quaternary structure is fundamental for a complete understanding of the cooperative interactions. With the development of techniques which permit an analysis of quaternary structure by measuring the concentration-dependent deaggregation of hemoglobin [see, for example (1)], the nature of the subunit interactions giving rise to cooperativity are in principle accessible to an experimental approach. However, for technical reasons experiments have rarely been carried out in which information concerning quaternary structure and function was obtained simultaneously and this situation has hampered adequate structure-function correlations. While definite concepts have appeared, their basis was simply the most reasonable interpretation of the available, though often inadequate, information. Central among these concepts are the primary role assigned to the αβ unit in generating cooperativity (2-5) and the designation of single chains as the smallest unit of normal hemoglobin in which cooperativity is absent (6).

In the work reported here, we describe ligand combination kinetics and molecular weight measurements performed concurrently on the same solutions of hemoglobin. The experiments provide several new conclusions: In contrast to the widely

*This research was supported in part by United States Public Health Service Grant GM 14276-03.

held notion that salt tends to dissociate hemoglobin essentially to dimers while not altering at all its cooperative properties (see review by Antonini, ref. 7), a detailed analysis of the experiments in strong salt solutions (in which hemoglobin is presumed to dissociate to $\alpha\beta$ units) has shown that a reinterpretation is necessary to take into account hydration parameters. The effect of hydration is to require a downward revision of the extent of subunit dissociation. Moreover, close examination of the kinetic properties of hemoglobin under the same conditions reveals that significant salt-dependent functional changes are indeed present, namely a transition from slow to rapid rate of reactivity of hemoglobin with CO after flash photolysis of solutions about $10^{-4}M$ in heme. A similar transition occurs in the absence of excess salt at substantially lower concentrations. The appearance of rapid hemoglobin under these conditions correlates closely with the appearance of dimeric units, not monomeric units as had been concluded (6).

Thus we reach the general conclusion that the $\alpha\beta$ dimer rather than being the major functional unit of hemoglobin, is actually devoid of cooperativity in ligand binding and is the concentration-dependent rapidly reacting species observed in hemoglobin solutions at neutral pH by flash photolysis (8). By direct implication tetrameric hemoglobin $(\alpha_2\beta_2)$ is the functional unit of cooperative interactions in ligand binding.

II. The Properties of Hemoglobin in Strong Salt Solutions.

a) The Apparent Hydration of Hemoglobin in 2 M NaCl as Determined by Sedimentation Equilibrium in Natural and Heavy Water.

Since the experiments on hemoglobin in strong salt solutions provide one of the fundamental arguments for the "cooperative dimer hypothesis" (9), the properties of hemoglobin in these solvents must be given careful consideration in any correlations of structure and function. In brief, the data frequently cited show a strong tendency toward deaggregation to two chain units in, for example, 2 M NaCl, although ligand binding parameters remain unaltered by the addition of salt. While it is difficult to imagine a major source of error in the ligand binding data, there is serious concern that the molecular weight measurements may have underestimated the size of hemoglobin and thereby overestimated the extent of deaggregation by ignoring preferential interactions between the protein and water in the three component systems. Apparent hydration of this type, while insignificant in dilute buffers, can lead to large errors in molecular weight in 2M NaCl, since the bound water displaces the denser salt solutions, thereby elevating the effective bouyancy of the protein.

All of the existing methods for measuring such preferential interactions (cf Ref. 1, 10, 11) are poorly suited to hemoglobin by virtue of either their duration in time (days) or their incompatibility with partial subunit deaggregation. Therefore, a new method was developed, based on performing simultaneous sedimentation equilibrium experiments on solutions of hemoglobin in 2M NaCl prepared with natural and heavy water. The method requires little more effort than for a single sedimentation equilibrium experiment and can also be applied to measuring partial specific volumes in two-component systems of proteins undergoing partial deaggregation.[1]

$$\bar{M}_w \ (1 - \bar{v} \ \rho) = \frac{2RT}{\omega 2} \ \frac{c_b - c_m}{c_o} \ \frac{1}{r^2_b - r^2_m} \tag{1}$$

where $\bar{M}_w$ is the weight average molecular weight of the protein throughout entire contents of the solution examined. For the material in D_2O the expression of the left equation (1) becomes

$$k\bar{M}_w \ (1 - \frac{\bar{v}}{k} \ \rho). \tag{2}$$

Where two solutions of natural and heavy water are examined simultaneously, at the same concentration of a homogeneous protein (c_o) and the same radial coordinates, any differences

[1]The following abbreviations are employed in this paper: Hb, hemoglobin; M_w, weight average molecular weight; $\bar{v}$, partial specific volume; ρ, density where the subscripts s and w refer to solution and water, respectively, and the additional subscripts 1 and 2 refer to natural and heavy water, respectively; R is the gas constant; T is the absolute temperature (kelvin); ω is the rotor velocity in radians per second, c is concentration; k, the deuteration factor equal to the mol wt in D_2O divided by the mol wt in H_2O (12), 2 the hydration factor in mo water per gm protein.

in the distribution of the material at sedimentation equilibrium will be reflected in the term, $\Delta c = c_b - c_m$, for each
solution.

For solutions prepared in natural (subscript 1) and
heavy (subscript 2) water in the absence of added salt, the
partial specific volume may be obtained from the relationship

$$\bar{v} = \frac{k - (\Delta c_2 / \Delta c_1)}{\rho_{s,2} - \rho_{s,1} \; (\Delta c_2 / \Delta c_1)} \tag{3}$$

where the solutions are at the same initial concentrations
and radial coordinates. Equation (3) resembles the treatment proposed by Edelstein and Schachman (12), but is more
simple in that only the differences, $\Delta c = c_b - c_m$, must be
determined for each solution rather than the entire distribution of protein as a function of distance in the centrifuge
cell. Moreover, the derivation leading to equation (3) is
directly applicable to aggregation-deaggregation systems.

When three component systems are considered, as is represented by protein-water-2M NaCl, equations similar to
equation (1) may be written to describe the behavior of the
natural and heavy water solutions. In this case the expression for the left side of equation (1) takes the form for
the natural water solution

$$\bar{M}_w \; (1 + q\rho_{w,1}) \quad (1 - (\frac{\bar{v} + q}{1 + q\rho_{w,1}}) \; \rho_{s,1}) \tag{4}$$

while the corresponding expression for the heavy water solution is

$$k\bar{M}_w \; (1 + q\rho_{w,2}) \quad (1 - (\frac{\bar{v} + q}{1 + q\rho_{w,2}}) \; \rho_{s,2}) \tag{5}$$

The equations contain the term, q, which expresses the preferential hydration of the protein in ml water/gm protein.
Since the partial specific volume of the anhydrous protein
is either known or measureable in a separate experiment without excess salt (using equation 3), q may be obtained directly by solving the appropriate simultaneous equations (where
the same radial coordinates exist) yielding

$$q = \frac{\bar{v} \; (\rho_{s,2} - \rho_{s,1}) \; (\Delta c_2 / \Delta c_1) + (\Delta c_2 / \Delta c_1) - k}{k \; (\rho_{w,2} - \rho_{s,2}) - (\rho_{w,1} - \rho_{s,1}) \; (\Delta c_2 / \Delta c_1)} \tag{6}$$

To apply equation (6) to hemoglobin in 2M NaCl several additional factors must be considered. Hemoglobin will manifest reversible aggregation-deaggregation of its subunits under the conditions of the experiment. At the concentrations of protein studied only dimers and tetramers will be present. However, in the centrifugal field applied to obtain sedimentation equilibrium, the variation of concentration of protein throughout the cell will cause a redistribution of the proportion of dimers and tetramers. As a consequence, $\bar{M}_w$ (and $c_b - c_m$) will no longer reflect the distribution of dimers and tetramers in the solution prior to the application of the centrifugal field. Moreover, the perturbation of the initial equilibrium will differ for the H_2O and D_2O solutions because of their densities. Finally because of the presence of multiple molecular species, the measures $\bar{M}_w$ will apply to the protein at the concentration $(c_b + c_m) /2$ (13). In order to take into account all of these factors, a computer program was written to correct the values of q obtained simply from equation (6). In practice the q calculated from equation (6) required a reduction by about 25% for the experiments with 2M NaCl.

One final consideration is the possibility that D_2O alters the deaggregation equilibrium of hemoglobin. The method described here implicitly assumes that the equilibrium is not altered by changing the solvent from H_2O to D_2O. To test this assumption, a control experiment was performed by omitting the NaCl and determining the $\bar{v}$ of hemoglobin taking into account the factor described above. The results (Table IA) give a value of 0.749 ± 0.007 cc/gm, in excellent agreement with values determined classically (0.749 cc/gm, cf. Ref. 13). Had the aggregation-deaggregation properties of hemoglobin varied in D_2O as compared to H_2O, an incorrect value for the $\bar{v}$ would have been obtained, since the concentration of the experiment was sufficiently low that considerable deaggregation occurred. Indeed, further analysis of the experiment showed that the actual molecular weights in the centrifuge cell varied from 45,300 at the meniscus to 62,200 at the bottom, while the weight average of the entire contents of the solution was 58,940. These molecular weights are in good agreement with the values for these concentrations predicted by the tetramer-dimer deaggregation constant measured previously (18), $K_{4,2} = 1.5$ μM. Since the quaternary structure parameters of hemoglobin appear to be unaltered by substituting D_2O for H_2O, experiments with 2M NaCl solutions were undertaken.

+The values of C are normalized to one unit of Δr^2 (in cm^2).

When data for the sedimentation equilibrium distribution of hemoglobin in 2 M NaCl (Table 1B) were examined, simple calculations with equation (3) indicated that appreciable

TABLE I

Hemoglobin in Solutions of Natural and Heavy Water with and without Salt at Sedimentation Equilibrium[*]

$\underline{A}$	No Added Salt		$\underline{B}$ Plus 2 M NaCl	
	(1) H_2O	(2) D_2O	(1) H_2O	(2) D_2O
ΔC^+	0.507 mm	0.385 mm	0.602 mm	0.394 mm
$\Delta C_2/ C_1$		0.760		0.654

*The experiments were performed with a Spinco Model E ultra centrifuge with interference optics. The solutions in H_2O and D_2O were examined simultaneously in two interference cells with the proper masking(15). THe solutions were prepared by 1:10 dilutions of a stock hemoglobin (human) solution into the appropriate buffers. The experiment with no added salt was conducted at a rotor speed of 18,000 rpm with an equilibrium time of 18 hours. Each cell contained 0.1 ml of solution at a protein concentration of 0.43 mg/ml with 0.1 M phosphate as the buffer (pH 6.8). The densities of the solutions were measured (by pycnometry) for used in the calculation of $\bar{v}$ (equation 3) and found to be 1.010 gm/cc for the H_2O solution and 1.108 for the D_2O solution (90%). The experiment with 2 M NaCl was conducted at a rotor speed of 22,000 rpm, an equilibrium time of 20 hours and 0.1 ml of each solution at a protein concentration of 0.64 mg/ml. In addition to the 2 M NaCl, 0.1 M phosphate (pH 6.8) was present. The densities of the solutions with salt were measured for use with equation (6) and found to be 1.087 gm/cc for the H_2O-salt solution and 1.1796 gm/cc for the D_2O-salt solution. The additional values needed for equation (6), $\rho_{w,1}$ and $\rho_{w,2}$ were placed at 0.9982 gm/cc and 1.0945 gm/cc, respectively. All experiments were conducted at 20°C and the value of k (the deuteration factor) was assumed to be 1.015 (12) for all calculations.

water binding was occurring. Therefore, the data was analysed completely and a value of q = 0.25 ml water/gm protein was obtained. In other words, hemoglobin is effectively hydrated to a dramatic extent, about 25% by weight. The hydration may be actually occurring in the absence of salt, but can only become manifested under high solution density conditions.

Hydration to this extent will have a marked influence on the estimated molecular weight. If the corrections for q are not applied, molecular weights will be erroneously low by about 12%. When this error is introduced into the usual equations for subunit deaggregation constants, errors of about 250% result. The deaggregation constant (tetramer-dimer) had been estimated at 190 μM by sedimentation velocity (17). The hydration correction should lower this number to $K_{4,2}$ = 70-80 μM. Indeed, Guidotti (4) has reported a value of the constant, in the predicted range, 75 μM on the basis of osmotic pressure measurements. Determinations of molecular weight by osmotic pressure methods in strong salt solutions or any multicomponent solutions are free of hydration corrections. Therefore, the diminished estimate of the extent of dissociation by osmotic pressure supports the argument for a necessary correction in sedimentation measurements in solvents such as 2M NaCl. Similar corrections for preferential hydration should also apply to the data for hemoglobin in strong salt solutions obtained by light scattering (3), as specified by Cassassa and Eisenberg (10). However, it should be noted that even for sedimentation experiments corrections for preferential interactions could be eliminated by selecting a salt with a partial specific volume very close to the value for water, such as the salt triethylamine hydrochloride. In this case, corrections for hydration automatically "disappear", just as they do in simple protein-water (or dilute buffer) mixtures.

As has been recently emphasized by Tanford (18), in a mixture such as hemoglobin, water and salt, although an apparent binding of water is observed, the salt may also be binding in appreciable quantities. The measured parameter (q) simply reflects the net increment of one or the other small molecules over their proportion in the bulk solution.

b) The Kinetics of Combination of Hemoglobin with CO in 2M NaCl Solutions.

With the experiments described above suggesting that the extent of deaggregation of hemoglobin in 2M NaCl may have been, in some cases, substantially overestimated (while nevertheless occurring to a significant extent) attempts

were made to determine if the revised estimate of the deaggregation could by correlated with some functional property
of hemoglobin. Therefore, flash photolysis experiments (see
section III for details), were conducted on Hb-CO in 2M NaCl.
The results of these experiments show that the proportion of
rapid material is dependent on the hemoglobin concentration.
The results are consistent with a deaggregation equilibrium
constant of about 70 μM, in good agreement with the estimates
of the tetramer-dimer dissociation equilibrium from osmotic
pressure or hydration-corrected sedimentation data.

III. Correlation of a Rapid Reacting Species with Dimeric
Units from CO-hemoglobin in Dilute Buffer.

Since the experiments with solutions of hemoglobin in
2M NaCl diminished considerably the acceptibility of the
cooperative dimer model, studies were undertaken to examine
closely the correlation of the appearance of "rapid" hemoglobin with the appearance of monomers or other subunits in
the absence of salt. Experiments were performed simultaneously
on the same samples of Hb-CO to determine the fraction of
rapid material by flash photolysis and the degree of deaggregation by sedimentation velocity. Deaggregation of hemoglobin could also be measured by sedimentation equilibrium
(12), but sedimentation velocity was preferred in this case
to permit a point-by-point comparison with the kinetic measurements.

When the data for the proportion rapid hemoglobin as a
function of protein concentration were examined, it was clear
that the fraction rapid increased with dilution in a range of
protein concentration where the sedimentation coefficient
varied in the range between the values generally assigned to
dimers and tetramers. To permit an exact comparison of the
fraction rapid and the fractional deaggregation, the sedimentation data were analysed assuming that the $s_{20,w}$ was a weight
average value representing dimers and tetramers. By assigning
the value $s_{20,w}$ (tetramer) = 4.65S and $s_{20,w}$ (dimer) = 2.85S,
α (the fraction dissociated to dimer) was computed from the
relation:

$$\alpha = \frac{s\ (20,w)\ -\ 2.85S}{4.65S\ -\ 2.85S}$$

It has been previously shown (16) that in the dissociation
process the dimer is a stable intermediate, excluding the
possibility that the observed reaction is a concerted tetramer-
monomer dissociation process.

Flash photolysis experiments were performed on the same
solutions as those used for the ultracentrifuge measurements.
The solutions were prepared by taking a strong solution (about
4 mM) of oxyhemoglobin and equilibrating it with one atmosphere of pure CO. This solution was diluted in a syringe
with 0.1 M phosphate buffer (containing 2M NaCl when required)
which had been deaerated by bubbling with argon and complete
absence of oxygen was assured by the addition of a little
(0.05%) solid sodium dithionite immediately before use. The
required amount of CO-Hb was injected into about 5 ml of
deaerated buffer contained in an anaerobic tube furnished
with a 1 cm spectrophotometric cell together with an amount
of CO saturated buffer sufficient to give at least a 5-fold
excess of free CO over heme.

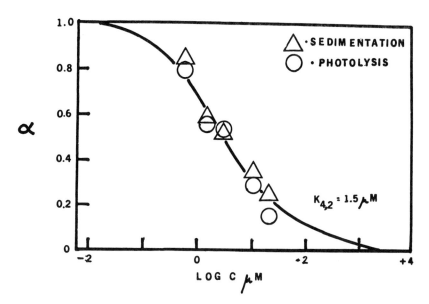

Figure 1. Simultaneous sedimentation velocity and flash
photolysis experiments on Hb CO. The sedimentation experiments were performed with a Spinco Model E ultracentrifuge
equipped with an automatic scanner. Traces were recorded
with light of 405 mμ and sedimentation coefficients were
calculated from the movement of the boundary at the height
corresponding to one-half of the plateau concentration. The
experiments were performed at 22° C with a rotor speed of
60,000 rpm. The flash photolysis experiments were carried
out as described in the text.

Photolysis was performed with a 100 J flash of duration (1/e) of 270 μsec, filtered through an aqueous auramine orange solution with absorbance 10 at 430 nm. The reaction was followed at 430 nm using a Bausch and Lomb 250 mm grating monochromator. The data were collected with an on-line computer, using 400 points to represent the reaction curve, as described by Gibson and DeSa (19). The proportion of slow and fast was obtained by fitting the reaction data by least squares, using the method of steepest descents. With rates differing by a factor of 30, as in the present case, the proportions could be estimated to within about $\pm$ 5%.

The kinetic and sedimentation data, when plotted together (Figure 1), give excellent agreement. Moreover, the line presented in Figure 1 was computed from a deaggregation equilibrium constant of 1.5×10^{-6} M, the value obtained earlier for the tetramer-dimer reaction by a more complete sedimentation equilibrium study (16). Thus there can be little doubt that it is the two chain unit, $\alpha CO\beta CO$ and CO-hemoglobin which gives rise to the concentration-dependent rapidly reacting form of hemoglobin.

DISCUSSION

While this communication is of a preliminary nature, it nevertheless presents strong evidence in favor of the view that the integrity of tetrameric hemoglobin is obligatory for cooperativity in ligand binding. Two experimental lines converge to fortify this conclusion--studies with hemoglobin in 2M NaCl and experiments on the photodissociation and quaternary structure of CO-hemoglobin.

The experiments in 2M NaCl have essentially eliminated the "salt paradox". This phrase has been used to describe the apparent contradiction of the dissociation of hemoglobin into 2 chain units in 2M NaCl with the maintenance of normal cooperativity (Hill coefficient - 2.7) and rate of reaction toward ligands (5). We now suggest that the "paradoxical" aspects of the situation were based on two incorrect conclusions--an over-estimation of the extent of dissociation and an under-estimation of the importance of the changes in the proportion of rapid hemoglobin on flash photolysis in 2M NaCl (6). Hemoglobin in 2M NaCl is only slightly dissociated to dimers at the concentrations previously examined. In fact, at the concentrations where many of the oxygen equilibrium measurements were performed (3) the average degree of dissociation of oxyhemoglobin is not quite at the three chain level. The over-estimation of the degree of dissociation stemmed from a failure to take into account the buoyancy effect of bound water inaccessible to salt. The concept of "bound water" is an old one in the hemoglobin literature (20) and indeed Perutz arrived at an estimate of the extent of preferential hydration in the same range as the value presented here, 33%. Similar values from proteins in general have recently been reported with an NMR technique (21) and Parkhurst and Gibson (22) reached qualitative conclusions of this nature from their studies of the reaction of horse hemoglobin with carbon monoxide in solutions and crystals.

Thus, when it is noted that oxyhemoglobin is dissociated to only a moderate extent and, at the same time, deoxyhemoglobin is recognized as being considerably more tightly assembled (4,23), there is no need to invoke a paradox to explain the behavior of hemoglobin in 2M NaCl.

Removal of the salt experiments from the paradoxical realm reopens the question of the extent to which the dimer of hemoglobin is functionally cooperative. Therefore, experiments were performed to evaluate this point. When solutions of CO-Hb in 2M NaCl were examined at only slightly lower concentrations than heretofore studied (5), a rapidly reacting species (characteristic of non-cooperative units such as myoglobin or single chains of hemoglobin) was observed and in just the range of protein concentration where the tetramer to dimer dissociation reaction is expected to occur on the basis of the data corrected for hydration. These experiments hence suggest that the $\alpha\beta$ dimer must be displaced from its position as the major cooperative unit to one substantially devoid of cooperativity.

The general conclusions from the experiments with salt were confirmed by experiments with solutions free of excess salt. Examination of CO-Hb in 0.1 M phosphate showed that the appearance of the rapidly reacting species in flash photolysis experiments correlates excellently with the deaggregation of tetramers into dimeric units as observed by ultracentrifugation. Centrifuge and kinetic experiments in parallel eliminate the possibility that the rapid behavior is caused simply by the loss or gain of an effector ligand, such as diphosphoglycerate (24).

The general conclusions presented here correlate satis-
factorily with the crystallographic changes in hemoglobin
accompanying ligand binding. Perutz (25) has reported that
most of the motion within hemoglobin upon ligand binding
occurs between $\alpha\beta$ units with very little occurring within
the $\alpha\beta$ unit itself. These findings are readily reconciled
with a model in which the $\alpha\beta$ unit is non-cooperative and all
of the cooperativity generating changes in quaternary structure occur between $\alpha\beta$ units.

References

1. Schachman, H.K. and S.J. Edelstein. Biochemistry, 5,
2681 (1966).

2. Wyman, J. Adv. Prot. Chem., 3, 407 (1948).

3. Rossi-Fanelli, A., E. Antonini, and A. Caputo. J. Biol.
Chem., 236, 391 (1964).

4. Guidotti, G. J. Biol. Chem., 242, 2685 (1967).

5. Antonini, E., E. Chiancone, and M. Brunori. J. Biol.
Chem., 242, 4360 (1967).

6. Antonini, E., M. Brunori, and S. Anderson. J. Biol. Chem.,
243, 1816 (1968).

7. Antonini, E. Science, 158, 1417 (1967).

8. Gibson, Q.H. and E. Antonini. J. Biol. Chem., 242, 4678
(1967).

9. Rossi-Fanelli, A., E. Antonini, and A. Caputo. J. Biol.
Chem., 236, 391 (1961).

10. Cassassa, F.F. and H. Eisenberg. Adv. Prot. Chem., 19,
287 (1964).

11. Kirby-Hade, E.P. and C. Tanford. J. Am. Chem. Soc.,
89, 5034 (1967).

12. Edelstein, S.J. and H.K. Schachman. J. Biol. Chem., 242,
306 (1966).

13. Van Holde, K.E. and R.L. Baldwin. J. Phys. Chem., 62,
734 (1958).

14. Svedberg, T. and K.O. Pedersen. The Ultracentrifuge,
Oxford University Press, New York, 1940.

15. Richards, E.G., D.C. Teller, and H.K. Schachman. Biochemistry., 7, 1054 (1968).

16. Edelstein, S.J.   Thesis., University of California, 1967.

17. Kirschner, A. and C. Tanford.   Biochemistry, 3, 291 (1964).

18. Tanford, C. J. Mol. Biol., 39, 539 (1969).

19. Gibson, Q.H. and R.J. DeSa.   Abst. Seventh Int'l. Conf. Biochem. Symp. III, 1967.

20. Perutz, M.F.   Trans. Faraday Soc., 42B, 187 (1946).

21. Kuntz, I.D., T.S. Brassfield, G.D. Law, and G.V. Purcell. Science, 163, 1329 (1969).

22. Parkhurst, L.J. and Q.H. Gibson. J. Biol. Chem., 242, 5762 (1967).

23. Benesch, R.E., R. Benesch, and M.E. Williamson.   Proc. Nat. Acad. Sci., Wash., 48, 2071 (1962).

24. Benesch, R., R. Benesch and C.I. Yu, Proc. Nat. Acad. Sci., U.S., 59, 526 (1968).

25. Perutz, M.F. Nature, 219, 131 (1968).

ON THE RELATION BETWEEN COOPERATIVITY, AMOUNT OF
FAST REACTING MATERIAL, AND DISSOCIATION IN HEMOGLOBIN

Ernaldo Antonini, Maurizio Brunori,
Emilia Chiancone and Jeffries Wyman

Institute of Biochemistry, University of Rome

Centre for Molecular Biology of the
Consiglio Nazionale delle Ricerche

and

The "Regina Elena" Institute for Cancer Research
Rome, Italy

From a correlation of their data on the concentration dependence of the amount of quickly reacting material observed on the flash photolysis of hemoglobin with data on the concentration dependence of molecular weight, Edelstein and Gibson argue that the (αβ) dimers are devoid of cooperativity. To this conclusion, reached by taking the presence of fast reacting material as characteristic of non-cooperativity, we raise the following objections resulting from a careful consideration of a wider body of information.

Facts

In connection with this problem there are three sets of facts to be taken into account and correlated; data on molecular weights of liganded and unliganded hemoglobin; data on cooperativity as measured in ligand equilibria; and data on the amount of fast and slow material seen in kinetic experiments both by rapid mixing and flash photolysis. The problem resolves itself into a comparison of the way in which all of these quantities vary with protein concentration under a variety of conditions which are known to influence the state of aggregation of the protein, namely, ionic strength, pH, and the specific nature of the buffer. We have assembled as much of the available data as we could find under three headings, according to conditions: a) neutral pH (6.8-7.2) in phosphate buffer and/or NaCl, to give a total ionic strength between 0.05 and 0.3 M; b) neutral pH (6.8-7.2) in 2 M NaCl, with or without phosphate buffer; c) in acetate buffer of pH 5.4 at an ionic strength of 0.25 M.

Under each of the three conditions, Figures 1, 2, and 3 show curves for the degrees of dissociation of tetramer to dimer based on the dissociation constants for liganded and unliganded hemoglobin given by various authors (1-9). In the case of the results obtained with the ultracentrifuge in high salt, the dissociation constants were corrected to take into account the water of hydration, in accordance with the work of Edelstein and Gibson (6). It should be noted that all these constants were calculated by the authors on the assumption that only tetramers and dimers were present, and that in some cases, where molecular weights were measured by osmotic pressure, the calculations involved the use of rather unexpectedly large virial coefficients. In the absence of any objective basis for selection, the curves shown embrace all available data without any attempt to reconcile discrepancies between different authors.

The equilibrium data present no problem from a factual point of view, being in all cases highly consistent. The horizontal arrows indicate the concentration region over which n, as observed in equilibrium experiments with oxygen, carbon monoxide, and ethyl isocyanide, is equal or greater than 2. It should not be assumed, however, that in high and low salt n is less than 2 at concentrations below those indicated by the arrow; it is only because data are lacking at lower concentrations that the left-hand end of the arrow is located as it is. In the case of the acetate measurements, there was a tendency for n to drop below 2 at lower concentrations.

The kinetic data on ligand binding were obtained either by stopped flow or flash photolysis and were analysed in terms of two components, one fast and the other slow.

Conclusions

1. Cooperativity, as shown by the values of n in the equilibrium measurements, is present when both oxy and deoxy forms of hemoglobin are largely dimers or tetramers or consist of some of each. We therefore reach the inevitable conclusion that cooperativity must be an inherent property of dimers, although of course, there may be some further cooperativity, and indeed must be whenever n is greater than two.

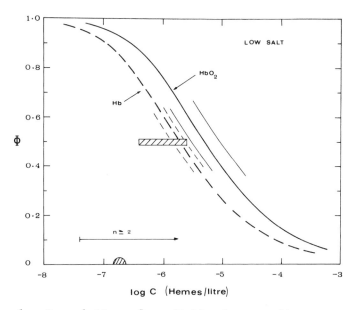

Figure 1. Correlation of available data on dissociation, co-operativity, and fraction of fast reacting material shown by hemoglobin in low salt buffer at neutral pH (6.8-7.2). The ordinates give the fractional dissociation, Φ, of tetramers into dimers on a weight or heme basis; the abscissae give the hemoglobin concentrations in hemes per litre. The heavy solid line corresponds to the average value of the dissociation constant for ligand-bound hemoglobin, and the lighter solid lines represent the upper and lower limits of the values reported by various authors. The heavy broken line corresponds to the average dissociation constant for unliganded hemoglobin, and the lighter broken lines represent the upper and lower limits of the values reported by various authors. The shaded rect-angle shows the range of concentrations over which half the material is quickly reacting when ligand-bound hemoglobin is totally photodissociated, as obtained in several experiments. The half-circle shows the lowest concentrations at which stop-ped flow measurements on the combination of hemoglobin with CO were carried out; at all these, and higher, concentrations, the amount of quickly reacting material was less than 5% of the total. Unpublished experiments in collaboration with N. Anderson; cf also References 10 and 11 for $\underline{n}$ values and Re-ferences 12 and 13 for rapid reaction methods.

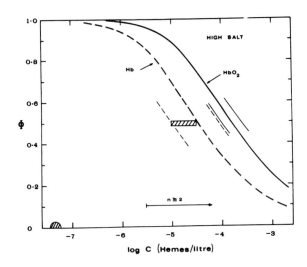

Figures 2 (above) and 3 (below). Correlation of available data on dissociation, cooperativity, and fraction of fast reacting material shown by hemoglobin in high salt buffer (Fig. 2) and acetate buffer, pH 5.4 (Fig. 3). Symbols as in Figure 1.

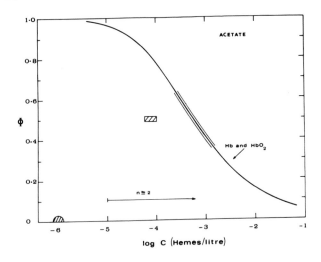

2. The quickly reacting form seen on flash photolysis of HbCO is not correlated in any simple way with the presence of dimers. In fact, in high salt and in acetate buffers the appearance of the dimers occurs well before that of the fast reacting material as the concentration goes down. In the case of unliganded material studied by flow, there is no fast reacting material even at concentrations where the dimers are predominant, although the α and β chains are fast reacting under all conditions. In connection with this it is suggestive that trout hemoglobin, which shows cooperativity in the oxygen equilibrium, does not show fast reacting material on rapid mixing or flash photolysis under conditions where it is dissociated to much the same extent as human hemoglobin.

3. Although reliable information about the dissociation of dimers into monomers is lacking, it seems clear that the quickly reacting form obtained on flash photolysis is not entirely restricted to monomers. Moreover, it is clear that the spectroscopic change observed when the isolated chains are brought together results from the formation of dimers, and not tetramers, since it occurs at concentrations where the deoxygenated material is essentially all dimeric (14).

In considering the flash photolysis results it should be remembered that three different kinds of quickly reacting material have been postulated: first, the form observed at alkaline pH, even at high concentrations; second, the form observed on partial photodissociation under all conditions; and, finally, the concentration-dependent form which appears on total photodissociation. It is exclusively with the last that we are concerned, although it is difficult to be sure that the other two forms have been completely excluded in all the observations.

In interpreting the change of observed molecular weight with protein concentration it is somewhat problematical whether we should completely ignore the possibility of an overlapping dissociation of the dimers into single chains, some of which (the β chains) are known to have a strong tendency to associate. Also, although there is no reliable information about the dissociation of either the oxy or deoxy material into constituent chains, nevertheless a maximum difference between the dissociation constants for the two forms is set, thermodynamically, by the difference in ligand affinity (given by the median ligand activity of the isolated chains and the tetramers).

In any attempt to rationalize these conclusions, it should be borne in mind that the hemoglobin tetramer can dissociate in two different ways, giving rise to two different types of dimer ($\alpha_1\beta_2$ or $\alpha_1\beta_1$, according to the Perutz notation); and it seems likely that neither the one nor the other of them is realized exclusively but that the relative proportions of the two depend on the special conditions prevailing (ligand state, ionic strength, pH, and nature of salt or buffer). By postulating the presence of two kinds of dimer with different functional properties as regards cooperativity and reaction rate and related by a ligand linked dismutation constant, as well, possibly, as the presence of some single chain molecules, there would seem to be a prospect of devising schemes capable of meeting the facts. Flash photolysis would be expected to liberate certain forms in amounts never realized in the course of equilibrium measurements.

References

1. Chiancone, E., L. M. Gilbert, G. A. Gilbert, and G. L. Kellett. J. Biol. Chem., **243**, 1212 (1968).

2. Kirschner, A. G. and C. Tanford. Biochem., **3**, 291 (1964).

3. Guidotti, G. J. Biol. Chem., **242**, 3685 (1967).

4. Ackers, G. K. and T. E. Thompson. Proc. Natl. Acad. Sci., **53**, 342 (1965).

5. Schachman, H. Personal communication.

6. Edelstein, S. and Q. H. Gibson. This symposium.

7. Antonini, E. and J. Wyman. Unpublished studies.

8. Chiancone, E. and G. A. Gilbert. J. Biol. Chem., 240, 3866 (1965).

9. Banerjee, R. and L. Sagaert. Biochim. Biophys. Acta., **140**, 266 (1967).

10. Anderson, S. and E. Antonini. J. Biol. Chem., **243**, 2918 (1968).

11. Rossi Fanelli, A., E. Antonini, and A. Caputo. Adv. Prot. Chem., **19**, 135 (1964).

12. Antonini, E., M. Brunori, and S. Anderson. J. Biol. Chem. **243**, 1816 (1968).

13. Antonini, E., E. Chiancone, and M. Brunori. J. Biol. Chem., _242_, 4360 (1967).

14. Brunori, M., E. Antonini, J. Wyman, and S.R. Anderson. J. Mol. Biol., _34_, 357 (1968).

GENERAL DISCUSSION
THE DIMER-TETRAMER CONVERSION AND INFORMATION TRANSFER

Eigen: I should say that a statement like "a certain model is right" or "wrong" do not seem to be the point. I think both the Monod and Koshland models are <u>per se</u> right; the question is whether they are fulfilled in one or the other case. It is quite clear that there are systems which come close to the Monod mechanism, and there may be others which come close to the Koshland mechanism. Both are meaningful, both are possible, and there are even many more mechanisms which are possible. I would just show two slides which demonstrate a number of possible intermediate mechanisms. It should be possible to come to some agreement among all the experimental evidence which has been presented here. The Monod and Koshland mechanisms apparently are the two possible extremes of such a scheme.

One possibility, shown in Figure 1, is that everything which Monod, Changeux and Wyman proposed is right for the dimer only. In other words, all the conformational changes within a dimer are all or none. For tetramers, this would mean that the reaction path comes already close to the diagonal which would represent the Koshland mechanism (Fig. 2)

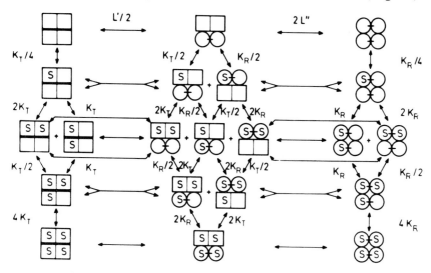

<u>Figure 1.</u> Possible mechanism of subunit cooperativity.

439

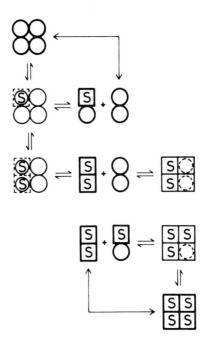

Figure 2. Possible mechanism of subunit cooperativity via tetramer-dimer dissociation.

A mechanism via dissociation of the tetramer into dimers, which then may only be able to undergo conformation changes, may be another alternative, as shown above, especially in the light of new experiments by Ilgenfritz and Schuster (cf pp. 399). I am wondering whether any such mechanism could not be in closer agreement with all the experiments presented here by the various speakers than the one extreme presented earlier.

Edelstein: What we measure in the experiments I showed is the dissociation of one state (liganded) to its corresponding dimer, and the dissociation of the unliganded state to its dimer. Now, if we assume for the moment that all of the cooperativity is between αβ dimers, which is certainly the direct conclusion to be drawn from our results, then if there is such a step as that described by L from the Monod

440

point of view, its value will be equal to the ratio of the subunit dissociation constants for these two states. We just do not find a high enough ratio to give the required L.

Eigen: There could be two L values expressing the additional cooperativity between the two dimers.

Edelstein: The only point I would make is that the overall L must be relatively large; even if you break it up, you must still set it to about 10^4. Even if there is intermediate isomerization, the overall partition will have to be maintained. The difference between the fully unliganded state and the oxygenated state, in terms of tetramer-dimer dissociation, will have to be even greater when you introduce isomerizations, or else you will lower the cooperativity.

Eigen: If there are two L-values, it would be hard to get all the information about them, and the different K_R and K_T values, from a certain type of experiment. In the general model, there must be a certain overall L-value, even if the reaction proceeds along a path for which L is not representative.

Shulman: The point about our results is that they prohibit the model in which the cooperativity is caused by a Monod type of tight coupling between the two subunits of a dimer, so long as the dimer is part of a tetramer. The $\alpha\beta$ dimer in the tetramer of our mixed state hemoglobin with α-ligated and β not, has high oxygen affinity. According to the Monod model, this high affinity is not going to come about because of the subsequent ligand-induced change. It is just intrinsically a change in affinity and the simplest view of this high affinity is that it is caused by some change around the heme groups. By NMR we look there, and we see no change in the structure or electronic properties of the heme. There is no evidence for the high affinity that the Monod model requires. While we have not been able completely to eliminate the possibility of a high affinity form, we do not have any evidence for it in the vicinity of the heme, where such evidence is most likely to be found. However, our experiments were all done on tetramers, and our concentrations were about $10^{-2}M$ in heme, so that we cannot say anything about the nature of the cooperativity in the free dimers.

Czerlinski: In the schemes shown in Dr. Edelstein's last slide (p. 417), dimerization has also been shown. How far has it been attempted to incorporate a "dimerizing" model with the (intrinsic) binding constants of the oxygen, which are different for the dimer as compared with those for the tetramer?

Edelstein: At usual concentrations, where the equilibria are measured, the population is essentially all tetramer, so we had to explain the behavior of the tetramer. One could also go to a lower concentration, at which the dimer would be the principal contributing factor, and derive the appropriate equations.

Czerlinski: Thus in the general case, one would have to consider both dimerization to and interconversion of tetramers for a full description of all data over the full concentration range of hemoglobin. It has been shown that polymerization alone can produce allosteric behavior (1); the polymerizing model is compared with Monod's interconversion model (2), the all-or-none effect is discussed, and some equations for chemical relaxation times are derived (n monomers to one polymer). Only the limiting case corresponding to Monod's derivation was considered. One might also treat a polymerizing model according to the limiting case of Koshland (3). The question then is, can one describe the data with some polymerizing model alone, not considering the interconversion model (at least in some restricted concentration range)?

Ilgenfritz: My first point is, that if hemoglobin dissociates markedly into dimers at pH 9, then this dissociation-association reaction must appear in the relaxation spectrum. We think, that we can identify the slowest relaxation phase as corresponding to this reaction (see paper), giving a recombination rate constant of about 10^6 $M^{-1}sec^{-1}$. We therefore conclude that hemoglobin is largely dissociated under the used conditions.

The relaxation phases which correspond to the oxygen binding steps of dimer or tetramer are not yet fully understood. There are however, still two relaxation times, which would have to be attributed to the dimer. But if two reaction steps appear in the oxygen binding of the dimer, you may say that the dimer has cooperativity.

The existence of an intermediate species, the dimer with one ligand, is also suggested by inspection of the fastest relaxation time. This fastest time is roughly linear when plotted against the oxygen concentration. Since the fastest time can always be approximated by $1/\tau = k_{Diss.} + k_{Recomb.}$ ([unliganded sites] + [O_2]), a linear dependence of $1/\tau$ with O_2 can only be observed if the unliganded protein sites, which correspond to the fastest step, are small compared to O_2. Such a small concentration, while the protein is largely dissociated can only represent an intermediate and one would again consider the dimer to bind O_2 cooperatively.

All these arguments, however, depend on whether there is so much unliganded and liganded dimer present at pH 9 or not.

Shulman: It does seem that we need cooperative dimers to explain Dr. Ilgenfritz's observations. However, the cooperativity could be sequential.

Gibson: Dr. Ilgenfritz has raised the question of identifying a rate of 20 sec^{-1} with the dissociation of the tetramer to the dimer. Here I must discuss kinetic measurements, where one can only argue rather than state facts. Nonetheless, there is a very simple experiment which Dr. Antonini and I carried out with human hemoglobin about three or four years ago, in which we took carbon monoxide hemoglobin and diluted this about 20-fold so that the final dilution was on the order of 5×10^{-7}, and we flashed at this photochemically at different time intervals after performing the dilution. If we flashed at it soon enough, most of the recombination was slow, but as we waited, the recombination reaction became progressively more rapid, and we identified this with some sort of dissociation step in the hemoglobin. By analogy with the ultra-centrifuge experiments which we have—and this applied to human hemoglobin—the slow phase was the tetramer and the rapid phase was, presumably, dimer. This was appearing however not at 25 sec^{-1}; the rate constant for this was about 0.5 sec^{-1}. For this reason, I am a little unhappy about accepting the identification of the 25 sec^{-1} step.

There is the additional point that if both rate const-

ants are correct--I think one was 1 x 10^5 and the other was
25 sec^{-1}--you are really going to have a great deal of dimer
around in the solutions all the time. I think also that we
must bear the species difference very strongly in mind; it
is most important.

However, to go back to the early oxygen experiments
which I did not mention this morning, if one mixes oxygen
with hemoglobin in such proportions so as to get about say
20% saturation, first, a reasonable ligand binding reaction
is seen and then over the course of about 1 sec, there is
a further reaction which I would again associate with the
appearance of the dimer in the solution. This is not 20
sec^{-1}, but at 1 sec^{-1}.

Eigen: In the light of previous discussions on models it
seems important to me to restate that both the "all or none"
and the "sequential" model are the extremes of one general
model in which each subunit can be present in two different
conformations. Thus, all states proposed by both models and
even some additional ones must be present, the question is
only, to which extent they are populated and which is the
main reaction path. When the work on GAPDH was done in our
laboratory, I tried to find out about the conditions under
which one might find one or the other extreme. The two figures represent two such cases.

Figure 3 shows a system in which the R- and the T-state
have almost the same free energy (L not far from 1) and also
similar affinities (expressed by K_R and K_T). If only one
slow first order time constant is observed, one can be pretty
sure to have an "all or non" change. A slow reaction of intermediate states will be high, the states will not be populated. If hybrid intermediates on the same energy level as
R_0 and T_0, etc., would be present, more than one time constant for the conformation change should be seen. Also, since
in the R- and T-state the affinity for ligands is similar,
there is no driving force for conformation change by ligand
binding. GAPDH belongs to this category. Even if L is 10
to 100, ΔF still is quite low.

If however, L becomes 10^4 to 10^6 the situation looks
quite different (Figure 4). If, in addition, the conformation changes are fast, the barriers can't be high and thus
the free energy for the hybrid intermediates must be in between the "all or none" R- and T-states. Whenever one takes

444

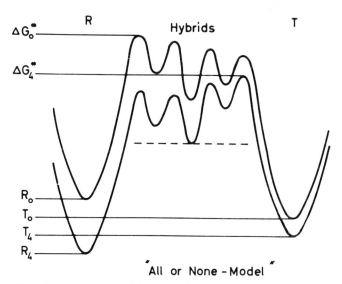

All or None - Model

Figure 3.   Free energy diagram for the Monod Changeux Wyman model.

into consideration both T_i and R_i, the hybrid intermediates are more populated than one of the symmetrical states.   It may be that only one of the states referring to the same degree of saturation is populated to a measurable extent. If now the model is symmetrical (i.e., $T_0/R_0 \quad R_4/T_4$) then one would be close to Koshland's sequential model.   Actually, it doesn't yet need to be cooperative.   Cooperativity comes in with any unsymmetrical partition of the free energy drop upon binding.   Hemoglobin--if it belongs to this general scheme--may be close to this sequential limit, since the parameters L and C (ratio of affinity in R- and T-state) differ appreciably from those for GAPDH in being far from one.   I must clearly state, however, that these consider- ations do not represent an exact theory but rather indicate certain tendencies.

Riggs: I would like to mention a useful procedure for look- ing at tetramers, dimers and monomers.   Dr. Bonaventura and I have been looking at human hemoglobin A with the electron microscope, and it turns out that if we prepare a solution which other data say is mostly tetramers, stain it and spray it on the grid (undoubtedly, complicated things happen dur--

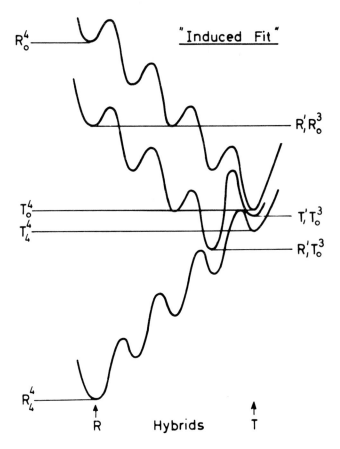

Figure 4. Free energy diagram for the Koshland model.

ing this process), then what one sees are objects the size
of tetramers. If one dilutes the solution sufficiently to
obtain what other data say are dimers, then what we see are
objects that contain mostly dimers, and a small number of
apparent tetramers, and some monomers. Now, if one takes a
solution of hemoglobin Kansas, which is known to be dimeric
at concentrations where hemoglobin A is tetrameric, what we
see are mostly dimers. This procedure, while wholly qualitative at present, may be quite useful because direct counting
of particles is possible.

Concerning the aggregation upon deoxygenation of various hemoglobins, it is well known that lamprey hemoglobin aggregates upon deoxygenation, and so does the human hemoglobin Kansas, which differs from normal human hemoglobin A only in one residue in the B chain which is very close to the contact. Bonaventura and Kitto have found that this aggregation is very widespread, and occurs in a number of invetebrate hemoglobins, including some of the echinoderms, the annelid work, Glycera and in the hemoglobin of the botfly, Gastrophilus. This wide range of phyla shows thay oxygenation-induced dissociation is very widespread and general.

Chance: One way to test the role of gross subunit movement in cooperativity such as that suggested by the 5.5 Å hemoglobin model is to determine whether physical constraints which supress this structural change interfere with the cooperativity. In horse hemoglobin, lattice forces seem insufficient to contain the tendency of the subunits to rotate in liganding reactions and shattering of the crystals results. This phenomenon is not observed in crystals of lamprey hemoglobin. In experiments in collaboration with Dr. Rumen, we have identified as high cooperativity in suspensions of crystalline lamprey hemoglobin as in the solution. Our results would suggest that the degree of subunit rotation that characterizes cooperativity in hemoglobin is not essential to the cooperativity that is observed in lamprey hemoglobin. One interpretation of this experiment is that the essential nature of cooperativity is a small scale or a tertiary structure change presumably related to the spin state changes of the iron atom as discussed elsewhere in this volume (6), and this may well be the prime mover in the larger quaternary structure changes which are so clearly observed in mammalian hemoglobins and are presumably responsible for the association-dissociation phenomena observed in lamprey hemoglobin (7).

Love: It is important that the crystal form of your deoxy lamprey hemoglobin be known, because the structural response occurring during exposure to CO is dependent upon crystal form and type (8). In some cases, the response is isomorphous" like that of myoglobin and Glycera hemoglobin, i.e., the lattice constants remain essentially the same and the diffraction patterns "before" and "after" are nearly indistinguishable, but none the less different. In other cases, the lattice constants change up to 14%, and the intensity distribution is seriously altered, in one case to the extent that the space group changed. The three crystal forms of lamprey deoxy hemoglobin A, B, and C, have respectively, 8, 10 and 8 molecules in the asymmetric unit. It is not known whether they are polymers or looser aggregates. In either case, the monomers appear to interact, or not interact, depending upon their arrangement and the salt employed for their precipitation. We would guess that crystals showing myoglobin-like behavior, with very little structural change, would follow a hyperbolic dissociation curve, but that the others would show cooperativity.

Chance:* Dr. G. Ogunmola has explored the suggestion in Dr. Love's comment to determine whether the crystalline species of lamprey hemoglobin designated A-6 and A-8 in which identifiable lattice constant changes are observed to change up to 2-9% on conversion of the carboxy form to the A-7 and A-9 forms respectively (9). The A-6 crystals show a cooperativity value of n=1.4 and the A-8 value is n=1.1. This suggests that the form A-6 showed high cooperativity in spite of a small change of lattice constants. In addition, Dr. Ogunmola measured with high sensitivity the birefringence of these crystals during reaction with CO and found there was no significant change. It seems that more data is necessary to establish a correspondence between the cooperativity of the crystalline hemoglobin and the change of lattice constants.

References

1. Czerlinski, G. Cur. in Mod. Biol., 2, 219 (1968).

2. Monod, J., J. Wyman and J.P. Changeux, J. Mol. Biol., 12, 88 (1965).

3. Koshland, D.E., G. Nemethy and D. Filmer. Biochemistry, 5, 365 (1966).

4. Guidotti, G., J. Biol. Chem., 242, 3673 (1967).

5. Rossi Fanelli, A., Antonini, E., and Caputo, A., Adv. Prot. Chem., 19, 73 (1964).

*Note added in proof November 1970.

6. Chance, B.   This volume, p. 321.

7. Rumen, N.M., Fed. Proc., 27, 2852 (1968).

8. Hendrickson, W.A., W.E. Love and G.C. Murray, J. Mol. Biol., 33, 829 (1968).

9. Hendrickson, W.A., Ph.D. Thesis, Johns Hopkins University, 1968.

CONFORMATION CHANGES IN CYTOCHROME c REACTIONS

Abel Schejter[*]

Department of Biochemistry and Molecular Biology
Cornell University, Ithaca, New York

There is one type of reaction of cytochrome c that requires a conformation change, and this is the binding of ligands to the iron. This is because the cytochrome c iron is held in a closed crevice structure, with the porphyrin nitrogens more or less in a plane, and two protein side chains above and below this plane. Thus, when a ligand binds the iron, one of the metal-protein bonds has to be broken. When this occurs, the side chain can be displaced by a small rotation, or it may actually swing away together with a considerable part of the protein chain.

The belief that the latter is true, namely, that a substantial conformation change accompanies ligand binding originated with the demonstration that the binding of cyanide is a very favorable process due to a large positive entropy change (1). When the cyanide binding constants of ferric cytochrome c are compared with those of a typical open crevice heme protein, such as ferrimyoglobin, it is possible to estimate the thermodynamic parameters of the crevice closing reaction in neutral ferricytochrome c (2,3). These are: ΔF = 3.5 kcal/mole, ΔH = -18.0 kcal/mole and ΔS = -48 eu. These values are not only a quantitative expression of the strength of the closed crevice structure, but also an indication of the nature of the forces at play.

The closing of the crevice is a very exothermic reaction. Part of this exothermicity is due, of course, to the formation of the iron-protein bond that satisfies the sixth coordination valence of the metal. However, this can only account for a fraction of the 18 kcal/mole involved. The rest must arise from the hydrophobic interactions that result from the very compact tertiary structure that X-ray studies have visualized.

[*] On leave, Tel Aviv University, Tel Aviv, Israel.

The large decrease in entropy that attends the formation of the crevice indicates that the closed crevice protein has frozen into a state with fewer degrees of freedom than its open crevice conformer.

The fact that the primary structure of cytochrome c shows great variability in two thirds of the amino acids (5) poses the question whether the thermodynamic parameters of crevice closing are dictated only by the portion of the protein nearest to the group involved in the closing of the crevice, or by interactions taking place without involvment of the heme group. The information available at present from chemistry, spectroscopy and crystallography of cytochrome c suggests rather strongly that the 70-80 portion of the sequence is in close proximity to the heme group, and that the methionyl residue at position 80 actually binds the iron through its sulfur atom (6). Since this sequence is invariant with evolution, if the thermodynamic parameters of ligand binding reactions depended only on the interactions between this portion of the chain and the heme group, they should not vary with the species considered.

This is, however, not the case. We have measured the heat and entropy changes for the reaction of cyanide with yeast iso-1 ferricytochrome c (7), in order to compare them with the corresponding data for the horse protein (1,8). The results are shown in Table I.

TABLE I

Thermodynamic Parameters for the Reaction of Horse and Yeast iso-1 Ferricytochrome c with Cyanide

Species	K	ΔF (kcal/mole)	ΔH (kcal/mole)	ΔS (e.u.)
Horse	1.2×10^6	-8.3	1.1	31.3
Yeast	3.8×10^5	-7.6	-12.9	-17.8

The enthalpy change for the yeast protein is very favorable, while that for the horse cytochrome c is slightly unfavorable. This suggests that the non-covalent interactions in the yeast cytochrome c are much weaker than those operating in the horse protein. Furthermore, the entropy change accompanying cyanide binding to horse cytochrome c is much more favorable than that observed for yeast cytochrome c. This suggests again that the movements required from the protein chain in order to reach the ligand bound conformation are smaller in the case of the yeast protein because its native conformation is less constrained.

These results, therefore, tend to indicate that yeast iso-1 cytochrome c is a less compact and rigid molecule than its horse homolog. This is in keeping with the fact that yeast cytochrome c is less stable to ionic strength and temperature changes, and more susceptible to proteolysis (3). Another important conclusion suggested by these results is that the fixation of amino acid changes during evolution is dictated by the free energy of formation of the native conformation of the protein. However, heats and entropies are allowed considerable margins of change, as long as their added contributions keep the resulting free energy within rather restricted limits.

A second type of reaction of cytochrome c that involves a change in its conformation is the ionization of the oxidized form in weakly alkaline solution. This ionization was first described by Theorell (9); its pK in the horse protein is 8.9 (10). There are two factors that indicate a conformation change linked to this ionization. One is the markedly low rate of the observed change, incompatible with that of a single proton dissociation. The other is the nature of the spectroscopic change: although the porphyrin bands show only a slight red shift, and the spectrum remains that of a low spin ferric heme compound, the band at 695 nm disappears entirely. This band, which was also described first by Bigwood in 1934 (11), has been shown to depend on the coordination of the heme iron with a methionine sulfur (12). Its disappearance accompanies invariably those processes that entail a conformation change, such as heating, ligand binding, denaturation with urea or guanidine, or polymerization (13). Heating and cooling, for example, is a reversible process with thermodynamic parameters very similar to those that govern the crevice closing reaction (14). Thus, in terms of the spirit of this symposium, it is fair to say that nature has provided cytochrome c with a conformation probe built-in in its native structure. We have used this property in order to follow the kinetic and equilibrium parameters of the heme-linked ionization with pK 8.9 (10).

Two types of experiments were conducted. In one case, the protein was brought by rapid mixing to pH's between

453

8.5 and 10, and the rates of disappearance of the 695 nm band were measured. Similarly, the appearance of the band was followed after jumping the pH from 10 to 7. It was also attempted to follow directly the release and uptake of protons working with indicators. In the alkaline range the indicators tried showed time dependent absorption changes, even in the absence of cytochrome c. This behavior was not observed with indicator chlorphenol red, which is suitable for use at pH 7. When cytochrome c at pH 10 was jumped down to pH 7, the change in the absorption of chlorphenol red indicated that a proton was taken up by the protein at the same rate at which the 695 nm band appeared, 0.04 sec^{-1}. Since protonation reactions are faster than this observed rate by several orders of magnitude, it follows that the rate limiting reaction of the observed process is the conformation change that restores the 695 nm band.

This, and considerations of microscopic reversibility, permits one to write the following scheme as the simplest one that represents the changes under consideration:

$$AH^+ \underset{}{\overset{K_i}{\rightleftharpoons}} A + H^+ \underset{}{\overset{K_c}{\rightleftharpoons}} B \qquad (1)$$

In Equation (1), A indicates the conformation possessing the 695 nm band, and B that is devoid of it. The fraction of component B at equilibrium β is a function of proton concentration H:

$$\beta = \frac{K_i K_c}{K_i(1+K_c) + H} \qquad (2)$$

Hence, from a plot of the inverse of β against H, it is possible to estimate $K_i K_c$. One could also in principle estimate K_c, but in this case it is markedly larger than unity. However, the kinetic experiments could be used for the evaluation of K_i and K_c. When the pH is jumped from neutrality to alkalinity, the ionization equilibrium is reached very fast, and the kinetics of disappearance of the 695 nm band measures the relaxation of the system between A and B. The observed rate for this reaction, k_{obs}, is pH dependent:

$$k_{obs} = k_f \frac{K_i}{H + K_i} + k_b \qquad (3)$$

This can be rearranged into:

$$\frac{1}{-k_{obs} - kb} = \frac{1}{k_f} + \frac{H}{k_f K_i} \qquad (4)$$

Using equation (4), the data can be conveniently analysed, because k_b can be determined independently. The experimental results at 21°, analysed in terms of equation (4), are shown in Figure 1. From the intercept in this figure, we evaluate $k_f = 2.9$ sec^{-1}, and from k_f and k_b, $K_c = 83$. From the overall equilibrium, $K_i K_c = 10^{-8.9}$ M/l. In addition to that, K_i can be obtained directly from the slope and intercept of Figure 1, resulting in this case $K_i = 1.5 \times 10^{-11}$ M/l; this provides an independent check on the consistency of the results and the proposed scheme.

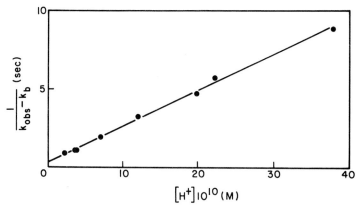

In brief, the alkaline ionization of ferricytochrome c can be described as a rapid ionization of a group with pK = 10.9, followed by a slow conformation change with an equilibrium constant of 83 and a forward rate of 2.9 sec^{-1}.

Two questions arise: first, what is the nature of the conformation change; and, second, can the ionizing group be identified? The conformation change may or may not involve the replacement of the methionyl sulfur by another protein ligand. The absence of the 695 nm band cannot be taken by itself as proof of such a replacement. According to one interpretation of the origin of this band, moving the sulfur closer to the iron or away from it will cause the band to shift to higher or lower energies, respectively. This problem will most probably be solved by resonance spectroscopy and by polarization studies.

It should be noticed that the ionization linked conformation change and the cold-hot equilibrium at neutral pH (14) are of the same order of magnitude, but in opposite directions, the latter having an equilibrium constant of 6×10^{-2} at 25°.

As to the nature of the ionizing group, judging from its

pK, 10.9, it can be tentatively identified as either a lysine or a tyrosine. Since complete guanydilation of the lysine residues causes no basic changes in the properties of cytochrome c (15), this would appear to rule out a lysine and leave a tyrosine as the responsible group. There is, finally, another result that appears to involve a tyrosine residue with conformation changes that affect the 695 nm band. When cytochrome c is nitrated with tetranitromethane, the appearance of one nitrotyrosine residue is paralleled in time with the disappearance of the 695 nm band (16), as shown in Table II. We hope that additional investigation of this new cytochrome c derivative will help to elucidate this matter.

TABLE II

Nitration of Ferricytochrome c with Tetranitromethane

Reaction time (min)	NO_2-Tyrosine (mol/mol cyt. c)	%695 nm band
0	0	100
5	0.24	80
10	0.50	53
15	0.74	30
20	0.90	2

References

1. George, P. and C.L. Tsou. Biochem. J., 50, 440 (1952).

2. George, P. and G.I.H. Hanania. Nature, 175, 1034 (1955).

3. Margoliash, E. and A. Schejter. Adv. Protein Chemistry, 21, 113 (1966).

4. Dickerson, R.E., M.L. Kopka, J.E. Weinzierl, J.C. Varnum, D. Eisenberg, and E. Margoliash. J. Biol. Chem., 242, 3015 (1967).

5. Nolan, C. and E. Margoliash. Ann. Rev. Biochem. 37, 727 (1968).

6. Margoliash, E.,W.M. Fitch, and R.E. Dickerson. Brookhaven Symp. Biol., 21, 259 (1969).

7. Aviram, I. and A. Schejter. J. Biol. Chem., 244, 3773 (1969)

8. George, P. S.C. Glauser, and A. Schejter. J. Biol. Chem., 242, 1690 (1967).

9. Theorell, H. and A. Akesson. J. Am. Chem. Soc., 63, 1818 (1941).

10. Schejter, A., L.A. Davis, and G.P. Hess. unpublished experiments.

11. Bigwood, E.J., J. Thomas, and D. Wofters. Compt. Rend. Soc. Biol., 117, 220 (1934).

12. Schechter, E. and P. Saludjian. Biopolymers, 5, 788 (1967).

13. Schejter, A. and I. Aviram. Biochemistry, 8, 149 (1969).

14. Schejter, A. and P. George. Biochemistry, 3, 1045 (1964).

15. Hettinger, T.P. and H.A. Harbury. Proc. Natl. Acad. Sci., U.S., 52, 1469 (1964).

16. Schejter, A. and M. Sokolowski. FEBS Letters, 4, 269 (1969).

SPIN-LABEL STUDIES OF CYTOCHROME C*

Henry R. Drott and Takashi Yonetani

Johnson Research Foundation
University of Pennsylvania
Philadelphia, Pennsylvania

The technique of spin-labeling having been successfully applied to numerous proteins (1,2), biological and artificial membranes (2,3), and polymers (4) was extended to cytochrome c (5). By labeling cytochrome c, the environmental surroundings of the label as well as the interaction with another protein were investigated. The bromoacetamide derivative of the five-membered nitroxyl ring (Figure 1) is the spin label of choice as it has been employed to alklate cysteine (6), methionine (7), and histidine (7,8) moieties in this and other proteins.

Attempts to label horse heart cytochrome c quantitatively were not successful, whereas baker's yeast cytochrome c appears to react stoichiometrically. Therefore attention is focused at first on results obtained with yeast cytochrome c and delayed until last for horse heart cytochrome c.

The pertinent conclusions of the alkylation studies of cytochrome c can be summarized in two statements. Carboxymethylation of histidine-33 and methionine-65 have been effected

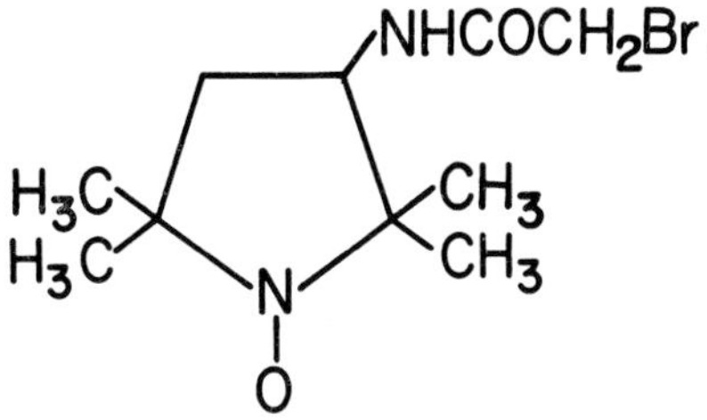

Figure 1. N-(1-oxyl-2,2,5,5-tetramethyl-3-pyrrolidinyl) bromoacetamide.

*Supported by research grants from the USPHS GM 12202, 1-F02-HE 39,533, GM 15435, 5-K3-GM35,331 and NSF GB 6974.

by other investigators (9-18) at neutral pH for vertebrates and non-vertebrates cytochrome c without any significant changes in the physical properties. At lower pH or in the presence of cyanide, methionine-80 was also alkylated causing gross changes in the physical properties.

In Figure 2, the absorption spectra for the native yeast cytochrome c in the oxidized and reduced forms are shown. There is very little difference between these absorption spectra of yeast cytochrome c and mammalian-type cytochromes (19). Thus, the carboxymethylation results were used as a guide for the spin-label studies of yeast cytochrome c.

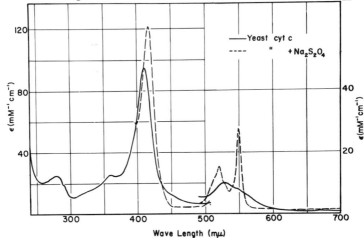

Figure 2. Absorption spectra of yeast cytochrome c. The spectra were recorded in 0.1M potassium phosphate buffer, pH 7.0 at 23°C.

Two different spin-labeled cytochrome c have been prepared. One product was made at room temperature and has at least one spin-label per cytochrome c. The other modified cytochrome c was prepared at 40°C and appears to have at least two spin-labels per cytchrome c. In Figure 3, the absorption spectra for the room temperature preparation is described. There are no spectral shifts, no change in extinction coefficients, nor appreciable binding of carbon monoxide. Furthermore, this preparation has 100% enzymic activity to the mitochondrial NADH oxidase system. In Figure 4, the optical spectra of the cytochrome c modified at 40°C is altered tremendously from the native material. Both the Soret bands of the oxidized and reduced forms of this cytochrome c undergo blue shifts and the alpha band appears to

460

have a hypochromic shift. Moreover, this preparation exhibits an appreciable binding of carbon monoxide and has only 60% enzymic activity to the NADH oxidase system.

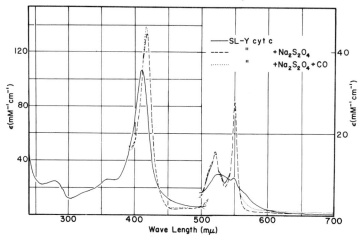

Figure 3. Absorption spectra of spin-labeled yeast cytochrome c prepared at room temperature. The experimental conditions were the same as in Figure 2.

An attractive feature of yeast cytochrome c is the single cysteine residue it possesses at position 103 next to the C-terminus (20). Although the native cytochrome c reacts with pCMB, both these samples do not react with pCMB, indicating that the sufhydryl has been blocked. From these optical data,

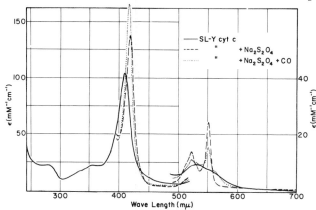

Figure 4. Absorption spectra of spin labeled yeast cytochrome c modified at 40°C. The experimental conditions were the same as in Figure 2.

461

it appears that the room temperature sample has been labeled at the sufhydryl and the 40°C preparation has been alkylated at both the sufhydryl and the methionine-80 positions. Alkylation of the histidine-33, -39 and methionine -64 is certainly a real possibility, but evidence for this requires amino acid compositions to be determined.

The EPR spectra of the different materials are shown in Figure 5. The upper spectrum corresponds to the room temperature sample, while the lower spectrum is for the high temperature material. In both of these spectra, the labels are weakly immobilized, indicating that alkylation has occurred on the surface of the cytochrome c. Moreover, both spectra appear to be homogeneous, suggesting that the surroudings of the different labels are very similar and that their rotational correlation times are of the same order. However, the degree

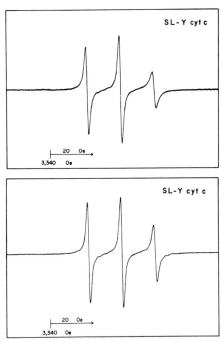

Figure 5. EPR spectra of the mono- and di-spin-labeled yeast cytochrome c. The top spectrum represents the mono-labeled material, while the bottom curve is for the di-labeled material. The spectra were recorded on a Varian 4502 spectrometer using the multipurpose cavity with the aqueous sample cell. The modified cytochrome c's were dissolved in 10 mM potassium phosphate buffer, pH 7 at 23°C.

of immobilization is much weaker in the lower spectrum. The
extra motional freedom is attributed to the label located at
methionine-80.

The spin-labeled cytochrome c was used to examine the
interaction with the enzyme cytochrome c peroxidase (CCP)
and with mitochondrial membrane fragments (3). In Figure 6,
the effect of adding an equal molar amount of CCP to the
spin-labeled cytochrome c is exhibited. Although the sample
is diluted 10%, the intensity of the center line of the mixture is reduced 23%. Along with the center line, the outer
lines have decreased in intensity and a complementary broad-
en accompanies all three lines, suggesting that the CCP
binds with spin-labeled cytochrome c. The binding does not
effect the rotational motion of the label except for the
imparting to it the motion of the cytochrome c-CCP complex.

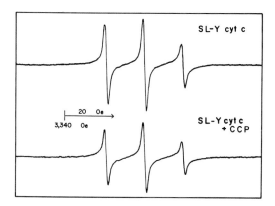

Figure 6. The interaction between spin-labeled yeast cytochrome c and cytochrome c peroxidase. An equal molar amount
of cytochrome c peroxidase (CCP) was mixed with spin-labeled
cytochrome c dissolved in 10mM potassium phosphate buffer,
pH 7.0, at 23°C.

In closing, experiments with spin-labeled horse heart
cytochrome c gave similar absorption and EPR spectra to
those just described for the room temperature spin-labeled
yeast cytochrome c. The binding of CCP exhibited no difference
from that described. In addition, reduction of spin-labeled
ferricytochrome c to spin-labeled ferrocytochrome c with
ascorbate failed to alter the EPR spectrum significantly.

References

1. Hamilton, C.L. and H.M. McConnell. Structural Chemistry and Molecular Biology, (A. Rich and N. Davidson, eds.) W.H. Freeman and Co., San Francisco, Calif., 1968, p. 115.

2. Griffith, O.H. and A.S. Waggoner. Accounts Chem. Res., 2, 17 (1969).

3. Lee, C.P. and H.R. Drott. Volume I p. 247

4. Griffith, O.H., J.F.W. Keana, S. Rottschaefer, and T.A. Warlick. J. A. C. S., 89, 5072 (1967).

5. Barratt, M.D., D.K. Green, and D. Chapman. Biochim. Biophys. Acta., 152, 20 (1968).

6. Ogawa, S. and H.M. McConnell. Proc. Nat'l. Acad. Sci., 58, 19 (1967).

7. Morrisett, J.D. Ph.D. Dissertation, University of North Carolina, Chapel Hill, 1968.

8. Smith, I.C.P. Biochemistry, 7, 745 (1968).

9. Tsai, H.J. and G.R. Williams. Can. J. Biochem., 43, 1409 (1965).

10. Tsai, H.J. and G.R. Williams. Can. J. Biochem., 43, 1995 (1965).

11. Schejter, A. and P. George. Nature, 206, 1150 (1965).

12. Motonaga, K., H. Katano, and K. Nakanishi. J. Biochem. Tokyo, 57, 29 (1965).

13. Ando, K., H. Matsubara, and K. Okunuki. Biochim. Biophys. Acta., 118, 240 (1966).

14. Ando, K., H. Matsubara, and K. Okunuki. Biochim. Biophys. Acta., 118, 256 (1966).

15. Stellwagen, E. Biochem. Biophys. Res. Comm., 23, 29 (1966).

16. Hartbury, H. Hemes and Hemoproteins (B. Chance, R. Estabrook, and T. Yonetani, eds.) Academic Press, New York, 1966, p. 391.

17. Fanger, M., T. Hettinger, and H. Harbury. Biochemistry, 6, 713 (1967).

18. Stellwager, E. Biochemistry, 7, 2496 (1968).

19. Armstrong, J., J. Coates, and R.K. Morton. Hematin Enzymes (J. Falk, R. Lemberg, and R.K. Morton, eds.) Pergamon Press, Oxford, England, p. 385.

20. Narita, K., K. Titani, Y. Yaoi, and H. Murakami. Biochim. Biophys. Acta, 77, 688 (1963).

HIGH RESOLUTION PROTON NMR STUDIES OF THE COORDINATION OF THE HEME IRON IN CYTOCHROME c

Kurt Wüthrich*

Bell Telephone Laboratories, Incorporated
Murray Hill, New Jersey

This paper describes the analysis of a series of high resolution proton nuclear magnetic resonance (NMR) spectra of cytochromes c. The NMR spectra yield data on the axial ligands of the heme iron, the complex formation of cytochrome c with cyanide ion and azide ion, and the electronic structure of the heme group. Based on these data on the coordination of the heme iron NMR spectroscopy can further provide information on the protein conformation in cytochrome c.

NMR studies of cytochrome c have been described previously. Kowalsky reported that hyperfine interactions with the heme iron give rise to large upfield and downfield shifts of three proton resonances of ferricytochrome c (1). More recently McDonald and Phillips studied the denaturation of cytochrome c, (2) and presented an interpretation of the NMR spectrum of ferrocytochrome c (3).

In a review presented to the third Johnson Foundation Colloquium Margoliash (4) concluded that the axial ligands of the heme iron in ferrocytochrome c were probably histidyl residue 18 and methionyl residue 80. It was left open whether the same residues are bound in both oxidation states. These conclusions were based mainly on comparisons of the amino acid sequences of cytochromes c from different species, (5) and on experiments with chemically modified cytochromes c (6,7). More recently X-ray studies of horse heart ferricytochrome c (8) showed that the axial ligands are histidyl residue 18, and an aliphatic amino acid residue. The present NMR data imply that the sixth ligand is methionyl in both oxidized and reduced cytochrome c.

*Present address: Institut für Molekularbiologie und Biophysik, Eidgenössische Technische Hochschule, 8049 Zürich, Switzerland.

Materials and Methods

Ferricytochrome c of guanaco, rabbit, and turkey was obtained from Dr. E. Margoliash, horse ferricytochrome c from Mann Laboratories, and Candida krusei ferricytochrome c from Sankyo Laboratories in Tokyo, Japan. For the NMR experiments ∿0.01 M solutions in 0.1 M deuterated phosphate buffer, pD 7.0, were prepared. The complexes of ferricytochrome c with cyanide ion and azide ion were prepared by addition of KCN and NaN₃, respectively (9,10). Ferrocytochrome c was prepared by reduction of ferricytochrome c with ascorbic acid or disodiumdithionite. The cyanide complex of ferrocytochrome c was obtained by reduction of cyanoferricytochrome c with dithionite (11).

High resolution proton NMR spectra were recorded on a Varian HR-220 spectrometer equipped with a standard Varian variable temperature control unit. The temperature regulation in the sample zone, which was determined from the chemical shifts of the resonances of ethylene glycol, was better than ± 1° under the conditions of the experiments described in this paper. Sodium 2,2-dimethyl-2-silapentane-5-sulfonate (DSS) was used as an internal standard. Chemical shifts are expressed in parts per million (ppm) from DSS, where shifts to low field are assigned negative values.

General Features of the NMR Spectra of Cytochrome c

The proton NMR spectrum at 220 MHz of ferricytochrome c is shown in Figure 1. The three parts of the spectrum contain all the observed lines. The strongly overlapping resonances between DDS and -9 ppm come from the bulk of the 650 protons of the polypeptide chain. The sharp lines between -4 and -6 ppm are the resonance of HDO and its side bands. The intensities of the resolved resonances observed in the regions from 2 to 7 and -10 to -35 ppm correspond to one to three protons. These resonances are shifted upfield or downfield by local magnetic fields arising both from aromatic ring currents (2, 13) and from the unpaired electron of the heme iron (1,14). The lines at -34.0, -31.4, and +23.2 ppm are the shifted resonances reported by Kowalsky (1).

(The structure of the heme group of cytochrome c and the axial ligands of the heme group are shown at the bottom of Fig. 1. Nielands (12) has pointed out that even though this is the most likely structure of porphyrin c, the possibility of β-attachment of the cysteinyl residues to one or both of the 2,4 positions has not been definitely ruled out. The NMR spectrum

might be taken to support the structure given below. In place
of the methionyl residue, histidyl, arginyl, and lysyl have
been suggested as well as the sixth ligand in at least one of
the two oxidation states of cytochrome <u>c</u> (4,5).)

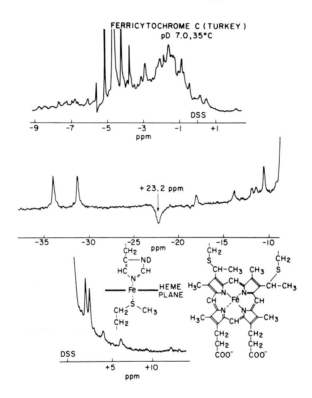

Figure 1. High resolution proton NMR spectrum at 220 MHz of
a 0.01 M solution of ferricytochrome c. No DSS was added to
this sample. The sharp lines between −4 and −6 ppm correspond to the HDO resonance and its first and second spinning
side bands. Different vertical and horizontal scales are
used for the regions from +1 to −9 ppm, and from 0 to +10
and −10 to −35 ppm. The high field line at +23.2 ppm is
observed as an inverted resonance of the center band of the
spectrum. (The HR−220 spectrometer operates with a 10 kHz
field modulation. Usually, one observes the first upfield
side band; if large hyperfine shifts occur, parts of the
center band and the different side bands of the spectrum
overlap.)

In a protein the proton resonances of the amino acid residues are shifted relative to their positions in the NMR spectra of the individual amino acids by the local magnetic fields of the neighboring groups (2). In diamagnetic proteins the largest upfield and downfield shifts come from the ring currents in aromatic rings (2,13,15,16). The resonances of protons located within a few angstroms above or below the plane of a phenylalanin ring can be shifted upfield by as much as 2 ppm (17). Considerably larger shifts may result if the protons are located near several aromatic amino acid residues, or near the plane of the heme group (15,16). Resonances of aliphatic amino acid residues can thus be shifted to positions several ppm upfield from DSS, and may be well resolved at 220 MHz. Ring current shifts are very sensitive to the relative positions of the observed protons and the aromatic rings in the three-dimensional arrangement of the polypeptide chain and hence to conformational changes in the protein (2,13). On the other hand, in the absence of conformational changes, ring current shifts are independent of temperature (2,14).

In the NMR spectra of paramagnetic heme proteins one observes hyperfine shifts in addition to the ring current shifted resonances (14). The unpaired electron of the iron in the low spin ferric hemes (Fe^{3+}, S=1/2) of ferricytochrome c and cyanoferricytochrome c is delocalized into the π-orbitals of the axial ligands and the porphyrin ring. Unpaired electron density is then transferred by spin polarization or hyperconjugation (18) from the carbon or sulfur atoms to the protons attached directly, or in methyl and methylene groups (Figure 1). The resulting contact shifts $\Delta\nu_c$ of the proton resonances are given by (19)

$$\Delta\nu_c = A \frac{\gamma_e}{\gamma_H} S(S+1) \frac{\nu}{3kt} \tag{1}$$

where A is the contact interaction constant of the proton γ_e, γ_H, and k have the usual meaning, S is the total electronic spin, ν the resonance frequency of the proton, and T the temperature in degrees Kelvin. Proton resonances could also be shifted by pseudo-contact coupling with the unpaired electron (20). However, it appears that for low spin porphyrin iron (III) complexes pseudo-contact shifts are small compared to contact shifts (1,21). Therefore unpaired spin densities on the atoms next to the observed protons can be derived from the hyperfine shifts. For a proton or a methyl group attached to an aromatic ring we have (22,23)

$$A = Q \rho_c^{\pi},$$

(2)

where A is the contact coupling constant in cps, and ρ^{π} the integrated spin density in percent of one electron centered on the π-orbital of the neighboring ring carbon atom. Q is a proportionality constant which is approximately the same for all aromatic protons, i.e., $Q = -6.3 \times 10^7$ Hz. For the protons of the ring methyls a value of $Q \approx 3 \times 10^7$ Hz seems to be in best agreement with the experimental observations (24).

Summarizing we then have that in addition to the strongly overlapping resonances of the polypeptide chain the NMR spectra of cytochromes c contain a number of resolved resonances, which will be the subject of the discussions in this paper. In diamagnetic ferrocytochrome c (Fe^{2+}, S=0) the resolved resonances have to come from amino acid residues located near the heme group or other aromatic rings in the molecule. The ferricytochrome c spectrum (Figure 1) contains further the hyperfine-shifted resonances of the ligands bound to the iron. Because of the very short electronic relaxation times in low spin ferric hemes (24) the line widths of these proton resonances are essentially unaffected by electron-proton interactions. Therefore the shifted resonances can easily be observed. Ring current shifts and hyperfine shifts in the ferricytochrome c-spectrum are distinguished by their temperature dependencies [Equation (1)] (14). Once the hyperfine-shifted resonances have been identified and assigned to specific protons of porphyrin c information can be derived on the unpaired electron distribution in the π-orbitals of the heme group.

Ferricytochrome c and Ferrocytochrome c

In the proton NMR spectrum of ferricytochrome c (Figure 1) the resonances between DDS and +1 ppm correspond to $\sim$20 protons. The temperature dependence of this part of the spectrum is shown in Figures 2 and 3. Two temperature-independent resonances of one and six protons are at -0.2 and -0.5 ppm. From Fig. 3, which shows a plot of the resonance positions of Fig. 2 vs. the reciprocal of temperature, it appears most likely that the resonances of intensities three and six protons observed at +0.5 ppm and +0.2 ppm (220 Hz = 1 ppm) are also shifted by ring current fields. Further

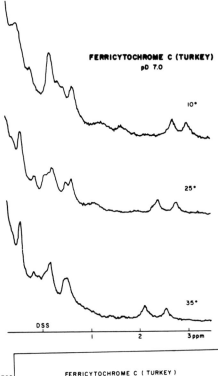

Figure 2. Temperature dependence of the NMR spectrum of ferricytochrome c between –1 and +3 ppm.

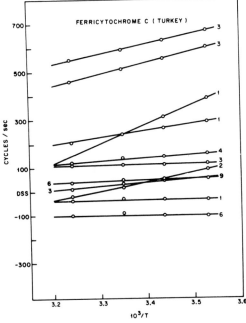

Figure 3. Dependence on the reciprocal of temperature of the positions of the resonance of ferricytochrome c in the range –1 to +3 ppm. The number of protons corresponding to the intensities of the resonances is indicated on the right-hand side.

evidence for the presence of ring current shifted lines at high fields from DDS comes from the observation that the total intensity of the resonances outside 0 to -9.5 ppm corresponds to a larger number of protons than are on the ligands bound to the iron (Fig. 1). From their temperature dependences, all the other resonances upfield from DDS (Figs. 3 and 4) are shifted by hyperfine interactions (Eq. 1). All the resonances at low fields from -10 ppm must come from hyperfine inter- actions, since the ring current fields could not possibly give rise to such large downfield shifts (14). Figure 4 shows that the positions of these resonances change with temperature. The intensities and temperature dependencies of all the re- solved hyperfine-shifted resonances are listed in Figure 5.

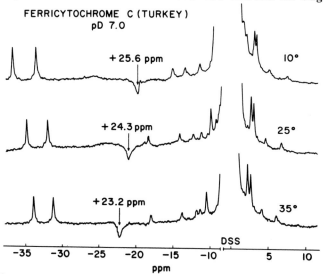

Figure 4. Temperature dependence of the NMR spectrum of ferri- cytochrome c in the regions -10 to -38 and +2 to +12 ppm. (See caption to Fig. 1)

The intensities and temperature dependences of all the re- solved hyperfine-shifted resonances are listed in Figure 5. Without counting the broad line at ca. -25 ppm, the total intensity of these resonances corresponds to 37 protons. In the structure of Fig. 1, one might expect 39 protons to ex- perience sizeable contact shifts, i.e., all the protons of porphyrin c except those originating from cysteine, the 2,4- imidazole protons of the axial histidine, and the protons of the methyl and methylene groups next to the sulfur in methionyl.

471

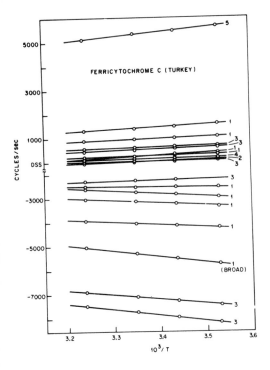

Figure 5. Dependence on the reciprocal of temperature of the positions of the hyperfine-shifted resonances of ferricytochrome c. The number of protons corresponding to the intensities of the resonances is indicated on the right-hand side.

It appears then that the resonances of only two of these protons are between 0 and -9 ppm, and therefore not resolved.

It was observed previously that the contact-shifted resonances of methyl and methylene groups in low spin ferric hemes (21) and heme proteins (14,25) are not usually split into single-proton resonances. The following assignments of the contact-shifted resonances of ferricytochrome c to specific protons of porphyrin c and of the axial ligands can then be made. The eight resonances of intensity one proton in Fig 5 correspond most likely to the six single protons of porphyrin c, and the 2,4 imidazole protons of the axial histidyl residue. From Fig. 1, at least five of the six methyl residues which are between -34 and +3 ppm at 35° C (Figs. 3-5) come from methyl groups of porphyrin c (Fig. 1). (In the methyl resonances at -34 and -31 ppm one observes a narrow and a broader component, which have approximately equal intensities. This arises because the dipole-dipole coupling between the methyl protons is modulated by two different rotational motions, a

very fast motion about the C-C bond from the porphyrin ring
to the methyl group, and the slower rotation of the entire
cytochrome c molecule. This effect was predicted from theoretical calculations by A. Redfield (30). To our knowledge
it has not been observed previously in a protein NMR spectrum.)
From the symmetry of the electronic wave functions of
the heme group, and because no large negative spin densities
would be expected on the carbon atoms of the porphyrin ring,
it appears then extremely unlikely that any of the methyl
or methylene resonances of the heme group could be shifted
to +23.2 ppm. Furthermore, no high field resonances of intensity two or more protons have been observed above 5 ppm
for any other low spin ferric hemes (21) or heme proteins,
(14,25,26) including cyanoferricytochrome c (Figure 7). This
implies that the resonance at +23.2 ppm comes from one of
the axial ligands. From previous work the sixth ligand is
a hemochrome-forming aliphatic amino acid residue (5,8). Of
these only methionyl (7) could conceivably give rise to a
contact-shifted line of 4 to 5 protons, i.e., if the resonances of the methyl and the methylene groups next to the
sulfur (Figure 1) were accidentally degenerate. The shape
of the resonance at 23.2 ppm indicates that it consists of
at least two almost degenerate lines. Thus the observation
of the high field resonance at +23.2 ppm implies that the
sixth ligand of the heme iron in ferricytochrome c is a
methionyl residue.

From Equation (1) the curves of Figure 5 should go to
the positions of the corresponding resonances in diamagnetic
porphyrins (15,16) when extrapolated to very high temperatures.
As in other low spin ferric hemes (21) and heme proteins
(14,25) most of the contact-shifted resonances in ferricytochrome c do not extrapolate to the predicted high temperature
positions. Most surprising is the temperature dependence of
the methyl resonance at ca. -11 ppm which moves to lower
fields when the temperature is raised (Figure 5). From these
temperature dependences detailed information on the electronic
structure of the heme group can be derived, as will be discussed in a future paper.

The high field part of the proton NMR spectrum of ferrocytochrome c, which was previously reported by McDonald and
Phillips, (2) is shown in Figure 6. At 35°C it contains
ring current shifted resonances of intensity three protons
at 3.3, 0.7, 0.6 and 0.6 ppm, a resonance of two protons at -0.1 ppm, and resonances of one proton at 3.7,
2.7, 1.9, 0.2, and -0.2 ppm. All but one of the line

positions are independent of temperature between 12 and 35°C. One methyl resonance moves from 0.7 ppm at 12° to 0.6 ppm at 35°, indicating localized changes in the protein conformation. However, the temperature dependence of all the ring current shifted resonances in ferrocytochrome c̱ is smaller than that of the lines which were identified from Figure 3 as hyperfine-shifted resonances of ferricytochrome c̱. McDonald and Phillips (3) suggested that the lines between 1.9 and 3.7 ppm come from the methyl group and three protons of the γ- and β-methylenes of an axial methionyl residue (Fig. 1) which would experience the strong ring current field of the porphyrin ring. As was discussed above, other explanations for the unusual positions of these resonances could be found. However, the NMR studies of the dissociation of cyanoferrocytochrome c̱ described below imply that the assignment to the methionyl protons is correct.

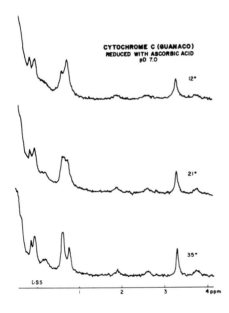

Figure 6. Temperature dependence of the NMR spectrum between −1 and +4 ppm of ferrocytochrome c̱.

Cyanoferricytochrome c and Azidoferricytochrome c

The NMR spectrum of cyanoferricytochrome c̱ in the region −30 to −9 and −1 to +12 ppm is presented in Figures 7 and 8. It is seen that the positions of most of the resonances are temperature dependent. At 35°, four resonances of intensity three protons are at −22.9, −21.1, −16.0 and −11.4 ppm, two

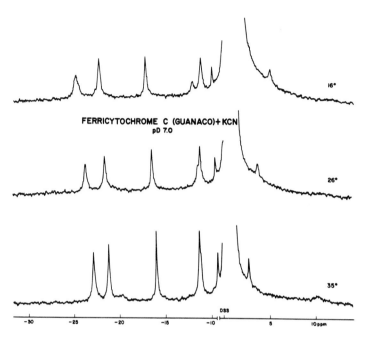

Figure 7. Temperature dependence of the NMR spectrum of cyano-ferricytochrome c in the regions −30 to −10 and 2 to 12 ppm.

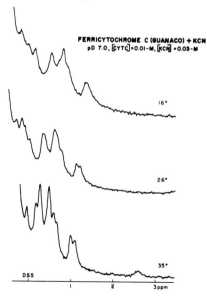

Figure 8. Temperature dependence of the NMR spectrum between −1 and +3 ppm of cyano-ferricytochrome c.

resonances of two protons at +1.0 and +1.1 ppm, and four resonances of one proton at -11.2, -9.3, +2.6, and +10.3 ppm. The
resonances between DSS and +0.9 ppm correspond to ca. 20 protons. Thus, the resonances observed outside the region from
DSS to -9 ppm correspond to more protons than are on the
ligands bound to the iron (replace methionyl by CN⁻ in Figure
1). From this and the temperature studies it appears that
three ring current shifted methyl resonances are between
DSS and +1 ppm, and one more is at ca. -0.3 ppm. The hyperfine-shifted resonances of cyanoferricytochrome c are quite
similar to those of cyanoferrimyoglobin (14,25). From comparisons with the latter and with various cyanoporphyrin iron
(III) complexes (21) it appears most likely that the resonances between -9 and -25 ppm correspond to the four ring
methyls of porphyrin c and the two imidazole protons of the
axial histidyl residue. Essentially all the remaining hyperfine-shifted lines would then be at high fields from DSS.
No resonance of intensity more than one proton is upfield
from 2 ppm.

Upon addition of sodium azide to a solution of ferricytochrome c, new hyperfine shifted resonances appear in the
NMR spectrum (Figure 9), indicating the formation of azido-
ferricytochrome c. Because the azide complex is rather unstable (10), its spectrum could only be observed in solutions
which contained an excess of uncomplexed ferricytochrome c.
Three resonances between -14 and -18 ppm (A) come from the
azide complex.

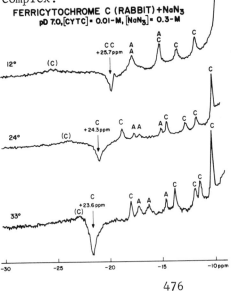

FERRICYTOCHROME C (RABBIT)+NaN₃
pD 7.0,[CYTC]= 0.01-M, [NaN₃]= 0.3-M

Figure 9. Temperature
dependence of the NMR
spectrum between −30
and −10 ppm of a solu-
tion of ferricytochrome
c and sodium azide. C,
resonances of ferricyto-
chrome c; A, azidoferri-
cytochrome c.

476

Formation and Dissociation of Cytochrome c
Complexes with Cyanide Ion

The difference between the NMR spectra of cyanoferricytochrome c and ferricytochrome c can be used to study the
reaction of the latter with cyanide ion. Figure 10 shows the
changes in the range -35 to -10 ppm of the ferricytochrome c
spectrum which arise from addition of variable amounts of
KCN. All the ferricytochrome c is complexed after addition
of a small excess of KCN, and no further changes of the NMR
spectrum are observed if more KCN is added. This shows that
a 1:1 complex is formed between ferricytochrome c and cyanide
ion. The complex formation does not occur instantaneously (9).
At 35°, the reaction was complete after about five minutes.

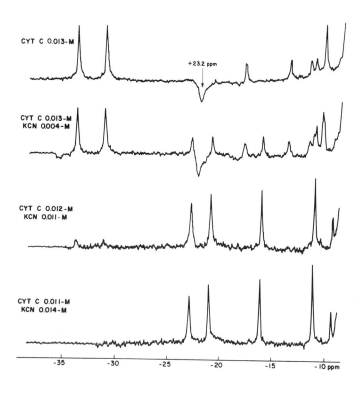

Figure 10. NMR study of the complex formation of ferricytochrome c with cyanide ion. The NMR spectra were recorded
approximately 15 minutes after the addition of KCN.

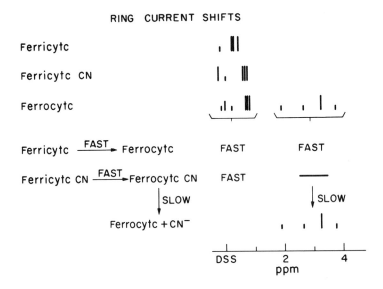

Figure 11. Comparison of the positions of the ring current shifted high field resonances of ferricytochrome c, cyanoferricytochrome c, and ferrocytochrome c. In the lower part of the figure, the rates of reduction with dithionite of the ferric compounds are compared with the rates at which the high field resonances of ferrocytochrome c appeared after addition of the reducing agent.

In Figure 11 the positions of the ring-current shifted high field resonances of ferricytochrome c, cyanoferricytochrome c, and ferrocytochrome c are compared. In each of the three spectra the region between -0.5 and +1.0 ppm contains a group of three methyl resonances which are at slightly different positions for the different compounds, and some other lines corresponding to one, two or three protons. Only ferrocytochrome c has lines between +1 and +4 ppm, i.e., one methyl resonance and three one-proton resonances. This can be used to study the dissociation of cyanoferrocytochrome c, which in turn yields information on the sixth ligand of the heme iron in ferrocytochrome c.

At 9° C the NMR spectrum of cyanoferricytochrome c contains hyperfine-shifted resonances of intensities one and four protons at +4.4 and +1.3 ppm. Figure 12 shows the changes of the NMR spectrum of cyanoferricytochrome c after addition of dithionite. As judged from the disappearance of the hyperfine-shifted resonances, the reduction was very fast at 9°C. The high field resonances of ferrocytochrome c between DSS and +1 ppm appeared immediately. On the other hand, the four resonances between 1.9 and 3.7 ppm of the ferrocytochrome c spectrum appeared very slowly. After 50 minutes the reaction was not complete, as is seen from a comparison of the last two spectra of Fig. 12. These observations agree with the reaction mechanism outlined at the bottom of Fig. 11 (11). The reduction to cyanoferrocytochrome c is very fast, and is followed by a slow dissociation of this unstable complex. Thus the axial amino acid residue which was displaced by cyanide goes back into its place in the native protein. The data of Fig. 12 show that the ferrocytochrome c resonances between 1.9 and 3.7 ppm come from this axial ligand, which then must be methionyl, since this is the only hemochrome-forming amino acid residue (4,7) that contains a methyl group.

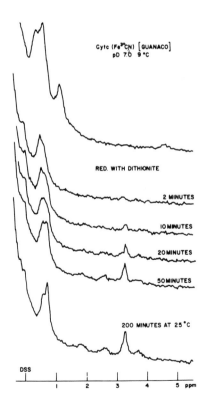

Cytc (Fe³CN) [GUANACO]
pD 7.0 9 °C

RED. WITH DITHIONITE

2 MINUTES

10 MINUTES

20 MINUTES

50 MINUTES

200 MINUTES AT 25°C

DSS

1   2   3   4   5 ppm

Figure 12. Proton NMR spectra of cyanoferricytochrome c and of the reaction products observed at variable times after the reduction of cyanoferricytochrome c with dithionite.

Discussion

The analysis of the ferricytochrome c spectrum (Figures 1-5) and the NMR studies of the dissociation of cyanoferrocytochrome c strongly imply that the sixth ligand of the heme iron is methionyl in both oxidized and reduced cytochrome c. This conclusion receives additional support from the observation that the ring current-shifted methionyl resonances of ferrocytochrome c were observed immediately after reduction of ferricytochrome c (Figure 11). In view of the data available on the dissociation of ferrocytochrome c complexes (5,11) it seems unlikely, though it is not impossible, that this would be so unless methionyl is the sixth ligand in both oxidation states.

Since the resonances of the different species are well resolved in the NMR spectra of solutions of ferricytochrome c and its complexes with cyanide ion and azide ion (Figures 9 and 10) these spectra show that the rate of a possible ligand exchange between different molecules of ferricytochrome c is too slow to be studied by NMR. This was to be expected from earlier data (5,9,10,11). On the other hand the present results might provide the basis for future applications of NMR spectroscopy to studies of the stability of ferricytochrome c complexes, and of the kinetics of dissociation of complexes with ferrocytochrome c.

Once the ring current-shifted resonances have been identified in the NMR spectra of the paramagnetic compounds (Figures 3 and 11) data on the protein conformation in cytochrome c can be derived. Since cyanide ion replaces an amino acid residue in the axial position of the heme iron, one would expect conformational changes upon formation and dissociation of the cyanide complexes. It has also been suggested that conformational changes occur upon interconversions between the ferrous and ferric oxidation states (5,27). A comparison of the NMR spectra shows that the ring current-shifted methyl resonances which are at 0.7, 0.6 and 0.6 ppm in ferrocytochrom c are at different positions in both ferricytochrome c and cyanoferricytochrome c (Figures 2, 6, 8, and 11). This indicates that conformational changes occur upon both, change of the oxidation state of the heme iron, and complex formation with cyanide ion. Furthermore, two ring current shifted methyl resonances at 0.6 and 0.7 ppm are at identical positions in cyanoferrocytochrome c and ferrocytochrome c, whereas one methyl resonance moves from 0.6 to 0.7 ppm upon dissociation of the cyanide complex (Figure 12). This latter resonance comes most likely from an amino acid residue near the axial methionyl, while the other two methyl resonances seem to come from positions along the polypeptide chain which are not affected by movements of the axial methionyl.

At the present state, the NMR data provide no evidence that the conformational changes upon interconversions between the ferrous and ferric oxidation states, or upon complex formation with cyanide ion, would involve major rearrangements of the molecule. All the major differences between the high field regions of the NMR spectra were found to come either from interactions with the unpaired electron in the ferric compounds, or from localized structural changes near the heme iron. For example, the ring current shifted methionyl resonances of ferrocytochrome c between +1 and +4 ppm (Figure 6) are not observed at these positions in the ferricytochrome c spectrum because they are shifted to even higher fields by hyperfine interactions (Figure 1). Direct evidence that the conformational changes are rather small seems to come from the similarity of the positions of the other ring-current shifted high-field resonances in the different compounds (Figure 11), and from the smallness of the changes in the spectral region -0.5 to +1.0 ppm during the dissociation of cyanoferrocytochrome c (Figure 12). A more detailed description of these conformational changes will result from the extension of the analysis to the entire NMR spectrum of the polypeptide chain (Figure 1).

The unpaired electron distribution in the heme group of ferricytochrome c differs greatly from that found in cyanoporphyrin iron (III) complexes (21). Thus, as was found in other heme proteins (14,25,26) the interactions with the polypeptide chain greatly influence the electronic structure of the heme group. The unpaired electron distribution in the heme group of ferricytochrome c differs from that found for cyanoferrimyoglobin (14,25) and cyanoferrihemoglobin (26) in that it deviates even more strongly from the fourfold symmetry of the atomic structure of the porphyrin ring.

The proton resonances at -34.0 and -31.4 ppm (Figure 1) appear to come from two of the ring methyls of porphyrin c. Otherwise one would have that Q of Equation (2) would be much larger for the methyl groups in the B-positions of the 2,4 substituents than for at least two of the ring methyls which are bound directly to the porphyrin ring (Figure 1).

The two remaining ring methyl resonances are then much closer to -3.5 ppm, which is the resonance position of the ring methyls in diamagnetic porphyrins (15,16). This shows that there are large positive unpaired electron densities on ring carbon atoms of two pyrrole rings of porphyrin c, and small positive or negative spin densities on the two other pyrroles. This seems interesting in view of the biological role of cytochrome c as an electron transporting enzyme. Since X-ray studies have shown that only one edge of porphyrin c is exposed to the solvent, (8) a possible path for the electron transfer would seem to be through this edge. Negative or small positive electron density at one of the exposed pyrrole rings might contribute towards a favorable free energy for the electron uptake at this position in ferricytochrome c.

A comparison with cyanoferricytochrome c implies that the coordination of the sixth ligand greatly affects the unpaired electron distribution in the heme group (Figures 1 and 7).[+] It is then of interest that there are at most very small differences between the electronic structures of cytochromes c from different species. From Figure 13 it is seen that the

---

[+]The exchange of the sixth ligand during formation of cyanoferricytochrome c involves two processes, $i.e.$, binding of different atoms to the heme-iron, and conformational changes of the protein near the heme group. From the present experiments, the relative contributions to the changes in the electronic structure of the heme group from these two processes cannot be distinguished. It should be pointed out, however, that cyanide ion and sulfur behave very differently in their complexes with metal ions (31). Therefore, appreciably different unpaired electron distributions in the heme group would be expected to arise from binding of cyanide ion and methionyl. In particular, from a qualitative molecular orbital scheme for the binding with methionyl, a structure where the unpaired electron of the low spin ferric iron would be localized essentially entirely in a molecular orbital derived from either the $3d_{xz}$ or $3d_{yz}$ atomic orbitals of the iron would seem to be energetically most favorable. In this structure, the unpaired electron distribution in the heme group would have essentially C_2 symmetry, $i.e.$, large positive electron densities would be observed on two diagonally opposite pyrrole rings, and small spin densities on the other two pyrroles (Figure 1). This wou be in good agreement with the experimental observations.

482

unpaired electron distributions in the heme groups of ferricytochromes c from rabbit, guanaco, turkey, and horse are essentially identical. For Candida crusei a typical ferricytochrome c spectrum is observed, even though the positions of the resonances at very high and very low fields are slightly different. There is only one methionyl residue in an identical position in the primary structures of these five species, i.e., methionyl in position 80 (28). Hence the data of Figure 13 might be taken as additional support for the assignment of the sixth ligand to position 80 (4,8). The invariance of the electronic structure of the heme group in different ferricytochromes c appears to parallel the invariance of the redox properties of cytochromes c from different species reported during this colloquium by Dr. E. Margoliash.

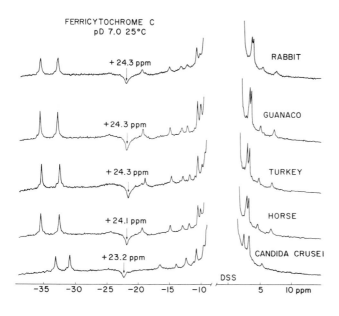

Figure 13. Proton NMR spectra in the regions −35 to −10 and +2 to +10 ppm of ferricytochrome c from different species.

Summary

High resolution proton nuclear magnetic resonance (NMR) spectroscopy was used to study oxidized and reduced cytochromes c, and the formation and dissociation of cytochrome c complexes with cyanide ion and azide ion. The NMR data provide strong evidence that the sixth ligand of the heme iron is a methionyl residue in both oxidized and reduced cytochrome c. It is further shown that in the formation of cyanoferricytochrome c cyanide ion replaces methionyl in the first coordination sphere of the heme iron. From the NMR spectra of ferricytochromes c from different species data on the electronic structure of the heme group are derived. As in other heme proteins, the protein environment greatly affects the unpaired electron distribution in the heme group of cytochromes c. In particular, from a comparison of ferricytochrome c and cyanoferricytochrome c, the importance of the coordination of the sixth ligand is apparent. On the other hand the unpaired electron distribution is essentially invariant for cytochromes c from different species, despite the differences in the primary structures of the polypeptide chains.

Acknowledgments

It is a pleasure to thank Dr. W. A. Eaton, Dr. J. B. Neilands, and Dr. A. Redfield for interesting discussions, Dr. E. Margoliash also for providing samples of cytochrome c, and Dr. R. G. Shulman for the hospitality extended to me in his laboratory.

References

1. Kowalsky, A., Biochem. 4, 2382 (1965).

2. McDonald, C. C., and W. D. Phillips, J. Am. Chem. Soc. 89, 6332 (1967).

3. McDonald, C. C., and W. D. Phillips, Proc. of the 3rd Int. Conf. on Magnetic Resonance in Biology, Warrenton, Virginia, October 1968.

4. Margoliash, E., in Hemes and Hemoproteins (B. Chance, R. W. Estabrook, and T. Yonetani, eds.) Academic Press, New York 1966, p. 371.

5. Margoliash, E. and A. Schejter, Adv. Protein Chem. 21, 113 (1966).

6. See references 14-29 of ref. 4.

7. Harbury, H. A., J. R. Cronin, M. W. Fanger, T. P. Hettinger, A. J. Murphy, Y. P. Myer, and S. N. Vinogradov, Proc. Nat. Acad. Sci. U. S. A. 54, 1658 (1965).

8. Dickerson, R. E., M. L. Kopka, J. Weinzierl, J. Varnum, D. Eisenberg, and E. Margoliash, J. Biol. Chem. 242, 3015 (1967).

9. Horecker, B. L., and A. Kornberg, J. Biol. Chem. 165, 11 (1946).

10. Horecker, B. L., and J. N. Stannard, J. Biol. Chem. 172, 589 (1948).

11. George, P., and A. Schejter, J. Biol. Chem. 239, 1504 (1964).

12. Neilands, J. B., and H. Tuppy, Biochim. Biophys. Acta 38, 351 (1960).

13. Kowalsky, A., J. Biol. Chem. 237, 1807 (1962).

14. Wüthrich, K., R. G. Shulman, and J. Peisach, Proc. Nat. Acad. Sci. U. S. A. 60, 373 (1968).

15. Becker, E. D., and R. B. Bradley, J. Chem. Phys. 31, 1413 (1959).

16. Caughey, W. S., and W. S. Koski, Biochem. 1, 923 (1962).

17. Johnson, C. E., and F. A. Bovey, J. Chem. Phys. 29, 1012 (1958).

18. Carrington, A., and A. D. McLachlan, in Introduction to Magnetic Resonance, Harper and Row, New York, 1967, pp. 81-85.

19. Bloembergen, N., J. Chem. Phys. 27, 595 (1957).

20. McConnell, H. M., and R. E. Robertson, J. Chem. Phys. 29, 1361 (1958).

21. Wüthrich, K., R. G. Shulman, B. J. Wyluda and W. S. Caughey, Proc. Nat. Acad. Sci. U. S. A. 62, 636 (1969).

22. McConnell, H. M., J. Chem. Phys. 24, 764 (1956).

23. Bersohn, R., J. Chem. Phys. 24, 1066 (1956).

24. Wüthrich, K., R. G. Shulman and T. Yamane, to be published.

25. Shulman, R. G., K. Wüthrich, T. Yamane, E. Antonini and M. Brunori, Proc. Nat. Acad. Sci., 63, 623 (1969).

26. Wüthrich, K., R. G. Shulman and T. Yamane, Proc. Nat. Acad. Sci. U. S. A. 61, 1199 (1968).

27. Dickerson, R. E., M. L. Kopka, J. E. Weinzierl, J. C. Varnum, D. Eisenberg and E. Margoliash, in Structure and Function of Cytochromes (K. Okunuki, M. D. Kamen, and I. Sekuzu, eds.) Univ. of Tokyo Press, Tokyo, 1968, p. 225-251.

28. Atlas of Protein Sequence and Structure 1967-1968 (M. O. Dayhoff and R. V. Eck, eds.) National Biomedical Research Foundation, Silver Springs, Md., 1968.

29. Kowalsky, A., Fed. Proc. 28, 603 (1969).

30. Redfield, A., private communication. See also Andrew, E. R. and R. Bersohn, J. Chem. Phys. 18, 159 (1950).

31. Cotton, F. A. and G. Wilkinson, Advanced Inorganic Chemistry, 2nd edition, J. Wiley and Sons, New York, 1966.

REVERSIBLE ALTERATIONS OF THE
TERTIARY STRUCTURE OF CYTOCHROME b_5[*]

Thomas E. Huntley, Philipp Strittmatter, and Juris Ozols

Department of Biochemistry, School of Medicine
University of Connecticut Health Center
Farmington, Connecticut

The primary structure of the heme protein isolated from bovine liver microsomes by pancreatic lipase extraction is shown in Figure 1. It contains a correction in the sequence previously reported (1,2), namely the addition of a missed, hydrophobic peptide containing the tryptophan and a fourth tyrosyl residue. This places the tryptophan residue in peptide T-12 rather than T-5, and results in a structure of 93 rather than 85 amino acids. The present experiments are concerned with a characterization of the tertiary structure which yields the precise heme binding as reflected in the spectral properties and enzymatic activity of this protein. The possibility of other interactions, covalent, ionic or hydrophobic, of this heme protein within the endoplasmic reticular structure is not questioned in this particular study.

The approach we used was to first simplify the protein by preparing the tryptic heme peptide (3). This removes peptides T-9, T-10, and T-11, and yields a heme peptide containing 82 amino acids, with spectral and catalytic properties identical to those of cytochrome b_5. The environment of the single tryptophan residue and the reactivity of the 4 tyrosyl residues were then examined under various conditions. Significantly, these five residues appear near the N-terminal sequence of the peptide. Concurrent heme binding studies have served to correlate alterations in this segment of the protein with the tertiary structure of the intact heme protein. This is of particular interest in view of rather indirect evidence, based upon a comparison of the primary structure of

*This investigation was supported by PHS GM-HE-15924 and research grants from the Connecticut Research Commission and the National Science Foundation.

cytochrome b to those of hemoglobins (4), suggesting that His-63 and H̄is-80, near the carboxyl terminus, may be involved in the heme iron coordination.

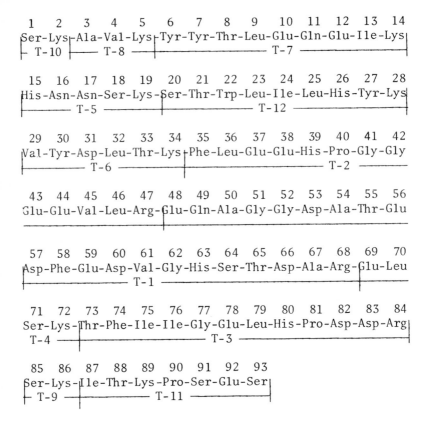

Figure 1. The amino acid sequence of bovine microsomal cyto-chrome $\underline{b}_5$. The tryptic peptides are shown.

Three of the four tyrosyl residues in apoprotein pre-parations can be detected spectrophotometrically by ionization at pH 13 or by iodination (Table I). The fourth tyrosine re-acts with iodine or ionizes at pH 13 only in the presence of urea. When heme is bound, only 3 tyrosyl residues are iodi-nated, and the heme protein spectrum is unchanged, from 360 to 600 nm.

TABLE I

Titration and Iodination of Tyrosyl Residues

Preparation	Urea M	Moles Tyrosyl/mole protein	
		Dissociated at pH 13	Diiodo derivative formed at pH 9.5
Apocytochrome $\underline{b}_5$	0	3.0	3.5
Apocytochrome $\underline{b}_5$	7.5	3.6	4.0
Apoheme peptide	0	3.0	-
Apoheme peptide	7.5	3.6	4.0
Cytochrome $\underline{b}_5$	0	-	3.0
Heme peptide	0	-	3.0
6 I-Apocytochrome $\underline{b}_5$ (diiodo-try)	7.5	0.7	4.0

The tryptic peptides, containing tyrosyl residues, can be separated by a simple procedure (Table II). Peptide T-12, the hydrophobic peptide which contains the tryptophan and one tyrosyl residue, is relatively insoluble. The tyrosyl residue in this peptide can only be ionized or iodinated in urea. It, therefore, contains the "buried" tyrosyl residue, and, moreover, retains, within this brief amino acid sequence, the structural requirements for an atypical tyrosine residue. The chromatographic procedure separates the other two tyrosyl peptides. On the basis of the stoichiometry and the net charge on the two peptides, low salt elution yields peptide T-6, and high salt, peptide T-7, the peptide with two tyrosyl residues.

The exposure of the heme peptide to limited amounts of iodine at pH 7.5 and 0°, results in the formation of partially iodinated, intact, heme protein derivatives. The iodine content of three preparations, which are electrophoretically distinguishable, is shown in Table III.

TABLE II

Separation of Tryptic Peptides of Apoheme Peptide

Fraction of tryptic hydrolysate	% of peptide		
	T-7	T-12*	T-6
Precipitate	15-20	65-75	15-20
Supernatant fluid	80-85	25-35	80-85
DEAE-A 50 chromatography			
0.01 M Tris-acetate pH 8.0 eluate	0	20-30	75-80
0.5 M NaCl eluate	75-80	0	0

*T-12 contains the tyr residue which dissociates only in urea.

Incorporation of two atoms of iodine results in two monoiodo-tyrosyl residues, and these are in peptide T-7. With slightly more iodine, peptide T-7 still contains all the iodine but now one monoiodo and one diiodo-tyrosyl residue as well. Peptide T-6 contains unmodified tyrosine in both preparations. Finally, excess reagent yields three diiodotyrosyl residues, and these are in peptides T-7 and T-6, since only the non-reactive tyrosyl residue in T-12 remains.

Alterations in the structure of the heme peptide derivatives can be detected in the circular dichroic spectra (CD).

TABLE III

Iodine Distribution in Heme Peptide Derivatives

Preparation	Apoheme peptide				Tyr content	
	Diiodo Tyr	Monoiodo Tyr	Tyr		T-7	T-6
			-Urea	+Urea		
2 I-Heme peptide	0.2	1.6	1.2	1.8	0.2	1.0
3 I-Heme peptide	1.0	1.0	1.0	1.6	0	1.0
6 I-Heme peptide	3.0	0	0	0.6	0	0

```
                  6    7
T-7 Tyr-Tyr-Thr-Leu-Glu-Gln-Glu-Ile-Lys
                 30
T-6 Val-Tyr-Asp-Leu-Thr-Lys        27
T-12 Ser-Thr-Trp-Leu-Ile-Leu-His-Tyr-Lys
```

Figure 2 shows the CD spectrum of heme peptide from 240 to 600 nm in which both the heme and aromatic amino acid bands occur.

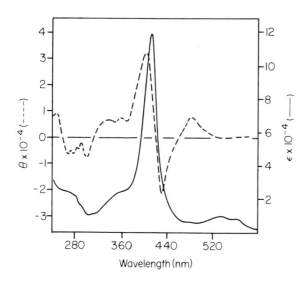

Figure 2. CD and absorption spectra of heme peptide.

The effects of iodination (Figure 3) are largely restricted to the area of the aromatic amino acid residues.

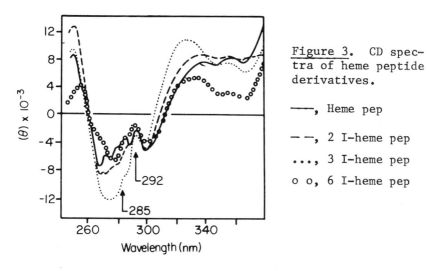

Figure 3. CD spectra of heme peptide derivatives.

———, Heme pep

— —, 2 I-heme pep

..., 3 I-heme pep

o o, 6 I-heme pep

The 292 nm band, presumably arising from the asymmetric environment of the single tryptophan residue, is retained in all of the iodinated heme proteins. The 2 I-heme peptide spectrum is very similar to the unmodified protein, but further iodination results in changes in the region of tyrosyl absorption at 260 and 290 nm, and in the case of the 3 I-derivative an apparent contribution from the 312 nm transition of the one diiodotyrosyl residue. These latter effects are seen more clearly with apoprotein solutions, where heme contributions to the CD spectra are eliminated (Figure 4). The diiodotyrosyl residue in the 3 I-derivative is then the only absorbing species above 300 nm.

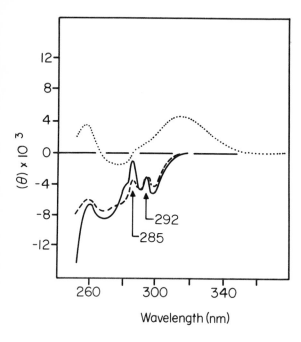

Figure 4. CD spectra of apoprotein derivatives.

——, apocyt. $\underline{b}_5$

— —, 2 I-apoheme peptide

•••, 3 I-apoheme peptide

Progressive iodination of the tyrosyl residues also alters the stability of the heme peptide in 1-propanol. Figure 5 shows that, at 25°, the dissociation of heme from the 2 I-derivative occurs at the same alcohol concentration as the heme peptide. Further iodine substitutions, however, increase the sensitivity of the protein to propanol. In each case, the dissociation, and therefore the alteration of the tertiary structure of the peptide chain, is reversible by dilution. In aqueous buffers, the fluorescence of the tryptophan residue

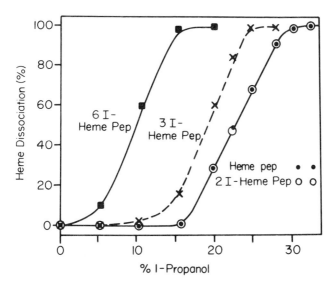

<u>Figure 5</u>. Heme peptide derivative dissociation in 1-propanol.

is completely quenched, but heme dissociation is accompanied by the appearance of fluorescence. Table IV compares the tryptophan fluorescence yields for the heme protein derivatives when heme dissociation is complete. The maximum fluorescence yield of the heme peptide actually slightly exceeds that of free tryptophan under the same conditions.

TABLE IV

Tryptophan Fluorescence of Heme Peptide Derivatives

Preparation	Relative Fluorescence	
	Tris-acetate pH 8.0	30% 1-propanol
Cytochrome b_5	0	100
Heme peptide	0	100
2 I-Heme peptide	0	75
3 I-Heme peptide	0	36
6 I-Heme peptide	0	9

The fluorescence changes permitted an examination of
the solvent effects on apoprotein unfolding. The apoprotein
does show some fluorescence in Tris buffer, pH 8.0, but
unfolding of the protein can be followed in each case by
the increase in fluorescence as shown in Figure 6. A comparison of these results with those obtained with the heme proteins (Figure 5) shows that apoprotein unfolding begins at
a lower alcohol concentration in each case, and that both
the 2 and 3 I-derivatives are less stable than the unmodified
peptide. The unfolding of proteins suggested by both the
heme dissociation and fluorescence data, is reinforced by
CD measurements. The circular dichroism of the apoprotein
preparations in the near ultraviolet region (Figure 3) is
not observed when these proteins are in 30% propanol solutions.

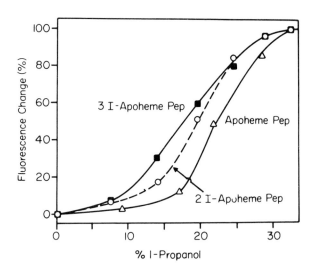

Figure 6. Apoheme peptide derivative fluorescence in 1-
propanol. Excitation, 290 nm; emission, 350 nm.

These experiments identify Tyr-6 and Tyr-7 in the
tryptic peptide T-7 as the most reactive tryosyl residues
in cytochrome b_5. These aromatic rings may contain single
iodine substitutions with little apparent alteration in
tertiary structure or stability. Conversion of one of

494

these residues to the diiodo-derivative results in less
stable structure. Moreover, formation of diiodotyrosine
at residues 6, 7, and 30 produces an extremely unstable heme
peptide. In view of the large ionic radius of iodine, the
substitution of 2 iodine atoms in a single tyrosine residue
may force considerable steric distortion. The tyrosine in
T-12 (Tyr-27) is remarkably unreactive, and iodination of
this aromatic ring requires denaturation by urea.

Two consequences of these experiments deserve further
comment. The 2 I-derivative results in the least apparent
distortion of the heme protein structure, and may therefore
serve to solve the difficult problem of obtaining an isomorphous derivative for a solution of the X-ray crystallographic data on cytochrome b_5 that has been obtained by Dr.
Scott Mathews. Previous efforts have utilized a derivative
we now know to have been largely the 3 I-cytochrome b_5, a
preparation containing diiodotyrosine that does not crystal-
lize in an isomorphous form. Finally, it should be emphasized
that this peptide contains only one tryptophan which provides
an excellent, constitutive "reporter" group for alterations
in the tertiary structure of the protein as illustrated in
the modifications of tyrosyl residues. This approach can
obviously be extended to studies on the roles of other amino
acid residues in the tertiary structure of cytochrome b_5.

References

1. Ozols, J. and P. Strittmatter. J. Biol. Chem., 243,
   3367 (1968).

2. Ozols, J. and P. Strittmatter. J. Biol. Chem., 243,
   3376 (1968).

3. Strittmatter, P. and J. Ozols. J. Biol. Chem., 241,
   4287 (1966).

4. Ozols, J. and P. Strittmatter. Proc. Nat. Acad. Sci.
   U.S.A., 58, 264 (1967).

OPTICAL ROTATORY DISPERSION OF CYTOCHROME c' FROM RHODOPSEUDOMONAS PALUSTRIS‡

F.C. Yong and Tsoo E. King

Department of Chemistry, State University of New York
Albany, New York

In the continuing effort to characterize RHP cytochromes by various physicochemical methods, we have studied the optical rotatory dispersion (ORD)[*] behavior of a monoheme and several diheme cytochromes[**] from photosynthetic bacteria. In this communication, we report the ORD spectra of cytochrome c' from Rhodopseudomonas palustris. The cytochrome, isolated by methods described elsewhere (2), has a covalently-bound mesoheme. Like all cc' cytochromes it exhibits a double Soret peak, at 435 and 426 nm (1,3). The amino acid composition of the cytochrome has been reported (3).

The heme concentration was determined from the pyridine hemochrome of the cytochrome using E_{mM} = 29 at 550 nm (1), and 14,000 was taken as the molecular weight. ORD measurements were performed in a Cary spectropolarimeter, model 60, programmed and calibrated by procedures previously described (4). The observed rotations were expressed as molar rotations in the unit of degrees per mole of heme per liter per decimeter without refractive index correction. Computer analysis of the ORD spectra were made by the method described previously (5-7).

Rps. palustris cytochrome c' at pH 9.5 exhibited large distinct Cotton effects in the Soret and the aromatic regions and small anomalous rotations at wave length above 500 nm. The ORD behavior in the oxidised, reduced and CO-reduced states followed a general pattern also observed in the other cc' cytochromes studied (8). The experimental ORD and the resolved Cotton effects are presented in Figures 1, 2 and 3 for the oxidised, reduced and CO-reduced forms,

[*]Abbreviations: ORD, optical rotatory dispersion; CD, circular dichroism; D.M., Debye magneton.
‡Supported by grants from NSF and USPHS.

respectively. Table I summarizes some parameters of the
Cotton effects. Figure 4 shows the computed circular dichroic
curves from fitted ORD spectra for the cytochrome.

Oxidised. The ORD for the oxidised form was resolved
into a set of four Cotton effects and the protein backbone
rotations. These resolved Cotton effects are depicted in
Figure 1 and detailed in Table I for their respective band
center, rotational strength, band width, and amplitude.

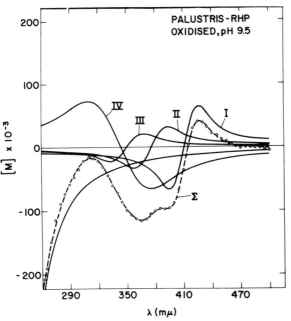

Figure 1. ORD of
Rps. palustris
cytochrome c' in
the oxidised form.
o—o—o, experimental
points; Σ, sum of
computed Cotton
effects; I, II,
III and IV, the
computed Cotton
effects. 0.05
M Tris, pH 9.5;
temp., 23°.

Both Cotton effect I, at 410.8 nm and Cotton effect II at
376.2 nm, were within the envelope of the Soret absorption
spectrum which showed a maximum at 405 nm and a shoulder at
about 370 nm (1). Cotton effect I had rotational strength
of 0.60 D.M. and Cotton effect II, 0.36 D.M. The computed
CD spectrum (cf. Figure 4) likewise demonstrated these characteristics. Cotton effects III at 350.6 nm and IV at 343.1
nm were evidently associated with transitions in the aromatic
region. The rotational strength for Cotton effect III decreased significantly at neutral pH and this transition was
not observed for any of the diheme proteins studied. Cotton
effect IV possessed large rotational strength (-1.58 D.M.)
and broad band width (35.9 nm). It was apparently a composite
of overlapping transitions.

TABLE I

Resolved Cotton Effects of Palustris
Cytochrome c' at pH 9.5[*]

	Oxidised	Reduced	CO-reduced
Cotton Effect I			
Center (nm)	410.8	428.5	416.8
Rot. str. (D.M)	0.60	0.53	0.16
Band width (nm)	16.9	18.5	3.8
Amplitude ($\times 10^{-3}$)	132.4	112.6	151.2
Cotton Effect II			
Center (nm)	376.2		414.8
Rot. str. (D.M.)	0.36		0.29
Band width (nm)	19.1		10.2
Amplitude ($\times 10^{-3}$)	64.4		106.5
Cotton Effect III			
Center (nm)	350.6	360.3	371.6
Rot. str. (D.M)	0.26	-0.11	-0.03
Band width (nm)	19.0	18.9	9.1
Amplitude ($\times 10^{-3}$)	42.0	18.7	11.1
Cotton Effect IV			
Center (nm)	343.1	307.2	327.2
Rot. str. (D.M.)	-1.58	-1.57	-1.06
Band width (nm)	35.9	39.2	28.8
Amplitude ($\times 10^{-3}$)	138.2	112.6	110.6
Abs. Max. (nm)	405,∿370	435, 426	417,∿396

[*]Rot. str. (D.M.), rotational strength in Debye magneton.

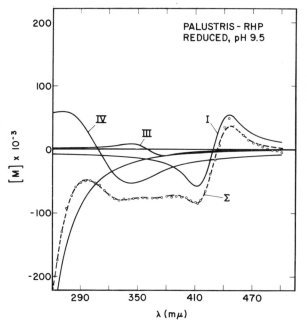

PALUSTRIS - RHP
REDUCED, pH 9.5

Figure 2. ORD of
Rps. palustris
cytochrome c' in
the reduced form.
Conditions same
as in Figure 1.

Reduced. Reduction of the cytochrome shifted the anomalous dispersion in the Soret region to longer wave lengths. The resolved Cotton effects, as given in Table I, likewise demonstrated a red shift. As shown in Figure 2, the ORD for the reduced cytochrome at pH 9.5 was resolved into three Cotton effects plus the protein backbone rotation. Cotton effect I, centered at 428.5 nm, had rotational strength of 0.53 D.M. and band width of 18.5 nm. In contrast to the oxidised form, a second Cotton effect was not detected in this region despite the fact that its absorption spectrum show double peaks, at 426 and 435 nm (1,3).

The electronic transition giving rise to the 426 nm absorption maximum is undoubtedly also responsible for the generation of Cotton effect I, although one cannot rule out minor contributions from the transition occurring at 435 nm. Inspection of the Soret absorption spectrum suggests that both the absorption peaks should be relatively sharp and have approximately similar oscillatory strength. This would imply that the transition occurring at 435 nm is, in all probability, a magnetically forbidden transition, or at most, only weakly allowed.

The rotational strength for Cotton effect I was significantly smaller than that observed for the oxidised (sum of
I and II). Upon reduction, Cotton effect III was shifted to
360.3 nm and the rotational strength became -0.11 D.M. Cotton effect IV at 307.2 nm possessed a large rotational
strength and broad band width. It was also evidently composed of multiple transitions. These results indicate that
reduction causes significant changes in the asymmetric perturbations of the heme chromophore.

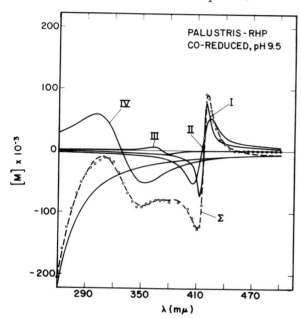

Figure 3. ORD
of Rps. palustris
cytochrome c' in
the CO-reacted
reduced form.
Conditions same
as in Figure 1.

CO-reduced. Reaction of the reduced cytochrome with CO
resulted in general sharpening of the Soret Cotton effects.
The experimental ORD as shown in Figure 3 was resolved into
four Cotton effects and the backbone rotation. Details of
the resolved Cotton effects are given in Table I. Cotton
effect I (416.8 nm) and Cotton effect II (414.8 nm) were located close together and almost coincided with the Soret
peak (415 nm). The band width for Cotton effect I was narrow
(3.8 nm) as compared to that of Cotton effect II (10.2 nm);
a characteristic feature likewise observed for other CO-
reacted cc' cytochromes (8). Cotton effect III at 371.6
nm showed a rotational strength of -0.03 D.M. These
three Cotton effects appeared to be situated within the
doubly degenerated Soret absorption band. Indeed, a summation

of their corresponding CD functions in the positive mode approximated the Soret absorption spectrum with maximum at 415 nm and a shoulder at about 390 nm. As observed in the oxidised and reduced forms, Cotton effect IV was apparently a composite of overlapping transitions.

A'parameters. The Schecter-Blout parameters (A'_{193} and A'_{225}) evaluated from the computer analysis in the manner described (6,9) reflect the effective backbone rotation of the protein molecule. The rotation encompasses, in general, contributions from the α-helix and other structures in the molecule, heme-protein interactions as well as heme transitions, if any, occurring in this region. The A' values at both wave lengths (193 nm and 225 nm) were significantly larger for the reduced (from 2178-1521) than either the oxidised (1370, -1104) or the CO-reduced (1695, -1228). The increase in the A' values together with the observed changes in the Soret transitions are indicative of conformational changes in the molecule upon reduction. Comparable results were observed for the cytochrome at neutral pH.

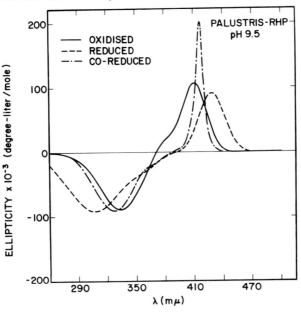

Figure 4. Computed CD spectra of Rps. palustris cytochrome c' in the oxidised, reduced and CO-reduced forms. ————, oxidised; – – – –, reduced and –.–·–·, CO-reduced.

ORD is a sensitive tool in probing conformational changes of protein in solution, particularly of cytochromes containing the highly conjugated heme chromophores. The results presented show that, in general, the anomalous rotations in the Soret region are resolved into positive Cotton effect(s), while those in the aromatic region show a negative Cotton effect of broad band width and large rotational strength. Very similar ORD behavior has been observed for the palustris cytochrome at pH 7.5. These characteristics appear to conform to a general pattern also observed in the ORD behavior of other cc' cytochromes (8).

Changes in the oxidation states and ligands in the cytochrome definitively alter the electronic configuration or geometry of the heme chromophore. The concomitant increase in the backbone rotation upon reduction of the molecule is likewise indicative of an altered conformation.

For Chromatium cytochrome cc' and Rubrum cytochrome cc', there is no indication of heme-heme interactions, despite the fact that they are diheme proteins. This is in distinct contrast to Chromatium cytochrome c552, which shows strong heme-heme interactions, as first reported by Bartsch et al. (10) and then by our own results (8).

Acknowledgment

We thank Prof. M.D. Kamen, University of California at San Diego, for the cytochrome samples.

References

1. Bartsch, R.G. in Bacterial Photosynthesis, (H. Gest, A. San Pietro, and L.0. Vernon, eds.) Antioch Press, Yellow Springs, 1963, p. 475.

2. Kamen, M.D., R.G. Bartsch, T. Horio, and H. De Klerk. in Methods in Enzymology, (S. Colowick and N.0. Kaplan, eds.), 6, Academic Press, New York, 1963, p. 391.

3. Dus, K., H. De Klerk, R.G. Bartsch, T. Horio, and M.D. Kamen. Proc. Natl. Acad. Sci. U.S., 57, 367 (1967).

4. Yong, F.C. and T.E. King. J. Biol. Chem., 244, 509 (1969).

5. Schellman, J.A. and T.E. King. in Hemes and Hemoproteins, (B. Chance, R. Estabrook, and T. Yonetani, eds.), Academic Press, New York, 1967, p. 507.

6. King, T.E., F.C. Yong, and P.M. Bayley. in Structure and Function of Cytochromes, (K. Okunuki, M.D. Kamen, and I. Sekuzu, eds.), University of Tokyo Press, Tokyo, 1968, p. 204.

7. King, T.E., P.M. Bayley, and B. Mackler. J. Biol. Chem., 244, 1890 (1969).

8. Yong, F.C. and T.E. King. J. Biol. Chem., 245, 1331, 2457 (1970).

9. Schecter, E. and E.R. Blout, Proc. Natl. Acad. Sci. U.S., 51, 695 (1964).

10. Bartsch, R.G., T.E. Myer, and A.B. Robinson. in Structure and Function of Cytochromes, (K. Okunuki, M.D. Kamen, and I. Sekuzu, eds.), University of Tokyo Press, Tokyo, 1968, p. 443.

X-RAY STUDY OF CALF LIVER CYTOCHROME b_5*

F. Scott Mathews, Michael Levine and Patrick Argos

Department of Physiology and Biophysics
Washington University School of Medicine
St. Louis, Missouri 63110

Cytochrome b_5 from calf liver microsomes has a molecular weight of 11,000 and contains 93 amino acid residues of known sequence (1). The three dimensional structure of this cytochrome is being investigated by X-ray diffraction methods to determine its mode of action and its relationship to other hemoproteins such as myoglobin and cytochrome c. This report describes progress made subsequent to the preliminary communication (2).

Cytochrome b_5 crystals were prepared from electrophoretically homogeneous protein by Dr. Philipp Strittmatter by adding a 4-5 fold excess of 4M phosphate buffer, pH 7.5, to 0.1 ml samples of a 1% protein solution. A uniform preparation of crystals is achieved by seeding. Crystals grown from buffered ammonium sulfate are badly split, but single crystals grown in phosphate can be transferred to ammonium sulfate without damage.

The crystals, which float in the mother liquor, are orthorhombic and form dark red prisms about 0.2 mm in size. Unit cell dimensions are a = 64.55 Å, b = 46.01 Å and c = 29.89 Å. The space group is $P2_12_12_1$. The volume of the unit cell, 88,800 Å , and the approximate crystal density, 1.28 g/cc, indicate one molecule in the asymmetric unit. Å 25 degree (1.9 Å resolution) precession photograph (2) of the hk0 zone shows strong reflections at high scattering angles, indicating that the protein is suitable for a high resolution structural investigation.

*This work is supported by USPHS Grant No. GM13925, NSF Grant No. GB5899 and Life Insurance Medical Research Fund No. G68-41. One of us (P. A.) is supported by a USPHS Postdoctoral fellowship.

We are attempting to obtain suitable derivatives of cytochrome b_5 for the isomorphous replacement method of phase determination. Unfortunately the protein contains no cysteine residues; therefore, sulfhydryl reagents containing heavy atoms will not be expected necessarily to bind to the protein, and empirical methods must be used. Thirty-five heavy atom reagents have so far been surveyed by soaking a few test crystals in a solution of reagent, usually 10^{-1} M, in 3.6 M phosphate buffer, pH 7.3. In some case the material is only sparingly soluble and solutions saturated with the heavy atom reagent have been used. In each test an 8 degree precession photograph using CuKα radiation was taken of the hk0 zone and compared visually with a corresponding photograph of the native crystal. We rejected possible derivatives if no intensity changes occurred or if changes were limited to a few low order reflections which probably are sensitive to the salt concentration or nonspecific binding of the reagent.

TABLE I

Possible Heavy Atom Derivatives of Cytochrome b_5

Mersalyl (sodium salyrganate)	Large differences; 5.5 Å hk0, h0ℓ, 0kℓ DPP's.*
Sodium mercuriated malate[1]	Large differences; 5.5 Å hk0, h0ℓ DPP's.
K_2HgI_4	Large differences; crystals sink; 5.5 Å hk0 DPP's.
Tris (5 nitrosalicyl-aldehyde) europium	Large differences; 5.5 Å and 2.8Å hk0 DPP's.
Dimercuriacetate[2]	Large differences; saturated solution; 5.5 Å and 2.8 Å hk0 DPP's.
bis(diphenylithio-carbazono) mercury	Large differences; saturated solution; 5.5 Å hk0 DPP.
0-chloro mercury phenol	Large differences; crystals sink; saturated solution; 5.5 Å hk0 DPP.
Methyl mercuric chloride	Crystals decompose; saturated solution.
Phenyl mercuric nitrate	Small differences; saturated solution.
Di p-tolyl mercury	Small differences; saturated solution.

*DPP, difference Patterson projection
[1]Gift of Dr. Leonard Banaszak   [2]Gift of Dr. Hunter Mermall

Our search has produced 10 potential derivatives worthy of further study which are summarized in Table I. A few of these cause the crystals to sink or otherwise break up, indicating that a specific interaction with the protein has taken place.

Difference Patterson projections have been calculated at 5.5 Å resolution for 7 of the 10 potential derivatives and at 2.8 Å resolution for 2 of the derivatives (cf Fig. 1). The data were recorded on precession camera films and measured on a Joyce-Loebl microdensitometer. A LINC computer was used to control the densitometer and process the data in the high resolution studies. Even though none of the maps indicates a single binding site, some have features in common. Progress on interpreting the maps is hampered by the high noise level

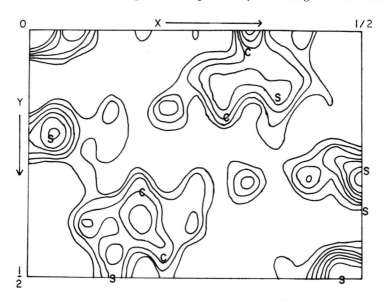

Figure 1. DMA difference Patterson at 2.8 Å resolution projected down the c axis. The coefficients used were $(kF_D - F_N)$ exp $(-2B \sin^2\theta/\lambda^2)$ where F_D and F_N are the scaled derivative and native structure factors, k was set to 1.1 to account for mercury and B was set to 25 to reduce the effects of series termination and non-isomorphism. Arbitrary contours at equal intervals are shown. The letters S and C indicate the positions of the self and cross vectors calculated from the two site solution described in the text.

and difficulties in reproducing the results. The latter
problem may arise from chemical degradation of the heavy atom
compounds or the protein crystals or both.

The dimercury acetate (DMA) derivative is an example
of our progress to date. Figure 1 shows a 2.8 Å difference
Patterson projection of DMA. The map has been artificially
smoothed, to reduce the effects of errors on the high order
terms, and agrees with an earlier 5.5 Å difference map. The
Harker peaks indicate a major site at x = (1/64)a, y = (5/48)b
in the plane group Pgg. A second site at x = (12/64)a, y =
(20/48)b is suggested by the difference Fourier synthesis,
shown in Figure 2, using signs calculated from the first site.

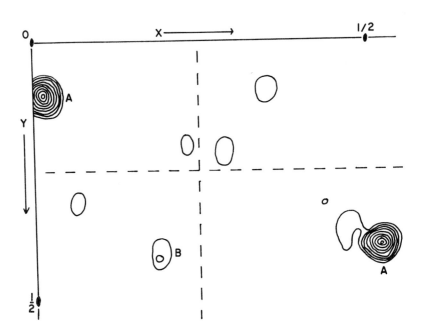

Figure 2. DMA difference Fourier projection along the c axis
based on the signs calculated from the site at peak A. About
25% of the terms, to which the contribution of site A was small
were omitted. The plane group is Pgg. Peak B is interpreted
as a second site. Arbitrary contours at equal intervals are
shown.

The vector set calculated from the two sites is indicated in Figure 1. The agreement between the calculated and observed structure factor differences is poor but may improve by least squares refinement. Final verification of the 2 site DMA solution must await a three dimensional study or another independent derivative.

The reduction of the heme iron from the ferric to the ferrous form has been studied in the crystalline state using the flow cell method (3). Visual comparison of 12 degree precession photographs of the h0ℓ zone before and after reduction with sodium dithionite showed no intensity changes despite a distinct color change in the crystal. Any conformational changes of cytochromes b_5 on reduction must be very small.

Calf liver cytochrome b_5 contains four tyrosine residues, three of which can be iodinated without affecting its interaction with heme (1). Attempts to iodinate crystalline cytochrome b_5 with KI_3 led to breaking up of the crystals. Limited iodination in solution produces a mixture of products containing 2, 3 and 6 moles of iodine per mole of protein (1). The intermediate 3:1 species containing a monoiodo- and a diiodotyrosyl residue at positions 6 and 7 was isolated and crystallized by Dr. Strittmatter. The crystals were much smaller and more elongated than the native but were sufficiently developed to produce 8 degree precession photographs of the hk0 and h0ℓ zones. The cell symmetry is now monoclinic, space group $P2_1$ with c the unique axis. The cell dimensions are a = 63.92 Å, b = 45.83 Å, c = 32.93 Å and β-90.48°. The a and b axes are very similar in length to the native orthorhombic crystals while the c axis is about 10% longer. The unit cell volume, 96,500 Å, (3) suggests two molecules per asymmetric unit, assuming no great change in density. The similarities between the native and iodinated crystal lattices suggest a simple relationship between the packing of molecules in the two cases despite the change in space group symmetry. Application of the Patterson rotation function (4) to the iodinated protein X-ray data might define the local symmetry elements within the asymmetric unit and establish its relationship to the native protein crystal.

Three dimensional low resolution studies of the derivatives under various soaking conditions are now in progress using the flow cell method (3).

509

References

1. Huntley, T. E., P. Strittmatter and J. Ozols, this volume, p. 491

2. Mathews, F. S. and P. Strittmatter, J. Mol. Biol. 41, 295 (1969).

3. Wycoff, H. W., M. Doscher, D. Tsernoglou, T. Inagami, L. N. Johnson, K. D. Hardman, N. M. Allewell, D. M. Kelly and F. M. Richards, J. Mol. Biol. 27, 563 (1967).

4. Rossmann, M. C. and D. W. Blow, Acta Cryst. 15, 24 (1962).

DISCUSSION

Mason: The results of Lucille Smith on the specificity of beef heart particles towards cytochrome c of various species are extraordinarily interesting from a comparative point of view. It seems reasonably evident that some selective evolutionary pressures must have been exerted on the structure of cytochrome c, otherwise, why do the differences exist? Now the question is, what can these have been? May I suggest the following possibilities:

1. Cytochrome c structure may influence the rate of respiration at the oxygen tensions existing in mitochondria _in vivo_, rather than at the high oxygen tensions of the assay system.

2. The structure of cytochrome c may be adapted to specific reduction by the b-c_1 system.

3. The structure of cytochrome c may be adapted to site-specific binding by "structural protein" or some other binding system.

Perham: I should just like to remind people that the nasty paradox to which Dr. Margolaish referred with respect to cytochrome c is found in other systems also. Thus, we know that _in vitro_ the turnover numbers of the mammalian glyceraldehyde-3-phosphate dehydrogenases vary little, yet the primary structures show small but real differences. Again, the question arises as to what stabilizes these alterations during evolution. One possibility is that it is due to necessity to interact with other molecules in which there are complementary structural changes imposed for other reasons. However, in this as in other explanations there exists an element of special pleading, and I would not wish to press it at this stage.

Nicholls: I think the simplest representation of the conformation change of cytochrome c at alkaline pH, a change which shows analogies with the cyanide induced conformational change and other conformation changes, is as in Scheme A:

Alkaline Forms of Cytochrome $\underline{c}$

"cyt. $\underline{c}$" indicates the ferric form that predominates at neutral pH. "(cyt. $\underline{c}$)'" indicates the altered conformational state that predominates above pH 8.5.

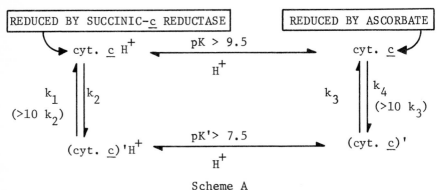

Scheme A

There are two conformational states and one ionizing group, with a pK in the normal form above 9.5. Dr. Schejter gave some particular values for these constants. The pK in the other form (pK' in Scheme A) has to be below 7.5. The ratio between these two hydrogen ion concentrations, which is at least 100, must equal the ratio of the other two equilibrium constants, i.e. antilog (pK-pK') = (k_1/k_2) x (k_4/k_3), so that under acid conditions the rate of formation of the normal conformer must be at least 10 times the rate of formation of the other conformer and vice versa under alkaline conditions. That is, k_4 must be greater than 10 times k_3. "cyt. $\underline{c}$ H$^+$" is now the form which is enzymatically reactive. For example, it is reducible by the succinate cytochrome c reductase complex. If these are the only changes which we allow, then "cyt. $\underline{c}$" (Scheme A) must be the ascorbate-reducible component, because the rate of $\underline{c}$ reduction by ascorbate increases with increasing pH despite the fact that the proportion of ascorbate-reducible material simultaneously goes down. This phenomenon has been documented by Greenwood and Palmer. "(Cyt. $\underline{c}$)'" is unreactive both with enzymes and with ascorbate. Two questions may now be posed: Firstly, if methionine-80 supplies the sixth ligand and yet is so easily disturbed that we get all these inactive forms of cytochrome $\underline{c}$, with no apparent functions, why has it been preserved so tenaciously throughout the evolutionary time scale? This is a problem for me. The

second question may be looking forward to tomorrow's session.
If the reduction route to the iron requires this methionine,
if there's a single electron transfer route involved, how is
it that cytochrome c attached to the mitochondrial membrane
can act both as an acceptor of electrons and as a donator of
electrons for two different kinds of enzyme, the reductase
and the oxidase?

Kowalsky: We have been interested in the mechanism of electron transfer into cytochrome c. Some years ago, I carried
out an experiment where I reduced cytochrome c with chromium.
This has the result of giving a cytochrome c which has chromium covalently bound to it. For various reasons, which I
won't go into, there is also phosphate on the compound.
This is a 1:1:1 complex of cytochrome c, chromium and phosphate. As it turns out, the results of the experiment indicate that there is more than one pathway of electron transfer
into cytochrome c. But this does represent one possible route
and the problem then becomes one of determining where the
chromium is on the cytochrome c. We have tried looking at
the NMR spectrum of this chromium-c complex and through the
courtesy of Varian, we observed the spectrum at 220 MHz.
Since the electron eventually winds up in the heme, one would
expect that the chromium would be some place close to the heme
group and therefore we concentrated on the resonances which
are associated with the heme. One does find changes in the
NMR spectra of cytochrome c which has been reduced with chromium. particularly the low field contact resonances seem to
be shifted slightly upfield by about two parts per million.
The widths are unchanged. The odd result is that the upfield
contact resonance which Dr. Wüthrich has so nicely shown
results from the methionine is affected by the chromium. This
methionine is buried in the protein. I find this result very
hard to explain, but the resonance is very markedly altered
by the presence of the chromium. Since the methionine did
seem to be affected by the chromium and since there are
other sulfur linkages in the protein, we looked for an interaction between model thioester compounds and chromous ion.
Although we have shown by continuous wave NMR that acetyl
methionine complexes through its sulfur with heme or with
the heme octapeptide we have not been able to find any interaction of the sulfur of methionine with chromous ion.

Wüthrich: Was the methionine resonance split?

513

Kowalsky: It is very difficult to say because it becomes very broad, and I don't know what happens to it. It is much broader than it was before.

Peisach: I would like to report the EPR data for cytochrome c and some of its derivatives which may simplify or actually may make the situation more complicated. Examine cytochrome c under various conditions and no changes in EPR spectra. So, for example, cytochrome c cyanide gives the same EPR as hemoglobin cyanide, indicating that the findings have been removed and cyanide has replaced it. Above pH 10, you can obtain a different type of low spin EPR spectrum attributable to hydroxide cytochrome c at the same EPR spectrum attributable to hydroxide cytochrome c at the same EPR spectrum as hydroxide hemoglobin. Below pH 4, cytochrome c changes from a low spin form to a high spin form. EPR taken in the range pH 2 to 4 indicate typical high spin EPR spectra except with a splitting around g = 6 or a spectrum indicating some sort of rhombic distortion very much like in the alpha chain that I mentioned before. At pH around 1 the EPR spectrum goes to axial, which means that the protein no longer effects the symmetry of the heme. Each of these steps is completely reversible, namely by bringing the protein down to pH 1 going to the rhombic axial spectrum, raising the pH again, of course, this returns back to normal cytochrome c EPR spectrum.

Czerlinski: Figure 1 shows a scheme which I presented originally two years ago and can be well utilized for discussing some of my own and other data. The scheme has been thoroughly discussed before. In the left bottom corner, reduced cytochrome c is shown. In our experiments, this compound was oxidized with ferrihexacyanide at various pH-values between 7 and 11, which allows one to follow directly the proton release upon production of oxidized cytochrome c. One could simultaneously observe the disappearance of the 695 mμ band, corresponding to the conversion time from 2 to 3 in the upper part of the scheme. pH-switching experiments were also conducted. The results presented by Dr. Schejter are somewhat different from our earlier measurements, which might be due, in part, to a difference in the method. If one compares the original kinetic data with those from chicken heart cytochrome c, one finds the conversion from 2 to 3 considerably slower in this latter case. Apparently, this indicates a tighter structure of the oxidized cytochrome

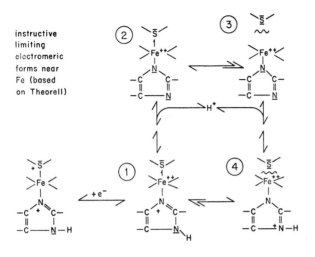

instructive
limiting
electromeric
forms near
Fe (based
on Theorell)

Figure 1.

c from chicken heart compared to that from horse heart.
The scheme in the figure also indicates that the conversion
from 2 to 3 is associated with an increase in the distance
between methionine and the iron; the question is then: what
does show up at the sixth ligand position of Fe around pH 10,
where component 3 is mainly present at equilibrium? Possibly
a water molecule appears, but one should then be able to detect it in NMR investigations. The scheme also extends somewhat the scheme of Micholls. With respect to ascorbate, I
feel, this reducing agent reacts with compound 1 in the form
of the di-enolate anion, producing the proper pH dependence.

Moss: I'm glad Dr. Yong mentioned cytochrome c-c' because
Drs. Michael Cusanovich, Anders Ehrenberg, and I have some
data on it which we would like to discuss. These are again
simple suspicion-confirming experiments, the suspicion being
that just as the cytochrome c-c' class is chemically neither
fish nor fowl, it is also magnetically ambiguous. The
magnetic experiments are being done in the long run in order
to find out if there are changes in iron electronic configuration which correspond to the changes in CO flash kinetics with

515

pH which Drs. Cusanovich and Gibson measure. There are, however, some early implications of the data to report immediately.

Figure 2 illustrates the susceptibility of <u>Chromatium</u> cytochrome <u>c-c</u>'. For pure axial high-spin heme proteins, with the coefficient D of the S_z splitting arbitrarily large, the moment per iron atom cannot fall below 19 = 4.36 Bohr magnetons. Both <u>Chromatium</u> and <u>R. rubrum</u> (3.4 and 3.8 Bohr magnetons, respectively) fall considerably below this.

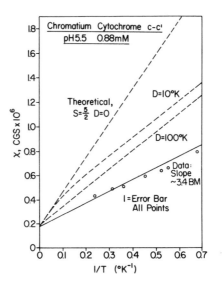

<u>Figure 2.</u> Susceptibility of Chromatium cytochrome <u>c-c</u>'. The dotted curve for D = 100°K represents the limit of arbitrarily large D for this temperature range, where D is the coefficient of S_z in the spin Hamiltonian.

There can't be any question of thermal mixtures of spin
states in interpreting this data because we are working as
low at 1.6° K. There is a temperature difference optical
spectrum at 630 mμ between 273 and 303° K, indicating a
possible thermal mixture of spin states in this temperature
range, but the molecule would be entirely in the ground
state at 4° K and below.

In the EPR spectra, the most obvious points are: first,
that the signal is highly temperature dependent. At 77° K,
the intensity of the derivative signal is very low, and it
increases by more than an order of magnitude as one goes down
to 30° K. Unlike other high spin iron proteins, the iron
spin appears in this case to be very strongly coupled to
lattice vibrations. The second point to note is that
there is a distinct splitting of the EPR g_1 signal at low
field (Figure 3). Further, the g values of the g_1 pair are
much lower than 6, more like 5.2 and 5.5. Drs. Ehrenberg and
Feher have also observed this splitting in R. rubrum cytochrome c-c'.

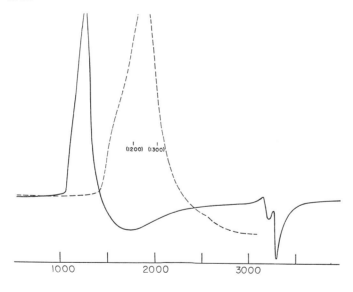

Figure 3. EPR spectrum of Chromatium cytochrome c-c' at
30° K. The dotted cruve shows the g_1 region on an expanded
field scale, illustrating the splitting of the g_1 peak. The
structure on the g = 2 signal is due to a small amount of
copper bound to the protein.

By chemically fish nor fowl I meant, for example, that while the cytochromes c-c' have optical spectra like myoglobin and hemoglobin, with weak peaks near 630 m , they do not bind simple ionic ligands like fluoride and azide. Their optical spectra are myoglobin-type but the chemical behavior is more like that of c-type cytochromes. Unlike c-type cytochromes, however, the reduced cytochromes c-c' do bind CO. By magnetically fish nor fowl, I mean, for example, that in coupling to the lattice vibrations, the iron configuration is more like the low- spin iron in c-type cytochromes, and yet the susceptibility and EPR g-values are nearly, but not quite, like those of the high-spin iron in myoglobin and hemoglobin. It may be that allowing a large rhombic distortion of the otherwise axial ligand field about the iron will enable us to account for both the g-values and the low susceptibility, and permit the spin-orbit coupling that is needed to explain the strong coupling of the spins to the lattice. Then we will see if this kind of distortion can be related to the ambiguous chemistry.

References

1. Kamen, M. Acta Chemica Scand., 17, 41 (1963).

2. Kotani, M. Prog. in Theoret. Phys., 17, 4 (1961).

X-RAY STUDIES ON CYTOCHROME c PEROXIDASE

Lars-Ove Hagman, Lars-Olor Larsson and Peder Kierkegaard

Institute of Inorganic and Physical Chemistry
University of Stockholm, Stockholm Sweden

An X-ray diffraction study of the enzyme cytochrome c peroxidase (CCP) has been started at this Institute in cooperation with Dr. Takashi Yonetani of the Johnson Research Foundation, School of Medicine, University of Pennsylvania. The crystalline CCP used in our work has been supplied by Dr. Yonetani.

Crystals suitable for X-ray investigations have been obtained by dialysis of 2% protein solutions at 4°C and pH 5.2-6.8 against a weak phosphate buffer (ionic strength 0.05 M) and using the alcohol 2-methyl-2,4-pentanediol as a precipitating agent. For the crystallization we use microdiffusion cells as described by Zeppezauer et al. (1) or the dialysis bag technique. The preparative work is continued in order to supply crystals suitable for X-ray studies as a matter of routine.

Parallel to the work with the native enzyme preparations of heavy-atom derivatives have been started. By soaking of enzyme crystals in an uranyl salt ($K_3UO_2F_5$) solution, a derivative has been obtained which has been found to be isomorphous with the native enzyme. A mercury derivative, containing one metal atom per enzyme molecule prepared by Yonetani (2), is another isomorph.

THe X-ray studies are being performed in a room kept at a constant temperature of +4°C. Precession film cameras and the standard techniques with the protein crystals mounted in sealed capillaries are used. The elimination of the β-radiation from the Cu X-ray tube is performed by means of a Ni filter or by a plane graphite monochromator mounted in a crystal holder.

Photographs obtained with the graphite monochromator method had much lower background than those obtained with the conventional Ni filter method.

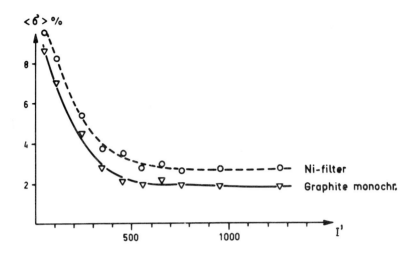

Figure 1. Mean deviation in the intensity measurements of X-ray reflections from films (hk0 layer) registered with Ni-filtered or graphite monochromatized CuK radiation of CCP crystals. The values

$$\sigma = \frac{100\Sigma|I_{obs} - I_{mean}|}{4 \cdot I_{mean}} \; ; \; I_{mean} = \frac{\Sigma I_{obs}}{4}$$

of four intensity measurements of symmetry-equivalent reflections were calculated. The values ($< \sigma >$) in different intervals of relative intensities were calculated with the formula

$$< \sigma > = \frac{\Sigma \sigma}{n}$$

(n = the number of σ-values within the interval). The $< \sigma >$ values calculated are plotted as a function of the relative intensities. $I' = 100$ corresponds approximately to $D' = 10 \log (T_b/T) = 0.09$ (T_b = background transmission and T = peak transmission) and $I' = 1500$ to $D' = 1.1$. THe total number of measurements from each film is 1224.

The diffraction intensities of the photographs are measured by an automatic film scanner (originally designed by Abrahamsson (3) and manufactured by a division of the Swedish Aeroplane Company: DATA SAAB AB, Gothenburg) controlled by an IBM 1800 computer. An integrated set of programs for calculation of structure factors from the film scanner process has been written by Werner (4). The programming languages used are the IBM 1800 Fortran and Assembler Language. The accuracy of the intensity measurements has been tested by comparison of symmetry-related reflections. Figure 1 shows the mean deviations of four measurements of symmetry-equivalent reflections in different intervals of relative intensities plotted as a function of the intensity on an arbitrary scale. From the figure it is obvious that the mean deviation values observed are lower for reflections registered by the graphite monochromator method than by the Ni-filter method.

The radiation damage of the protein crystals seem to be negligible for periods of exposure up to at least 90 hours when graphite monochromatized radiation is used. However, when the Ni-filter method is applied an appreciable radiation damage is observed after an exposure up to 50 hours.

The CCP crystals are orthorhombic, space group $P2_12_12_1$ and the unit cell dimensions are

$$a = 107.6 \text{ Å}, \ b = 76.8 \text{ Å and } c = 51.4 \text{ Å}$$

There are four molecules per unit cell.

Preliminary X-ray investigations have also been performed on iron-free crystals (supplied by Dr. Yonetani). The compound was found to be isomorphous with the native enzyme with the unit cell dimensions:

$$a = 107.2 \text{ Å}, \ b = 76.7 \text{ Å and } c = 51.4 \text{ Å}$$

In order to search for heavy-atom derivatives we use the flow-cell technique. The cell is easy to handle and good diffraction photographs are obtained in this way.

References

1. Zeppezauer, M., H. Eklund, and E. Zeppezauer. Arch. Biochem. Biophys., 126, 564 (1968).

2. Yonetani, T. Private communication.

3. Abrahamsson, S. J. Sci. Instr., <u>43</u>, 931 (1966).

4. Werner, P.-E. Arkiv Kemi, 1969, in press.

RECOMBINATION OF APOCYTOCHROME c PEROXIDASE WITH MODIFIED HEME DERIVATIVES

Toshio Asakura, Henry Drott and Takashi Yonetani

Johnson Research Foundation
University of Pennsylvania School of Medicine
Philadelphia, Pennsylvania

In spite of their identical heme structure, hemoglobin, myoglobin, cytochrome c peroxidase and catalase exhibit different types of enzymic activities. These variations might be caused by the different interactions between heme and the apoproteins.

One of the known methods for investigating the relation between heme and protein is the recombination study of the apoprotein with various heme derivatives, as has been reported for various hemoproteins except for catalase (1-25).

Cytochrome c peroxidase (CCP) has several advantages for the recombination experiment, as follows:
1) CCP and apo-CCP are obtained in pure crystalline form (24).
2) Apo-CCP is stable compared with other apoproteins (21).
3) Apo-CCP recombines with protoheme to form a holoenzyme indistinguishable from the native enzyme (24).
4) The recombination is investigated not only by spectral changes but also by the peroxidase activity.

Apocytochrome c peroxidase has been recombined with various heme derivatives. The structure of the heme derivatives used in our experiments is shown in Fig. 1. These modified hemes can be divided into three large groups: (1) modifications at the 2 and 4 positions, (2) modifications of the central metal, (3) modifications at the 6 and 7 positions of the porphyrin ring.

523

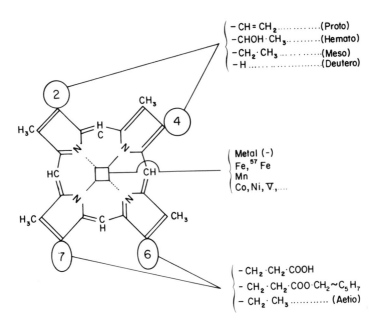

Figure 1. The structure of heme derivatives.

The results of recombination experiments of these modified hemes with apo-CCP are summarized in Table I. The side chains at the 2 and 4 positions of the porphyrin ring are considered to be not important for the peroxidase activity, since the synthetic enzymes containing proto-, hemato-, meso-, and deuterohemes showed almost the same activity as the native enzyme (20).

As shown in the second line of Table I, the metal-free porphyrins also recombined with apo-CCP to form well-defined porphyrin-protein complexes (21). The absorption spectra of these complexes were very close to those of porphyrins dissolved in organic solvents, suggesting that the porphyrins were located in a non-polar pocket of the protein. These complexes, however, had no peroxidase activity; therefore it

TABLE I

Recombination of apo-CCP with various heme derivatives

Position Modified	Name	Metal	Enzymic activity Cyto. c++	Ferrocyan
2,4	Proto-, Hemato Meso-, Deutero-H	Fe	96 100	88 100
2,4	Proto-, Hemato-, Meso, Deutero-P	-	0	0
Metal	Proto-	Mn	10	10
		Co	0	0
		V	0	0
6,7	Etio-H	Fe	1	50
		Fe	1	40
6,7	Etio-P	-	No binding	
	Proto-P methylester	-	No binding	

*Abbreviations: H, heme; P, porphyrin; Cyt. c++; ferrocyto-chrome c.

** The peroxidase activities relative to those of the native enzyme containing iron-protoheme.

was concluded that iron atom is essential for the enzymic activity (21).

The iron atom can be replaced by other metals such as manganese, cobalt, and vanadium. Among them, the manganese-containing enzyme exhibited a peroxidase activity of about 10% of the native enzyme (19).

The side chains at the 6 and 7 positions of the porphyrin ring appear to be especially important for the peroxidase activity of CCP. As seen in the fourth line of Table I, the synthetic enzymes containing etioheme or the dimethylester of protoheme showed a peroxidase activity of less than 1% of the native enzyme when ferrocytochrome c was used as the substrate (25). On the other hand, the enzymic activity toward ferrocyanide was at about half the original activity. This activity was found to decrease by elongation of the esters, as will be shown later.

As can be seen in the fifth line of Table I, metal-free etioporphyrin and dimethylester protoporphyrin did not combine with the apo-enzyme, suggesting that either the iron atom or free carboxyl groups at positions 6 and 7 are necessary for the heme binding.

Furthermore, the 6 and 7 positions of the porphyrin ring are shown to have a special meaning in the peroxidase activity of CCP. The effect of the length of the side chains at positions 6 and 7 was investigated by preparing the series of heme diesters having 1 through 5 carbon atoms. The affinity of these dimethyl through dipentylester protohemes for the apoprotein was found to decrease by the elongation of the side chains.

In Fig. 2, the enzymic activities of these diester-CCPs are compared with those of the native enzyme. As seen in the figure, all the ester-CCPs showed extremely low activity when ferrocytochrome c was used as the substrate. On the other hand, the enzyme activities toward ferrocyanide were dependent on the length of the side chains and decreased with the elongation of the chains at positions 6 and 7.

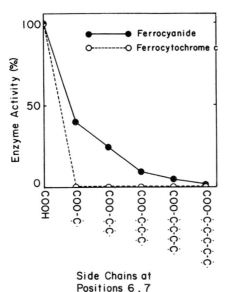

Figure 2. The effect of side chains at positions 6 and 7 of the porphyrin ring on the enzymic activities of CCP. Assay conditions as described in Ref. 20.

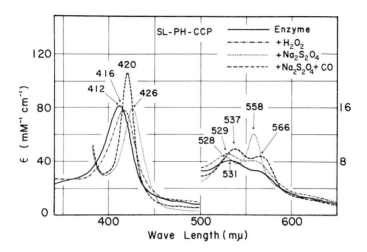

Figure 3. Light absorption spectra of the synthetic enzyme containing spin labeled protoheme. The measurements were made in 0.1 M potassium phosphate buffer, pH 6.0 at 23°.

The importance of positions 6 and 7 of the porphyrin ring was further confirmed by using a spin labeling technique on the heme. By reacting protohemin with nitroxide free radicals, we could obtain spin labeled protohemin, of which 6 and 7 positions were labeled. This spin labeled protohemin (SL-protohemin) was recombined with apo-CCP. The absorption spectra of the synthetic enzyme containing SL-protohemin are shown in Fig. 3. It is very interesting that the peaks of each of the curves are almost the same as those of ester-CCP, especially to the enzymes containing protoheme butyl-or pentylesters (25). The SL-protoheme-CCP reacted with hydrogen peroxide to form Complex ES. However, the enzymic activity of this enzyme with both ferrocytochrome c and ferrocyanide was very small. These results were also very close to those of butyl and pentylester-CCP.

The EPR spectra of the unbound spin labels and SL-protohemin in dimethylsulfoxide and tetrahydrofuran are shown in Fig. 4. The upper curves show the spectra of free spin labels

527

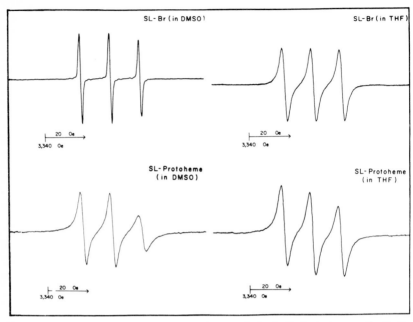

Figure 4. EPR spectra of N-(1-oxyl-2,2,5,5-tetramethyl-3-pyrrolidinyl)-bromoacetamide (SL-Br) and spin-labeled protoheme (SL-protoheme).

and the lower curves are those of SL-protohemin. The labels were only slightly immobilized by binding with protohemin.

However, by recombination with apo-CCP, the labels were strongly immobilized, as shown in Fig. 5 , indicating that the molecular motion of the labels attached to the 6 and 7 positions of the porphyrin ring is strongly hindered by binding to the protein. On the other hand, as shown in Fig. 6 , when the 2 and 4 positions of the porphyrin ring are labeled with nitroxide free radicals, the recombined enzyme exhibits an EPR spectrum having only moderately immobilized labels.

As has been established by X-ray crystallography studies (26,27), the heme in cytochrome c and myoglobin is schematically shown in Fig. 7 to be oriented in a particular fashion in the protein pocket. In cytochrome c, the 4 and 6 positions of the heme point toward the outside of the pocket (26), while

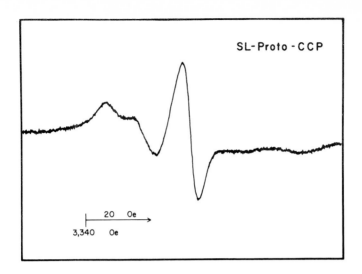

Figure 5. EPR spectra of synthetic enzyme containing spin labeled protoheme. Protohemin was labeled with 2,2,5,5-tetramethyl-3-aminopyrrolidine-1-oxyl at positions 6 and 7, and recombined with apo-CCP. The synthetic enzyme was dissolved in potassium phosphate buffer, pH 7.0 and measured with Varian EPR spectrometer (V-4502) at 23°.

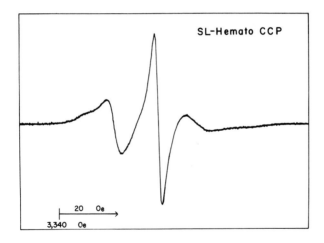

Figure 6. EPR spectra of synthetic enzyme containing spin-labeled hematoheme (2 and 4 positions labeled as above).

in myoglobin the 6 and 7 positions are oriented toward the outside of the pocket (27). In the case of CCP, spin labeling studies showed that the 6 and 7 positions of the heme side chains interact more strongly with the apoprotein than do positions 2 and 4. This result, and also the fact that the affinity of the heme for the apoprotein is dependent on the length of the side chains at positions 6 and 7, lead us to speculate the heme orientation shown in Fig. 7A. Of course, other possibilities such as Fig. 7B cannot be neglected.

The variations of enzymic activity found in hemoproteins containing protoheme may be interpreted by different heme orientations in the protein, but the size and flexibility of the pocket may also influence these variations.

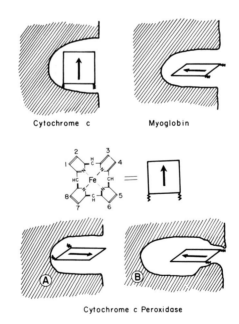

Cytochrome c   Myoglobin

Cytochrome c Peroxidase

Figure 7. Scheme of heme orientation in the protein molecule.

References

1. Hill, R., and H.F. Holden. Biochem. J., 20, 1326 (1926).

2. Rossi Fanelli, A., and E. Antonini. Arch. Biochem. Biophys., 80, 308 (1959).

3. Antonini, E., M. Brunori, A. Caputo, E. Chiancone, A. Rossi Fanelli, and J. Wyman. Biochem. Biophys. Acta, 79, 284 (1964).

4. Haurowitz, F., and H. Waelsch. Z. Physiol. Chem., 182, 82 (1929).

5. O'Hagan, J.E. Biochem. J., 74, 417 (1960).

6. O'Hagan, J.E., and P. George. Biochem. J., 74, 424 (1960).

7. Warburg, O., and E. Negelein. Biochem. Z., 244, 9 (1932).

8. Smith, M.H., and Q.H. Gibson. Biochem. J., 73, 101 (1959).

9. Rossi Fanelli, A., and E. Antonini. Arch. Biochem. Biophys., 72, 243 (1957).

10. Gibson, Q.H., and E. Antonini. J. Biol. Chem., 238, 1384 (1963).

11. Theorell, H., and A.C. Maehly. Acta Chem. Scand., 4, 422 (1950).

12. Theorell, H., S. Bergstrom, and A. Akeson. Arkiv Kemi Mineral Geol. 16A, no. 13 (1942).

13. Paul, K.G. Acta Chem. Scand., 13, 1239 (1959).

14. Paul, K.G., H.S. Gewitz, and W. Volker. Acta Chem. Scand., 13, 1240 (1959).

15. Gjessing, E.C., and J.B. Sumner. Arch. Biochem., 1, 1 (1943).

16. Oxols, J., and P. Strittmatter. J. Biol. Chem., 239, 1018 (1964).

17. Asakura, T., S. Minakami, Y. Yoneyama, and H. Yoshikawa J. Biochem. (Tokyo), 56, 594 (1964).

18. Asakura, T., H. Yoshikawa, and K. Imahori. J. Biochem. (Tokyo), 64, 515 (1968).

19. Yonetani, T., and T. Asakura. J. Biol. Chem., 243, 3996 (1968).

20. Yonetani, T., and T. Asakura. J. Biol. Chem., 243, 4715 (1968).

21. Asakura, T., and T. Yonetani. J. Biol. Chem., 244, 537 (1969).

22. Breslow, E., R. Kohler, and A.W. Girotti. J. Biol. Chem., 242, 4149 (1967).

23. Fabry, T.L., C. Simo, and K. Javaherian. Biochem. Biophys. Acta, 160, 118 (1968).

24. Yonetani, T. J. Biol. Chem., 242, 5008 (1967).

25. Asakura, T., and T. Yonetani. J. Biol. Chem., 244, 4573 (1969).

26. Dickerson, R.E., M.L. Kopka, J. Weinzierl, J. Varnm, D. Eisenberg, E. Margoliash. J. Biol. Chem., 242, 3015 (1967).

27. Kendrew, J.C., H.C. Watson, B.E. Strandberg, R.E. Dickerson, D.C. Phillips, and V.C. Shore. Nature, 190, 660 (1961).

PEROXIDASE: SOME PROPERTIES OF THE SITE NEAR IRON SIXTH COORDINATION POSITION*

Gregory R. Schonbaum, Karen Welinder
and Lawrence B. Smillie

Department of Biochemistry, University of Alberta
Edmonton, Canada

Confronted with the functional diversity among hemoproteins which contain ferriprotoporphyrin as the prosthetic group, it must be assumed that apoenzyme determines the catalytic properties of a given system (1). For myoglobin and hemoglobin, this generality has been translated into more specific terms, thanks to outstanding physical and chemical investigations of the past decade (2,3,4); in the case of peroxidases and catalases, however, the same degree of detailed knowledge remains to be achieved.

But, already, it would appear that the catalytic properties of these enzymes are not neccessarily governed by a single dominant feature, such as the difference in the nature of coordinating ligands, but could stem from an interplay of other factors bearing on: conformational freedom of the porphyrin in different redox states of the enzyme (5); diverse dielectric properties of hemin environment or changes in the orientation of the apoenzyme ligands with respect to the prosthetic group (6).

Perhaps the case in point could be illustrated by considering some properties of horseradish peroxidase (HRP). This enzyme contains three histidines (7,8); of these, one was placed as in myoglobin at the proximal, or 5th, coordination position to the prosthetic group (9,10); and another histidine, again in analogy to the myoglobin, appears to be located in the distal region. This assignment and the proposed role of the latter group, as a general acid-base catalyst

*This work was supported by Research Grants MT 1270 and G-64-36 from the Medical Research Council of Canada and the Life Insurance Medical Research Fund.

are based on the results of reversible inhibition of HRP following enzyme acetylation. Specifically, the inhibition is attributed to acetylation of the imidazole group of distal histidine, a residue which, rather surprisingly, was not shown to be subject to acetylation in previous studies of any protein (11). Since this report apparently constitutes the first description of such a modification, some limitations of the acylation procedure should be briefly reviewed.

Acetylation of HRP. Given the gross reactivity of most common reagents used so frequently in protein labeling, it is only seldom that a selective modification becomes possible (11). And thus, laborious, but unsuccessful attempts were made to devise specific reagents capable of interacting with a basic residue which -- from ligand substitution studies (6) -- was suspected to be present at the distal site of HRP. It was therefore rather unexpected that the desired modification was achieved using anhydrides. Typically, these reagents are used for random acylations, particularly of the ε-amino group of lysine, the reaction being generally carried out at approximately pH 8.0 in the presence of acetate and a large excess of anhydride (12,13). Under these conditions, acetylation of the peptide-linked imidazole group of histidine has not been observed (11) although in model systems rapid formation of acetyl imidazole can be readily demonstrated (14,15). At least two factors could be responsible for this behavior; one, having its origin in some steric restrictions, the other representing rapid hydrolysis of any N-acyl imidazole which might be formed. Therefore, it seemed that, if histidine were to be at all accessible to modification, the resulting N-acyl imidazole derivative might show some stability providing the reaction were carried out at lower pH, between 5.5 and 7.0, in the absence of acetate, which is known to catalyze hydrolysis of such compounds (16). Moreover, to facilitate trapping of any intermediate the reaction was studied at high, but equivalent concentraitons of HRP and acetic anhydride. This approach proved successful. For example at 0°, pH 6 using 3 mM enzyme and anhydride, approximately 32±4% of HRP was converted, within 100 sec., into catalytically inactive derivative. Since anhydrides readily hydrolyse (17,18,19) the observed substantial degree of inhibition suggests a marked nucleophilicity of the group undergoing acetylation.

Properties of acetylated HRP. An important characteristic
of the activated enzyme is the transient nature of the inhibition; and, what is more significant, the rate of reversion
to the fully active form occurs faster under acidic, or basic
conditions. Indeed, the cycle of inhibition and enzyme reactivation can be demonstrated even under unfavorable conditions,
at pH 5, in the presence of acetate and using a large excess
of anhydride over HRP (fig. 1). The limited stability of the

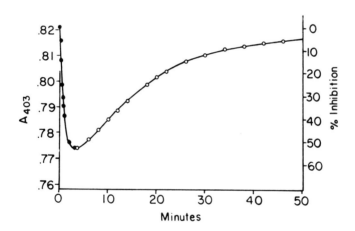

Figure 1. Absorbancy changes and per cent inhibition accompanying a) acetylation of HRP with acetic anhydride (●) and
b) regeneration of active enzyme (o). HRP 8.0 x 10^{-6} M;
$(CH_3CO)_2O$ 8.0 x 10^{-3}; 0.05 M acetate, pH 5.1, 25°.

acetyl derivative and particularly its susceptibility to acid catalyzed hydrolysis suggest that the modification occurred at the imidazole ring of histidine rather than at an amino or hydroxyl moiety. The same conclusion was reached from the analysis of the difference spectrum of acetylated vs. native enzyme. The spectrum, presented in fig. 2, shows distinctive features of N-acetyl imidazole both in regard to the wavelength of maximum absorption and the magnitude of the extinction coefficient (16,20). These spectroscopic differences between the native and inhibited HRP are eliminated upon regeneration of the active enzyme; and furthermore, the same rate constant characterizes the reactivation process irrespective of the method used in its evaluation, namely by following absorbancy changes in the ultra-violet and in the Soret region or by determination of the capacity to form enzyme derivatives with hydrogen peroxide. The inhibition and reactivation of horseradish peroxidase, which can be demonstrated repeatedly with the same sample, the unique difference spectrum in the ultra-violet (fig. 2) and the outlined kinetic features indicate that acetylation of HRP entails reaction at a single, functionally important residue. Of course, simultaneously with the inhibitory reaction other groups could undergo acetylation. If this is the case, then such modifications do not alter enzyme catalytic properties and are not reflected in its spectroscopic features.

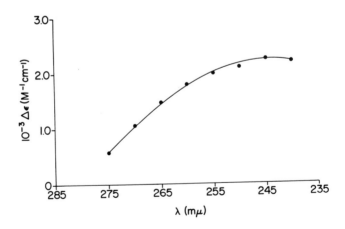

Figure 2. Difference spectrum of 'acetylated HRP' minus 'HRP', pH 7, 25°.

It is noteworthy also that in its inhibited form the enzyme shows, apart from characteristic differences in the ultra-violet, relatively small absorbancy changes even in the Soret region (fig. 3). Since the modification of a proximal ligand might be expected to cause a major perturbation of the ferriprotoporphyrin environment, the locus of the modified group was tentatively assigned to the distal region. Other evidence, in accord with this suggestion, comes from the studies of enzyme reactivation.

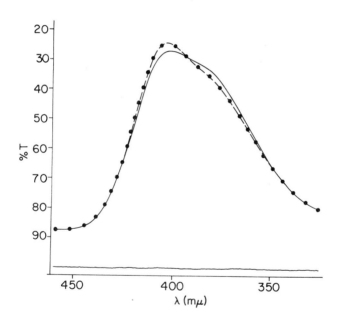

Figure 3. Spectra of HRP (---) and its acetylated derivative (—) 70% inhibition. The spectrum of regenerated HRP is essentially identical to that of the native enzyme (••••). HRP_T 6.0 x 10^{-6} M; 0.05 M phosphate, pH 6.8, 25°.

<u>Catalysis of HRP deacetylation.</u> The acetylated enzyme does not form spectroscopically identifiable complexes with various ligands which combine with the native HRP. And yet, hydrogen fluoride, formic acid and acetic acid catalyze reversion of the inhibited enzyme to its active form -- as illustrated in the case of HF (fig. 4). It would appear, therefore, that

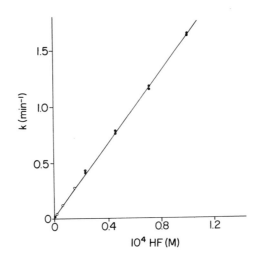

Figure 4. Reactivation of peroxidase catalyzed by hydrogen fluoride at pH 6 (●) and at pH 7 (○); 0.05 M phosphate; 25°.

inhibition of HRP on acetylation does not necessarily stem from hindrance of ligand approach to the distal site, but reflects decreased proton affinity of a functional group situated near the iron sixth coordination position. Indeed, from the kinetic analysis of enzyme regeneration, we estimate pK_a of the modified group to be approximately 3.0 at 25°; a value within the range expected for N-acetyl imidazolium ion which, in solutions of 0.2 ionic strength and at 25°, is characterized by pK_a 3.6 (16). But apart from decreasing basicity, acetylation will alter also another property of the imidazole group -- namely, its capacity to mediate proton transfers (21,22). The implied role of distal histidine, as a general base-general acid catalyst in ligand substitutions could be represented as:

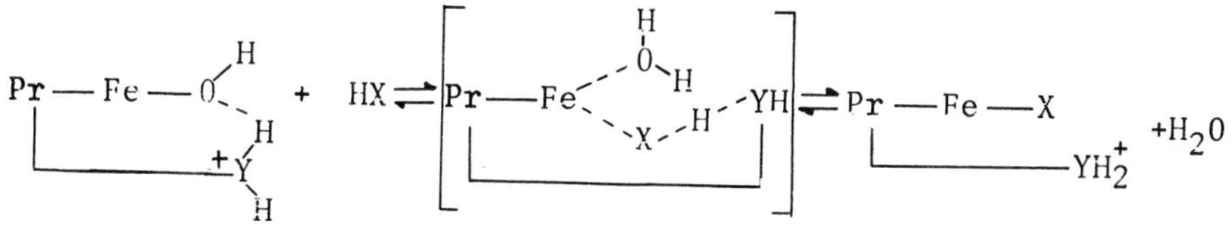

(HX = HF, HCO_2H, CH_3CO_2H)

The acetylation of histidine blocks such "cyclic" proton transfers with concomitant enzyme inhibition. Similarly, on association between the acetylated enzyme and a general acid, HX, the catalysis of enzyme reactivation can be attributed to a partial proton transfer from XH to $Y-COCH_3$:

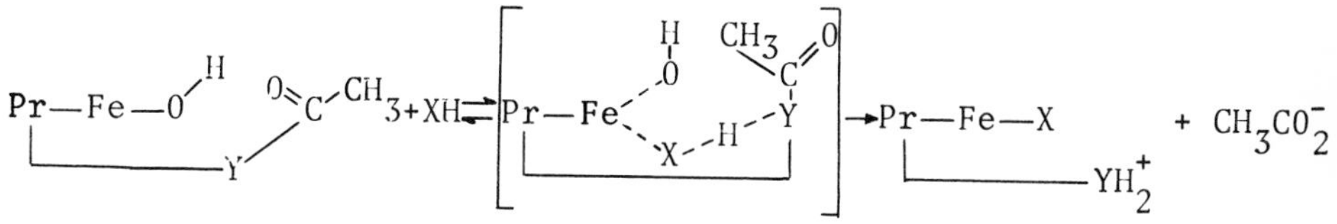

The catalytic effect of a given acid in HRP deacetylation would then depend not only on the proton affinity of its conjugate base but also on the complex forming capacity of HX with the inactive enzyme. Providing then that steric factors do not play a dominant role, the expected effectiveness of HX in enzyme reactivation should be: $HF > HCO_2H > CH_3CO_2H >> HN_3$. This was found to be the case. The participation of added anions in the displacements at the carbonyl group is also rendered less likely since azide, a relatively strong nucleophile, has only a marginal effect (6).

Undoubtedly, these proposals will have to be further developed before the mechanism of ligand interchange is fully understood. Already, evidence is at hand implicating -- apart from histidine -- another group, possibly a carboxylate residue, in proton transfers and transacylations outlined above. Nor are these ideas solely limited to ligand substitutions, but apply equally to the rearrangements attending the formation of compound I -- the primary peroxide derivative of HRP (6).

In this context, it may be also asked whether the contribution of a general acid-base catalysis to the overall enzyme function constitutes a differentiating, or even dominant, factor governing the properties of hemoproteins containing the same coenzyme. Further examination of this novel hypothesis will center on studies of horseradish peroxidase and demands a detailed knowledge of its structure. Some progress has been made in this direction, particularly with regard to the primary strucutre of the enzyme, and the preliminary results are summarized in Figure 5 and Table I.

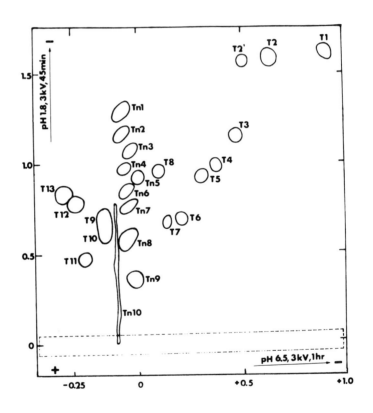

MOBILITY RELATIVE   MOBILITY RELATIVE
TO ASPARTIC ACID   TO LYSINE

Figure 5. Peptide "map" of tryptic digest of apo horse radish peroxidase. The amino acid composition or sequence of the peptides is shown in Table I.

TABLE I

Amino acid composition or sequence of tryptic peptides from horse radish peroxidase.

Peptide	Amino acid composition or sequence
T1	Arg
T2	Met-Lys
T2'	MetSO-Lys
T3	Val-Pro-Leu-Gly-Arg
T4	Ile-Ala-Ala-Ser-Ile-Leu-Arg
T5	Phe-Thr-Leu-Pro-Gln-Leu-Lys
T6	Met-Gly-Asn-Ile-Thr-Pro-Leu-Thr-Gly-Thr-Gln-Gly-Gln-Ile-Arg
T7	Asn-Val-Gly-Leu-Asn-Arg
T8	Ser-Ser-Asp-Leu-Val-Ala-Leu-Ser-Gly-Gly-His-Thr-Phe-Gly-Lys
T9	Composition unknown. His-positive.
T10	Ala-Ala-Val-Glu-Ser-Ala-Cys-Pro-Arg (Glx,Thr$_2$,Pro,Leu,Tyr,Phe)Asx(Cys,Asx$_3$,Ser$_2$,Pro,Val$_2$, Ile)Arg$^+$)
T11	Gly-Leu-Ile-Gln-Ser-Asp-Gln-Glu-Leu-Phe-Ser-Ser-Ile-Asn-Ala-Thr-Asp-Thr-Pro-Pro-Leu-Val-Arg
T12	Tyr-Tyr-Val-Asn-Leu-Glu-Glu-Gln-Lys
T13	Asp-Thr-Ile-Val-Asn-Glu-Leu-Arg
Tn1	Thr-Glu-Lys
Tn2	Ser-Asp-Pro-Arg
Tn3	Asp-Ser-Phe-Arg
Tn4	Phe-Ile-Met-Asp-Arg
Tn5	Gly-Phe-Pro-Val-Ile-Asp-Arg
Tn6	Thr-Pro-Thr-Ile-Phe-Asp-Asn-Lys
Tn7	Asp-Ala-Phe-Gly-Asn-Ala-Asn-Ser-Ala-Arg
Tn8	Val-Val-Asn-Ser-Asn-Ser-Asn
Tn9	Leu-Tyr-Asn-Phe(Asn,Thr$_2$,Ser,Leu),(Asn-Thr),(Gly-Leu-Pro-Asp-Pro-Thr),Tyr-Leu-Gln-Thr-Leu-Arg
Tn10	Composition unknown. Trp-positive

On nomenclature: Asx = Asp or Asn

Glx = Glu or Gln

+) T10 is a disulfide bridge peptide.

References

1. Theorell, H., Arkiv Kemi 16A, 3, 1 (1942).

2. Braunitzer, G., K. Hilse, V. Rudloff and N. Hilschmann, Adv. in Protein Chem. 19, 1 (1964).

3. Rossi Fanelli A., E. Antonini and A. Caputo, Adv. in Protein Chem. 19, 74 (1964).

4. Caughey, W. S., Ann. Rev. of Biochem. 36, 611 (1967).

5. Willick, G., G. R. Schonbaum and C. M. Kay, Biochemistry, 1969, in press.

6. Schonbaum, G. R., 1968, unpublished observations.

7. Klapper, M. H. and D. P. Hackett, Biochim. Biophys. Acta 96, 272 (1965).

8. Brill, A. S. and I. Weinryb, Biochemistry 6, 3528 (1967).

9. Brill, A. S. and H. E. Sandberg, Proc. Natl. Acad. Sci. U. S. A. 57, 136 (1967).

10. Brill, A. S. and H. E. Sandberg, Biochemistry 7, 4254 (1968).

11. Cohen, L. A., Ann. Rev. of Biochem. 37, 695 (1968).

12. Riordan, J. F. and B. L. Vallee, in Methods in Enzymology, Vol. XI (C. H. W. Hirs, ed.) Academic Press, N. Y., 1967, p. 565.

13. Fraenkel-Conrad, H., in Methods in Enzymology, Vol. IV, (S. P. Colowick and N. O. Kaplan, eds.) Academic Press, New York, 1957, p. 247.

14. Kirsch, J. F. and W. P. Jencks, J. Am. Chem. Soc. 86, 837 (1964).

15. Brouwer, D. M., M. J. Van der Vlugt and E. Havinga, Proc. Koninkl. Ned. Akad. Wetenschap. B60, 275 (1957).

16. Jencks, W. P. and J. Carriuolo, J. Biol. Chem. 234, 1272 (1959).

17. Bunton, C. A., N. A. Fuller, S. G. Perry and V. J. Shiner, J. Chem. Soc. 2918 (1963).

18. Gold, V., Trans. Faraday Soc. 44, 506 (1948).

19. Kilpatrick, M., J. Am Chem. Soc. 50, 2891 (1928).

20. Stadtman, E. R., in The Mechanism of Enzyme Action (W. D. McElroy and B. Glass, eds.) Johns Hopkins Press, Baltimore, 1954, p. 581.

21. Eigen, M. and G. G. Hammes, Adv. in Enzym. 25, 1 (1963).

22. Eigen, M., G. G. Hammes and K. J. Kustin, J. Am. Chem. Soc. 82, 3482 (1960).

TEMPERATURE-DEPENDENT SPIN TRANSITIONS
IN CYTOCHROME c PEROXIDASE*

Takashi Yonetani**

Johnson Research Foundation, School of Medicine
University of Pennsylvania, Philadelphia, Pennsylvania 19104

Several years ago Yonetani and Ehrenberg (1) reported that cytochrome c peroxidase has a magnetic moment of 5.2 Bohr magnetons at 20°C, indicating that the heme iron of this enzyme is essentially in a high spin state at room temperature. However, the EPR spectrum of cytochrome c peroxidase at liquid nitrogen temperature shown in Figure 1 suggests that cytochrome c peroxidase is a mixture of high and low spin ferric compounds at -196°C. Since EPR can distinguish clearly high spin and low spin states of ferric hemoproteins (2,3), a number of investigators have attempted to analyze the temperature-dependent transition of the spin state of ferric hemoproteins by

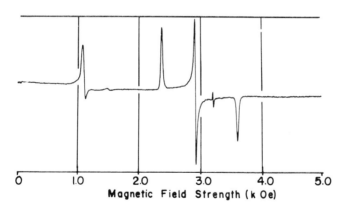

Magnetic Field Strength (k Oe)

<u>Figure 1</u>. EPR spectrum of cytochrome c peroxidase, pH 7.0, at -196°C.

*These investigations have been carried out in collaboration with Drs. T. Iizuka and M. Kotani, Faculty of Engineering Science, Osaka University and have been supported by NSF GB 6974 and PHS GM 15,435 and GM 12,202.
**The author is a Career Development Awardee (5-K3-GM 35,331).

EPR techniques. However, as shown in Figure 2, when the temperature is increased, the intensity of EPR signals of the
heme iron decreases drastically. In addition, the linewidth
of EPR signals is considerably broadened. Thus the quantitative analysis of the spin concentrations of high and low spin
components becomes difficult above -100°C.

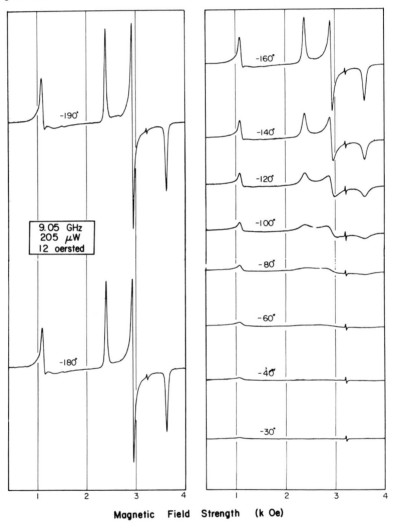

Magnetic Field Strength (k Oe)

Figure 2. The effect of temperature on EPR spectra of cytochrome c peroxidase.

Certain correlations between optical spectra and spin states of various hemoproteins were mentioned by several investigators (4-8). A high spin ferric hemoprotein has a Soret band at 405 to 410 mμ, intense charge transfer (CT) bands at about 500 and 600 to 650 mμ, and weak α and β bands. In contrast, a low spin ferric hemoprotein has a Soret band at longer wavelengths at 415 to 425 mμ, intense α and β bands, and weak CT bands, as schematically indicated in Figure 3. Typical patterns of optical spectra of high and low spin ferric hemoproteins are represented here by those of fluoride and cyanide complexes of ferric myoglobin. The interpretation of the origin of these absorption bands is theoretically difficult. However, these correlations between optical spectra and spin states of ferric hemoproteins have been empirically well established today.

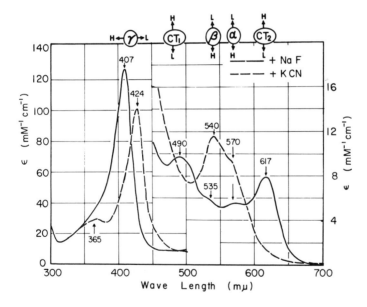

Figure 3. Absorption spectra of fluoride and cyanide complexes of ferric myoglobin. Changes in spectral characteristics accompanying the transition between high (H) and low (L) spin states are illustrated schematically.

In 1961 George, Beetlestone and Griffith (7) compared
optical spectra and magnetic susceptibilities of several hemoprotein hydroxides at different temperatures and demonstrated
that the temperature-dependent transition of the spin state
of hemoproteins, which was predicted by Griffith (9), could
be measured spectrophotometrically. Figure 4 illustrates
optical spectra of leghemoglobin at several ambient temperatures. The magnitude of these spectral changes is relatively
small, since the range of temperature is limited from 0°C to
about 50°C. The ratio of the intensities of CT to β bands
is used to monitor the spin transition. As shown in the inserted graph in Figure 4, this ratio decreases more steeply at
lower temperatures, indicating that the spin transition may
continue below 0°C. Measurements of optical spectra of cytochrome c peroxidase at various cryogenic temperatures (cf.
Fig. 5) (10) show clearly such spin transitions to take place
even in frozen samples. At -196°C, the spectrum of this enzyme is essentially a low spin type, which is characterized
by intense α and β bands and weak CT bands. At higher temperatures, the intensity of the α and β bands decreased, which
was accompanied by an increase in the intensity of the CT bands.

Temperature-Dependent Transition of Optical Spectra of LegHb.

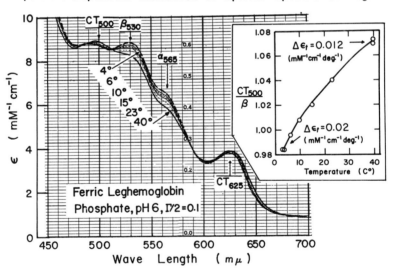

Figure 4. The effect of temperature on absorption spectra
of ferric leghemoglobin.

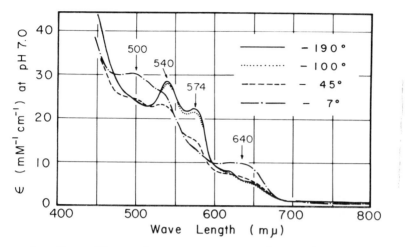

Figure 5. The effect of temperature on absorption spectra of cytochrome c peroxidase.

In Figure 6 the change in the intensity of the α band is plotted against the inverse absolute temperature (11), in order to indicate the range of temperature in which the spin transition occurred. The transition appears to take place in a narrow range of temperature from -100°C to -30°C. It should be pointed out that the kinetics of the spin transition may be investigated by the use of the temperature-jump technique. Judging from linewidths of Mössbauer spectra, it is estimated that the transition time between high and low spin states has

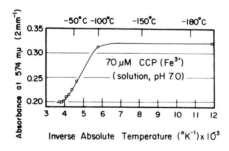

Figure 6. The temperature dependence of the intensity of the β band of cytochrome c peroxidase, which characterizes the low spin compound. The intensity of the β band is temperature-independent below -100°C. The spin change takes place above -100°C.

549

a lower limit of several times 10^{-8} seconds below $-78°C$ (12). At higher temperatures, it may well be of the order of nano seconds.

Iizuka, Kotani and Yonetani (11) measured the temperature dependence of the magnetic susceptibility of cytochrome c peroxidase at cryogenic temperatures (cf. Fig. 7). In agreement with the above-mentioned optical results, the susceptibility of cytochrome c peroxidase was found to change anomalously above $-100°C$. Below $-100°C$ the susceptibility of this enzyme obeyed Curie's law, indicating that the heme iron of this enzyme is in the ground state below $-100°C$. The effective Bohr magneton numbers (n_{eff}) of the enzyme at pH 5 and pH 7 below $-100°C$ were calculated to be 5.20 and 3.91, respectively. Assuming that these intermediate values of n_{eff} were derived from a mixture of two compounds, one with a high spin ground state and the other with a low spin ground state and that n_{eff} values of purely high and low spin compounds were 5.91 and 2.24, respectively, relative proportions of these two chemical species of this enzyme at pH 5 and pH 7 were estimated: this enzyme was found to contain 73% and 34% high spin compounds at pH 5 and pH 7, respectively, in the ground state. On the basis of the pH dependence, the compounds with high and low spin ground states were tentatively called acidic [CCP(FE^{III}-H$_2$O)] and alkaline [CCP(Fe^{III}-OH$^-$)] forms of cytochrome c peroxidase, respectively.

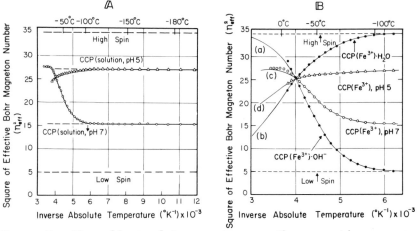

Figure 7. The effect of temperature on the magnetic susceptibility of cytochrome c peroxidase at pH 5 and pH 7. A, from $-196°C$ to $0°C$. B, above $-100°C$, where the spin change was observed.

By assuming the relative proportions of these two forms of cytochrome c peroxidase to be constant in frozen states, the anomalous temperature-dependence of the magnetic susceptibility of the enzyme (cf. Fig. 7) was analyzed by the method of George et al. (7). CCP(Fe^{III}-H$_2$0) was found to have a transition temperature at +1°C with an excitation energy of -1,230 cm^{-1}, whereas CCP(Fe^{III}-OH$^-$) was found to have a transition temperature at -41°C with an excitation energy of +1,830 cm^{-1}. On the basis of these observations, it is concluded that the EPR spectrum of cytochrome c peroxidase shown in Fig. 1 represents a mixture of two chemically different compounds having high and low spin ground states rather than a thermally equilibrated mixture of high and low spin isomers of one chemical compound.

This conclusion was further supported by the oxidative titration of the enzyme with hydroperoxide, as shown in Figure 8. Cytochrome c peroxidase was titrated with increments of ethyl hydrogen peroxide and frozen in liquid nitrogen within 1 minute after the addition of hydroperoxide. Frozen samples were examined by EPR spectroscopy to determine which portion of the enzyme was converted to the peroxide compound (Complex ES) upon the addition of hydroperoxide. As shown in Fig. 8 the hydroperoxide preferentially reacted with CCP(FE^{III}-OH$^-$) having a low spin ground state. CCP(Fe^{III}-H$_2$0) was converted to Complex ES only after all the CCP(Fe^{III}-OH$^-$) reacted with hydroperoxide.

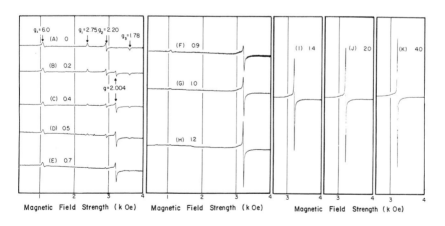

Figure 8. EPR titration of cytochrome c peroxidase with ethyl hydroperoxide.

The acidic to alkaline transition in hemoproteins is generally considered to occur at a rapid rate. If cytochrome c peroxidase is a mixture of so called acidic and alkaline forms, the relative proportions of these two forms after each addition of hydroperoxide should not change, since an original equilibrium between acidic and alkaline forms would be rapidly established. However, the results of Fig. 8 show that no transition of CCP(Fe^{III}-H$_2$0) to CCP(Fe^{III}-OH-) took place in 1 minute after the addition of hydroperoxide. Therefore, two forms of cytochrome c peroxidase, which were tentatively identified as CCP(Fe^{III}-$\bar{H}_2$0) and CCP(Fe^{III}-OH-) on the basis of their pH dependence, may not be so called acidic and alkaline forms of the enzyme. The chemical nature of these two forms of cytochrome c peroxidase is currently being investigated.

References

1. Yonetani, T. and A. Ehrenberg, in Magnetic Resonance in Biological Systems (A. Ehrenberg, B. G. Malmström and T. Vänngård, eds.) Pergamon Press, London, 1967, p. 155.

2. Gibson, J., D. J. E. Ingram and D. Schonland, Faraday Soc. Discussions 26, 72 (1958).

3. Ehrenberg, A., Arkiv Kemi 19, 119 (1962).

4. Theorell, A. and A. Ehrenberg, Acta Chem. Scand. 5, 823 (1952).

5. Scheler, W., Schoffa, G. and F. Jung, Biochem. Z. 329, 232 (1957).

6. Brill, A. S. and R. J. P. Williams, Biochem. J. 78, 246 (1961).

7. George, P., J. Beetlestone and J. S. Griffith, in Haematin Enzymes (J. E. Falk, R. Lemberg and R. K. Morton, eds.) Pergamon Press, London, 1961, p. 105.

8. Smith, D. W. and R. J. P. Williams, Biochem. J. 110, 297 (1968).

9. Griffith, J. S., Proc. Royal Soc. (London), Ser. A, 235, 23 (1956).

10. Yonetani, T., D. F. Wilson and B. Seamonds, J. Biol. Chem. 241, 5347 (1966).

11. Iizuka, T., M. Kotani and T. Yonetani, Biochim. Biophys. Acta 167, 257 (1968).

EFFECT OF FeOFe BRIDGING ON NMR AND EPR SPECTRA
OF IRON(III) PORPHYRIN SYSTEMS*

George A. Smythe[+], Ronald A. Bayne and Winslow S. Caughey[+]

Department of Chemistry, University of South Florida

We have recently converted several monomeric 2,4-substi-
tuted deuterohemins into dimers with an FeOFe linkage (1-4).
NMR and EPR spectra of these μ-oxobis (porphyrin iron [III])
compounds are of interest for comparison with other iron(III)
porphyrins and hemeproteins. The properties of these dimers
may be particularly relevant to studies of cytochrome oxidase
where bridging among the heme and copper components may occur
(cf. 5).

A typical proton NMR spectrum is shown by Figure 1.
All the resonances were found within the range of chemical
shifts normally found for diamagnetic metal porphyrins (2,6),
however for certain protons chemical shifts were significantly
different and line widths much greater, consistent with paramagnetic species. Assignments and chemical shifts (in ppm
downfield from TMS) are: α-CH_2 (of propionic acid esters)
6.15; CH_3(1,3,5,8) 5.24; OCH_3 3.57; acetyl-CH_3 3.40; β-CH_2
2.78; meso-protons ca. 1.8. These assignments may be compared with those found for the dipyridine iron(II) derivative
of this porphyrin (3): α-CH_2 4.30; CH_3(1,3) 3.75, 3.76;
CH_3(5,8) 3.42, 3.44; OCH_3 3.46; acetyl-CH_3 3.13, 3.17; β-CH_2
3.34; meso-protons 9.88(2), 10.96(1), 11.26(1). The rela-
tive magnitude and field direction of the shifts in the dimer
compared with the iron(II) compounds were similar to the
paramagnetic shifts observed for the paramagnetic nickel(II)
derivative (7), monomeric high-spin iron(III) hemins in
chloroform-d (8), and low spin cyano iron(III) porphyrins
(9). However, the absolute magnitude of the shifts were
generally far less; for example, the ring methyl protons
experienced downfield shifts of ca. 1.5, ca. 11 and ca. 45
ppm for dimer, cyanide and iron(III) chloride (at $\sim$30°C)

*Supported by USPHS HE-11807.

[+]Present address: Department of Chemistry, Arizona State
University, Tempe, Arizona 85281.

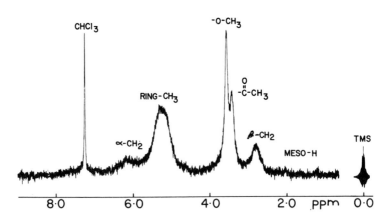

Figure 1. Proton NMR spectrum of μ-oxo-bis (2,4-deacetyldeuteroporphyrin IX dimethyl ester iron [III]) in CDCl$_3$ at 30°C obtained with a Jeolco JNM 4H-100 spectrometer operating at 100 MHz.

respectively.

An EPR spectrum of the diacetyldeuterohemin dimer preparations as a solid is shown in Figure 2. This spectrum, typical for the FeOFe dimers, was characterized by a small g = 6 absorption and a large g = 2 absorption. The possibility that at least a portion of the EPR signal arose from a monomeric iron (III) derivative present as an impurity cannot be unequivocally excluded.* It is noteworthy that the g = 6 and g = 2 signals decreased in intensity as the temperature was increased and were essentially non-detectable at room temperature. It is of further interest to note the striking similarity of the EPR signal obtained for the dimer (Figure 2) to the signal obtained (10) when cytochrome c peroxidase is treated with hydroperoxide to give the so-called "complex ES."

*Note added in proof: November 1970. More recently μ-oxohemin dimer preparations have been obtained which do not exhibit significant EPR signals (ag g = .6 or 2). Spectra such as in Figure 2 must, therefore, be due to the presence of compounds other than FeOFe dimers. The nature of the compounds, particularly those with narrow g = 2 absorptions, are of continued interest.

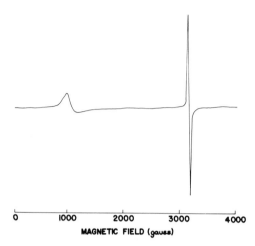

MAGNETIC FIELD (gauss)

<u>Figure 2</u>. EPR spectrum of μ-oxo-bis (2,4-diacetyldeutero-porphyrin IX dimethyl ester iron [III]) at 103°K obtained with a Jeolco JES ME spectrometer with 100 KHz field modulation unit. Conditions were: modulation amplitude, 4 gauss; micro-wave power, 4 mW; frequency, 9.24 GHz.

References

1. Alben, J.O., W.H. Fuchsman, C.A. Beaudreau and W.S. Caughey. Biochemistry, <u>7</u>, 624 (1968).

2. Sadasivan, N., H.I. Eberspaecher, W.H. Fuchsman and W.S. Caughey. Biochemistry, <u>8</u>, 534 (1969).

3. Smythe, G.A. and W.S. Caughey. Fed. Proc., <u>29</u>, 464 (1970).

4. Caughey, W.S. Advan. Chem. Ser., in press.

5. Caughey, W.S., J.L. Davies, W.H. Fuchsman and S. McCoy. In <u>Structure and Function of Cytochromes</u> (K. Okunuki, M.D. Kamen and I. Sukuzu, eds.), Tokyo Univ. Press, Tokyo, 1968, p. 20.

6. Caughey, W.S. and W.S. Kosai. Biochemistry, <u>1</u>, 923 (1962).

7. McLees, B.D. and W.S. Caughey. Biochemistry, <u>7</u>, 642 (1968).

8. Caughey, W.S. and L.F. Johnson. Chem. Commun., 1362 (1969).

9. Wüthrich, K., R.G. Shulman, B.J. Wyluda and W.S. Caughey. Proc. Nat. Acad. Sci. U.S.A., <u>62</u>, 635 (1969).

10. Yonetani, T., H. Schleyer and A. Ehrenberg. J. Biol. Chem., <u>241</u>, 3240 (1966).

DISCUSSION

Nicholls: I would like, as far as possible, to concentrate
the discussion to two central points of concern to us.
First, is the question as to what is the environment of the
heme iron of peroxidase. In particular, what are the heme-
linked groups and what other groups are nearby? I think most
of our speakers have addressed themselves to this point in
one way or another. The second question is one which is
raised by the very provocative slide of Dr. Schoenborn this
morning. He showed that in the metmyoglobin ethyl hydrogen
peroxide complex, the ethyl group appears to be retained, so
that the old chestnut of the structure of the red complex
(complex II) is a live one again today. I would like to ask
Dr. Blumberg who presented us with the grand scheme of all
heme proteins this morning how peroxidase fits into this
scheme.

Blumberg: There was some data about the peroxidases and
catalases on the truth diagrams 1 and 2 that I would like to
summarize here. Truth diagram 1 contains all of those proteins which were consistent with the same structure as hemoglobin and myoglobin, and catalase is one of these. Thus
the proximal ligand in catalase is histidine, as it forms a
normal azide and normal cyanide compound. Now the question,
does it have a distal histidine as well? To answer that you
have to ask whether it forms di-histidine compound. In the
Japanese literature, there is what appears to be a di-histidine compound, so I will leave the question of a distal histidine as a probability. Now as regards the peroxidases, and
I include Japanese radish peroxidase as well as horseradish
peroxidase and cytochrome c peroxidase, these three certainly
have the same proximal ligand, and it is not the same ligand
as in myoglobin and catalase. It might, however, be the
negative ion form of histidine. We have no information whatsoever on what the distal ligands of these three peroxidases
might be.

Theorell: When I heard about these substitutions of the hemins in cytochrome c peroxidase, it came to my mind that some
25 years ago, when we had crystallized for the first time
horseradish peroxidase, we did exactly the same. We took away
the protoheme and introduced some analogs with different side

557

chains. As far as I remember, "meso"-peroxidase, with saturated 2-,4- side chains was quite active; deuterohemin without
these side chains also worked, whereas hematohemin and some
others did not! It would be nice to know how these data compare with the cytochrome c peroxidase. From what we have
heard from Dr. Asakura, it seems that the results are quite
similar.

Asakura: Substitutions at positions 2 and 4 do not affect
the enzymatic activity of cytochrome c peroxidase. With deuteroheme, it has the same activity as the native enzyme.

Nicholls: Earlier Dr. Schoenborn showed an X-ray picture of
the ethyl hydrogen peroxide complex of metmyoglobin with an
ethyl group still there. The conventional structures for
metmyoglobin peroxide, indeed, for all the Fe^{4+} states of
heme proteins, including the cytochrome c peroxidase intermediate (when you forget about the free radical), peroxidase
Complex II and catalase Complex II--are essentially two in
number: FeO^{2+} (ferryl) or $FeOH^{3+}$ (ferric hydroxyl radical
complex). Dr. Schoenborn's data would suggest a third one
might be possible, namely, $FeOR^{3+}$, where the R group is retained with the oxidizing equivalent on the iron. I wonder
now whether we have any preference for any of these or other
pictures.

Chance: I am very happy that Dr. Schoenborn has obtained,
with reasonable resolution, the difference Fourier of the
peroxide compound of ferrimyoglobin. Herman Watson and I
tried this previously but used pH 8.5 in order to obtain a
more stable compound, and got a mixture of the alkaline and
the peroxide forms. I think there is no such objection to
Dr. Schoenborn's data; it is apparent that there is something
in addition to the peroxide oxygen at the heme of the compound.

Peisach: I would like to extend some of the correlations between optical and magnetic properties of high-spin ferric
heme. Based on data shown today by Dr. Yonetani for cytochrome
c peroxidase studied at low temperature, and also from published data for sulfite reductase (1), I would like to suggest
that there exists a correlation between the percent of ferric
heme with the splitting of the optical peak near 640 nm. If
you consider that peaks as representing an energy transition
to a doubly degenerate state, this degeneracy is partially
lifted by imparting rhombicity to the system. With large

rhombicity, as with sulfite reductase, the normal spectral peak at 640 nm is split even at room temperature, and the splitting is large.

Moss: There are in the literature experiments with Japanese radish peroxidase (2) and with horseradish peroxidase (3) in which the Mössbauer spectra of Compounds I and II are nearly identical. There does not appear to be a change in the iron configuration in going from Compound I to II, although there is a big change in going from either to the resting enzyme. An Fe(IV) state is compatible with the data for Compound II, but a further oxidation to Fe(V) is not compatible with the data for Compound I.

References

1. Siegel, L.M., and M. Kamin. In Flavins and Flavo-proteins (K. Yagi, ed.), Tokyo, Tokyo University Press, 1968, p. 15.

2. Maeda, Y. J. Phys. Soc. (Japan), 24, 151 (1968).

3. Bearden, A.J., T.H. Msoo, and A. Ehrenberg. Symposium on Cytochromes, Osaka, August 1967 (M. Kamen and K. Okonuki, eds.), Tokyo, Tokyo University Press, 1968.

HIGH-SPIN LOW-SPIN EQUILIBRIA IN DERIVATIVES OF BACTERIAL CATALASE[*]

Gunilla Heimbürger and Anders Ehrenberg

Biofysiska institutionen, Stockholms Universitet
Medicinska Nobelinstitutet, Karolinska Institutet
Stockholm, Sweden

Catalase (EC 1.11.1.6) was prepared from Micrococcus lysodeikticus (1). The purified enzyme had a ratio A_{406}/A_{280} of 0.90 and a Kat. F. (activity) of 140,000 (2). Before use the enzyme was dialyzed against 0.1 M EDTA (ethylenediamine tetra acetic acid) at pH 8.0 to remove Cu^{2+} ions, which often appear as an impurity in catalase preparations (3).

Magnetic susceptibility measurements were carried out at 20°C with a horizontal Gouy balance (4) on catalase, its azide and cyanide complexes, all at pH 8.0. The effective magnetic moments 5.6, 5.5 and 2.3 Bohr magnetons were obtained for the three derivatives, respectively. This shows that the catalase itself and its azide complex are in a high-spin form, while the cyanide complex is in a low-spin form at room temperature, which is in agreement with the findings for erythrocyte catalase (5).

ESR (electron spin resonance) spectra were recorded for the same compounds with a Varian X-band spectrometer with Fieldial, 100 kc field modulation and a variable temperature dewar insert. At -185°C the ESR spectrum of catalase (Figure 1) shows that it is a high-spin compound with g values of 6.0 and 5.2, and that of cyanide catalase shows that it is a low-spin compound with g values at 2.9, 2.2 and 1.6. This is in agreement with the susceptibility measurements at 20°C and demonstrates that the spin states of these two derivatives does not change noticeably with temperature. The spectrum of the azide complex, on the other hand, turned out to be of

[*]This investigation was supported by grants-in aid from the Swedish Statens Medicinska Forskningsråd and Statens Naturventenskapliga Forskningsråd and the U.S. Public Health Service (AM-05895).

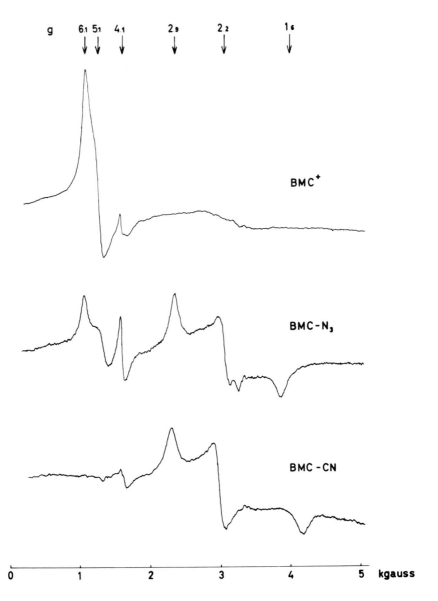

Figure 1. ESR spectra of catalase (BMC, Bacterium <u>Micrococcus lysodeikticus</u> Catalase), azide catalase and cyanide catalase at –185°C.

a mixed spin type at -185°C with g values of 6.2, 5.0, 2.8, 2.15 and 1.7. In all the ESR spectra an absorption with g = 4.15 was observed, but it could not be established whether this was due to the enzyme or an impurity.

The change from low-spin to high-spin form with increasing temperature was further investigated by taking ESR spectra at different temperatures. In Figure 2 the logarithm of the derivative amplitude is plotted against temperature. The

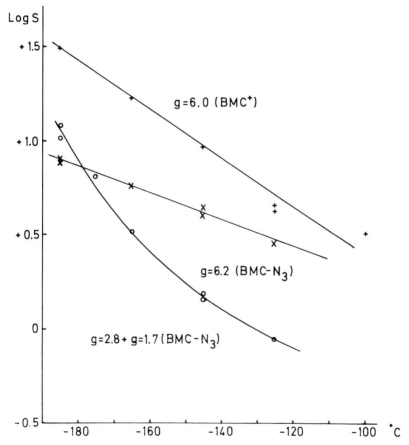

<u>Figure 2.</u> The logarithm of the derivative amplitude, S, of the ESR absorption of catalase (BMC$^+$) and the catalase-azide comlex (BMC-N$_3$) as a function of temperature (BMC$^+$:1.36 mM Fe, BMC-N$_3$:1.36 mM Fe). +:g=6.0 (BMC$^+$); x:g=6.2 (BMC-N$_3$); o:g= 2.8 + g = 1.7 (BMC-N$_3$).

line representing the high-spin signal at g = 6.2 decreases only slowly while the line representing the low-spin signal (g = 2.8 + g = 1.7) is much steeper and slightly curved. This may be interpreted in terms of a faster transition from low-spin to high-spin form with increasing temperature at the lower temperatures of the interval investigated. The same phenomenon was observed by Torii and Ogura (6) for the azide derivative of horse erythrocyte catalase. Since line broadening effects will also influence the curves of Figure 2, independent data are necessary for a more firm conclusion

The change from low-spin to high-spin form was also investigated in the visible region of the spectrum by the use of a Cary 14 R recording spectrophotometer, which was changed to record spectra at low temperatures by inserted dewars in the light paths. Specially made cuvettes of plexiglass in aluminum frames with a light path of 1 mm were used. When the bottom part of the cuvette was dipping into liquid nitrogen the temperature of the cuvette was about -190°C. The temperature was measured by a Pt-resistor placed in a cavity in the plastic part of the cuvette wall close to the sample. The reference cuvette contained 2% albumin solution in order to make the light scattering conditions comparable in both cuvettes. In the variable temperature experiments, the first measurement was taken at -190°C; then, the liquid nitrogen was removed and a slow flow of dry nitrogen gas was introduced, and spectra were taken at suitable temperature intervals.

The catalase azide complex was investigated by this method. The transition from low-spin to high-spin form with increasing temperature is indeed faster, up to about -140°C, which is in agreement with the ESR findings; but the pure high-spin form spectrum is not obtained until the temperature is raised to about -20°C.

The complexes of catalase and ammonia, methylamine, or ethylamine were investigated by low temperature spectrophotometry. Ehrenberg and Estabrook have found a low-spin cata-lase-ammonia complex at -196°C, and Scheler low-spin hemoglobin-ammonia and amino complexes at room temperature(+20°C) (7). The catalase complexes are in a low-spin form below -50°C. In contrast to the catalase azide complex, the transition from low-spin to high-spin form occurs within a fairly narrow temperature interval between -50° and 0°C (Figure 3). When the sample is thawed, the transition is complete, and as there is no difference in either wavelengths or

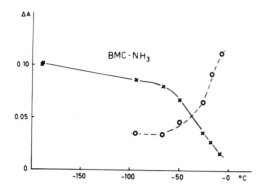

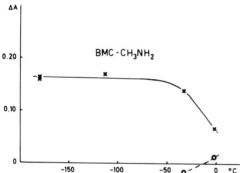

Figure 3. The temperature dependence of the high-spin and low-spin forms of the comlexes of catalase and ammonia methylamine and ethylamine. x:low-spin form $(A_{540}-A_{512})$; o: high-spin form $(A_{620}-A_{600})$.

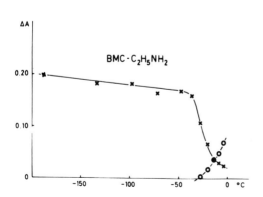

absorptions of maxima and minima, it is doubtful whether the catalase forms complexes with ammonia or amines at all at room temperature.

The proportion of low-spin compound formed below -50°C was dependent on the concentration of the added amine. Apparent dissociation constants were calculated for the complexes of catalase and the non-ionized base of ammonia (K=0.2mM), methylamine (K =0.2mM), and ethylamine (K = 8mM). The trend of these values is similar to that of the activities with hydrogen peroxide and its methyl and ethyl substituted analogs (8). Our results are thus consistent with the earlier suggestion (3) that catalase heme to amine complexes are formed. Since in all three cases the reversible high-spin to low-spin transition takes place in the same narrow temperature range between 0 and -50°C, it is likely that the species showing low-spin at low temperature are preformed already in the liquid or just upon freezing. In these species, the amine interacts with the protein in such a way that in the temperature range 0 to -50°C the changes of the lattice of ice will cause a small conformational change of the protein, so that the amine itself or an amino acid side chain, e.g. histidine imidazole, can coordinate to the heme iron in the sixth position.

The temperature dependent formation of the catalase amine complex is thus caused by a small conformational change taking place within a narrow temperature range. The spin state of the catalase-azide complex is, however, changing continuously from room temperature down to below -180°C and is best described as a temperature dependent high-spin low-spin equilibrium within the heme.

References

1. Beers, R.F. Science, 122, 1016 (1955).

2. Bonnichsen, R.K., B. Chance, and H. Theorell. Acta Chem. Scand., 1, 685 (1947).

3. Ehrenberg, A. and R.W. Estabrook. Acta Chem. Scand., 20, 1667 (1966).

4. Theorell, H. and A. Ehrenberg. Arkiv Fysik, 3, 299 (1951).

5. Deutsch, H.F., and A. Ehrenberg. Acta Chem. Scand., 6, 1552 (1952).

6. Torii, K. and Y. Ogura. J. Biochem., 64, 171 (1968).

7. Scheler, W. Biochem. Z., 330, 538 (1968).

8. Chance, B. and D. Herbert. Biochem. J., 4, 402 (1950).

DISCUSSION

Chance: I'd like to ask whether EPR data gives any information on the distance between this site and the heme group.

Ehrenberg: We have no information on that point at the moment.

Peisach: I should like to report briefly on a distinct difference between the EPR of heme iron of horse erythrocyte
catalase compared to the heme iron of a commercial preparation of beef liver catalase. The EPR of erythrocyte catalase
is clearly that of a single high-spin species in which the
heme is rhombically distorted from axial. This was reported
by Torii and Ogura (1). However, in addition to this particular resonance which one may also observe in beef liver
catalase, there is another EPR spectrum from another heme
which is in a totally different environment, also high-spin
but with a larger rhombic distortion.

Thomas: I want to make a few comments about one of the lesser
known peroxidases, chloroperoxidase, which seems to have
rather intermediate activities, in some respects, between
peroxidase and catalase. Physically the enzyme is similar
to horseradish peroxidase, with one heme, a molecular weight
of 45,000, about 20% carbohydrate and a largely acidic amino
acid composition. It manifests classical peroxidase activity
with respect to its ability to react with ascorbate and
guaiacol. In this respect, it has a turnover number of about
10% that of horseradish peroxidase. In terms of its catalytic
activity, the second order rate constant appears to be 10^4
$M^{-1}sec^{-1}$ as compared to 10^7 $M^{-1}sec^{-1}$ for catalase. The
catalytic activity of chloroperoxidase is stimulated by the
presence of chloride; this chloride stimulation occurs at
low pH and seems to be related to the ability of the enzyme to
bind chloride. Also, at low pH the enzyme acquires an ability
to catalyze halogenation reactions. Again, the ability of the
enzyme to catalyze halogenation reactions appears to be related to its ability to bind halides. With respect to active
sites, at the low pH and in the presence of chloride and peroxide, the enzyme denatures or kills itself, losing its

567

enzymatic activity. The rate of low inactivity has been shown to be first-order with respect to enzyme, indicating that the enzyme is killing itself rather than another enzyme molecule. Studies on the amino acid composition of the killed enzyme have indicated that one tyrosine becomes chlorinated. Otherwise there does not seem to be any important change in the amino acid composition. This indicates that a tyrosine residue plays an important role in the active site of chloroperoxidase.

Chance: Would Dr. Peisach care to describe the spectrum of the high-spin heme in beef liver catalase?

Peisach: Figure 1 is a trace of the spectrum taken at 1.4°K. In addition to the catalase EPR spectrum as seen with the erythrocyte protein with g values of 6.59 and 5.44, there is another resonance with g values of 6.95 and 5.07.

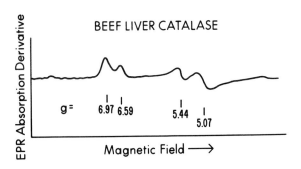

Figure 1.

Chance: The relationship might be suitable for assaying the number of hemes which are intact or "verdo"; can you specify the ratio of the two types of hemes:

Peisach: Approximately 2:1 in this sample.

Schonbaum: It might prove interesting to compare EPR spectra of bacterial and erythrocyte catalases which have no "verdo" heme.

Blumberg: Dr. Ehrenberg, do you know the g values of the low spin compound for any of the ammonia or the ammonia analogue complexes of catalase?

Ehrenberg: For the ammonia catalase compound, we reported 2.9, 2.3 and 1.8.

Blumberg: That is a dihistidine low spin form and thus catalase has a distal histidine as well as a proximal one.

Nicholls: When I see all the postulated histidines, both distal and proximal, I begin to wonder what causes the differences between catalase, peroxidase, hemoglobin and myoglobin. In particular, in the case of catalase, the distal histidine cannot be linked by a water molecule to the iron in the same way that it is in metmyoglobin and, presumably, methemoglobin. For there are no heme-linked ionizations throughout the range from pH 3 to pH 10 or so, and also because catalase binds the undissociated forms of acid ligands and not the anionic forms. I presume this might indicate that the histidine has become a much stronger acid, and never picks up a hydrogen ion in the normal course of events. It therefore has to be hooked up to the iron in some different way.

Margoliash: I would be delighted to be in a position to answer Dr. Nicholl's question in detail, but as we are all aware, we are still a fairly long way from such knowledge. May I just point out that in the discussion of the aminotriazole effect on catalase-H_2O_2,I, no mention whatsoever was made of a hypothetical proximal histidine. If the inhibitor had incorporated into such a histidine, one would expect a much larger spectral change in going from catalase to aminotriazolylcatalase than the very minor difference actually observed. Since the inhibitor nevertheless does react with a histidine specially rendered chemically highly reactive in Complex I, a reasonable position for it would be as a "distal" histidine. To extrapolate from this tentative idea to the conclusion that catalases, peroxidase, hemoglobins and myoglobins are all the same at the active center seems somewhat far fetched. If anything, the aminotriazole evidence would indicate that catalases and peroxidases differ in a very important way, since the inhibitor does not react with the latter except as an ordinary hydrogen donor, and one cannot invoke the usual difference in donor specificity between the two types of enzyme since aminotriazole does react with the Complex I of catalase.

One should also not forget, as amply discussed this morning, that the exact position of the iron may make a large difference in the reactivities of heme proteins and is quite capable of explaining the differences we are discussing.

Chance: The specificity difference is exhibited in the electron-accepting reactions of the intermediate complexes.

Schonbaum: Some ideas seem so plausible that we do not often reexamine the premises on which they are based. In the case of catalase, such an assumption is that water is hydrogen bonded to a distal group, at the sixth coordination position (3).

   To gain some insight into this problem, Mildvan and I have now compared proton relaxation rates of water, using solutions of catalase and its various derivatives. Our results do not invalidate current concepts; equally, we have no evidence bearing on hydrogen bonding between a presumed distal acceptor and water. However, the presence of water in the first coordination sphere is suggested by a nearly twofold decrease in the paramagnetic contribution to the proton relaxation rate (over the temperature range 1-36°) upon conversion of catalase into its azide complex.

   Another issue which has been raised concerns the interaction of catalase with ligands. I think that Dr. Chance's data--to which Dr. Nicholls referred--shows only that an undissociated form of entering ligand is involved in the rate limiting step of the reaction. Because, if non-rate limiting proton transfer were to follow the interaction of catalase and the acid ligand, then the state of ionization of the distal group would not be necessarily reflected in the kinetics of ligand exchange.

Moss: Does aminotriazole inhibit any of the peroxidases?

Margoliash: Aminotriazole has no effect on the activity of peroxidase. The only thing that appears to happen is that in the presence of H_2O_2 aminotriazole gets oxidized by the enzyme. This presumably indicates that the structures of the active sites of peroxidases and catalases are different.

Theorell: Does this work with hemoglobin as indicated?

Margoliash: We have not tried to react aminotriazole with hemoglobin. Since hemoglobin does not form with H_2O_2, a compound corresponding to Complex I, I would not expect the inhibitor to be incorporated in it, as it is in catalase. It certainly does not react with a fairly large variety of other proteins.

Schonbaum: The observation that aminotriazole does not affect HRP activity could be taken to indicate the absence of histidine at the distal site of the enzyme. Equally, these results could merely imply that peroxidase compound I, like the compound II of catalase, is reduced by 3-amino-1:2:4-triazole to give non-inhibitory oxidation products.

Chance: Can you give us more details on the atom activated by the peroxide?

Margolaish: It is easy enough to produce a number of hypothetical mechanisms for the possible addition of aminotriazole to the imidazole side-chain of histidyl residue 74 of catalase. One would require a strong electron withdrawl effect, making one of the imidazole nitrogens susceptible to nucleophilic attack. This can be obtained either if the H_2O_2 in complex I bridges between the heme iron and this imidazole group, or if the iron in complex I is in the Fe^{+5} form and is placed in an appropriate position with regard to the imidazole. Whatever the mechanism, it is clear that this particular imidazole is abnormally reactive. Aminotriazole does not react with any of the other 19 histidyl residues in the catalase subunit chain, as it does not with any residues in the other proteins investigated.

Schonbaum: Let me present a different interpretation of catalase-compound I aminotriazole reaction.

In short, my argument will rest on the premise that catalase inhibition by cyanogen bromide gives a product with strikingly similar properties to those shown by aminotriazole inhibited enzyme (4). To mention but a few points, you may recall that inhibited catalases obtained by either pathway: (a) have only one residue modified per enzyme subunit; (b) show similarities in their optical spectra; (c) retain small residual catalytic activity; and (d) have equally decreased affinities for various ligands.

Dr. Margoliash has now demonstrated that reaction with aminotriazole occurs at histidine 74. Again, it may be noteworthy that the chemical properties of the derivative formed in cyanogen bromide reactions parallel, in many respects, those of the cyanated imidazole. (We have failed to make a definitive structural assignment due to a limited stability of our compounds following proteolytic digestion of the enzyme.) It seems therefore that the inhibition of the enzyme, wehther with cyanogen bromide or aminotriazole, occurs at the same site.

We know also that the inhibition of catalase by cyanogen bromide occurs according to (4):

$$YH \text{ (Histidine ?)} + BrCN \longrightarrow Y\text{-}CN + Br^- + H^+$$

An analogous reaction could take place upon oxidation of 3-amino-1:2:4-triazole; the resulting intermediate, particularly if it were formed in two electron oxidation of aminotriazole, would react as readily as BrCN does with various nucleophiles, e.g. imidazole moiety of histidine.

Dr. Margoliash, would you comment on this suggestion?

Margoliash: I do not think I can answer that question if you are asking whether an aminotriazole molecule that is oxidized is the one that reacts, presumably invoking local concentration effects. The evidence at hand is that oxidized products of aminotriazole prepared separately, such as the 3,3'-azo-1, 2,4-triazole, have no effect on catalase (5) and that when the product of the reaction with catalase is treated with HI, one molecule of ordinary aminotriazole is recovered quantitatively for each molecule of reacted histidine, not any oxidized or decomposition products of it. Thus, I find it difficult to believe that an oxidation product of aminotraizole can be involved in the reaction. In this regard, it may be pertinent to recall that the kinetics of inhibition merely indicate the reaction to be a second order reaction between the inhibitor and Complex I (5). Whether this bears unequivocally on the question of a hypothetical intermediate form of aminotriazole would take too long to discuss.

Schonbaum: None of these remarks weaken my suggestion. And since the reaction is envisaged to proceed according to the generalized scheme (equation 1), the intramolecular oxidation

$$Catalase + aminotriazole \rightleftharpoons$$

$$[Complex] \longrightarrow Inhibited\ enzyme \tag{1}$$

of aminotriazole and the apparent second order kinetics of
inhibition are not necessarily mutually exclusive.
Furthermore, the product of such reaction could well regenerate aminotriazole upon reductive cleavage with HI--as seems
to be the case. Even more to the point is the explanation
which the oxidative scheme provides for the reactivity of
other compounds as catalase inhibitors -- compounds structurally related to 3-amino-1:2:4-triazole and like it readily
subject to oxidation.

This specificity would be surprizing if the only
necessary condition were to render, as we are told ". . . one
of the imidazole nitrogens susceptible to nucleophilic
attack." Indeed, the very concept of the reaction, as outlined by Dr. Margoliash, remains unclear to me since it would
seem that either aminotriazole, or imidazole moiety of histidine, or some intermediate formed from these compounds, must
be oxidized to give the inhibited derivative.

References

1. Peisach, J. In Developments of Magnetic Resonance in
   Biological System (S. Fujwara and L.H. Piette, eds.),
   Hirokowa Publ. Co., Tokyo, 1968, p. 101.

2. Nicholls, P. and G. Schonbaum. In The Enzymes (P.D.
   Boyer, H. Lardy and K. Myrback, eds.), Academic Press,
   Vol. 8.

3. Schonbaum, G. and F. Jajczay. In Hemes and Hemoproteins
   (B. Chance, R. Estabrook and T. Yonetani, eds.),
   Academic Press, New York, 1966, p. 327.

4. Margoliash, E., A. Novogrodsky and A. Schejter. Biochem.
   J., 74, 339 (1960).

5. Nicholls, P. Biochim. Biophys. Acta, 59, 414 (1962).

SOME PROBLEMS OF OXIDATION-REDUCTION OF CYTOCHROME c OXIDASE AND THE NATURE OF ITS "OXYGENATED" FORM*

H. Beinert, C.R. Hartzell and W.H. Orme-Johnson

Institute for Enzyme Research, University of Wisconsin
Madison, Wisconsin 53706

Cytochrome c oxidase had been known as an enzyme of complicated properties and behavior (1-6) even before the advent of EPR spectroscopy as a tool of inquiry. This technique of observation, although allowing a deeper insight in some respects, has added its share toward increasing the complexity of the picture. We propose in the following to pursue some lines of thought--initiated by our EPR spectroscopic observations--on the events during accumulation of electrons in and discharge from this enzyme and on the various forms in which one might expect to find this enzyme at the stages it goes through during these events. Obviously, all forms of the enzyme known to date are of interest in such a discussion and with some apprehension we are aware that we enter the "lions den" and will also try to explore how far our EPR observations may be useful in solving the mystery of the so-called "oxygenated" form of cytochrome c oxidase.

The basis for our approach was the observation--communicated in 1967 by van Gelder, Orme-Johnson, Hansen and Beinert (7,8) and documented in more detail elsewhere (9)--that during stepwise reduction of cytochrome c oxidase by any reducing agent applied, intermediate forms of the enzyme appeared which had characteristics found neither in the oxidized nor in the reduced form of the enzyme (see Figure 1). In other words, at states of partial reduction, there was not present a simple

*Supported by the Institute of General Medical Sciences, USPHS GM-12394, Research Career Award (GM-KG-18-442) to H.B., Postdoctoral Fellowship Award (1-F2-GM-33,198) to C.R.H., and a Research Career Development Award (5-K03-GM-10,236) to W.H. O.-J.

575

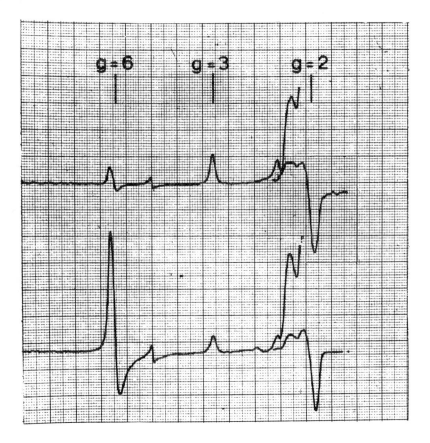

Figure 1. Low field portion of EPR spectra of cytochrome c oxidase in the oxidized (upper curve) and partly reduced (lower curve) state. Cytochrome c oxidase was prepared according to (9) up to the dialysis step. Further purification was achieved by two precipitations with ammonium sulfate at 20-30% saturation. The enzyme, 50 μM, was dissolved in 90 mM potassium phosphate, pH 7.3, containing 0.9% Tween 80. The enzyme was approximately 50% reduced with sodium dithionite. The conditions of EPR spectroscopy were: microwave power, 27 mw; modulation amplitude, 6 gauss; scanning rate, 1000 gauss per min; time constant, 0.5 sec; and temperature 39°K. The copper signal at g = 2 was recorded at one fifth the amplification used for the rest of the spectrum. The spectrum of the completely reduced enzyme, which is not shown, has only a small residual signal of "inactive" copper (10-13).

mixture of oxidized and reduced forms nor were there present forms simply combining properties of the fully oxidized and fully reduced forms.

This can be best explained and visualized if we introduce some schematic representation of the active center of cytochrome c oxidase. It is currently accepted by most workers in the field that the smallest enzymatically functional unit of cytochrome c oxidase incorporates four potential electron carriers, namely two hemes and two copper atoms (3,5,6). Both the hemes as well as the copper atoms, when in their proper location in the enzyme, differ among themselves in their more detailed properties. We will, therefore, use the following designation for the minimal functional unit of cytochrome c oxidase:

$$\underline{a} \qquad \underline{a}_3$$
$$Cu_d \qquad Cu_u$$

indicating the presence of cytochrome $\underline{a}$ and $\underline{a}_3$ and the EPR-detectable copper, Cu_d, and the EPR undetectable copper, Cu_u (10-14). For the oxidized form of the enzyme, we write the components in their oxidized forms $\underline{a}^{3+}$, $\underline{a}_3^{3+}$ and Cu_d^{2+}. There is some uncertainty as to the state of Cu_u in the oxidized form of the enzyme. From reductive titrations (14,15), it is known that the enzyme takes up at least 4 e^- before optical and EPR spectra indicate complete reduction. It was also shown that on denaturation with urea in the presence of PCMS and with careful exclusion of oxygen (12), all the copper of the enzyme appears in EPR-detectable form as Cu^{2+}. Neither of these observations is, however, conclusive, as Cu_u may, nevertheless, be present as Cu^+ (diamagnetic, i.e., no EPR signal) and in addition there may also be an unknown electron acceptor which is titrated on reduction or accepts an electron from Cu^+ on denaturation. A third possibility emerges from new experimental results revealed at this meeting (16), namely that Cu_d and Cu_u may, in fact, not be different, but that interaction with a neighboring paramagnetic component may diminish the intensity of the Cu^{2+} EPR signal (now representing both Cu atoms) without significant broadening. For the purposes of our discussion, we will assume that Cu_d and Cu_u are different and that both occur as Cu^{2+} in the oxidized form of cytochrome c oxidase. We can then write this form as

$$\underline{a}^{3+} \quad \underline{a}_3^{3+}$$
$$Cu_d^{2+} \quad Cu_u^{2+} \qquad (A)$$

and its fully reduced counterpart as

$$\underline{a}^{2+} \quad \underline{a}_3^{2+}$$
$$Cu_d^{+} \quad Cu_u^{+} \qquad (B)$$

Spectrophotometry detects $\underline{a}^{3+}$, $\underline{a}^{2+}$, $\underline{a}_3^{3+}$, and probably Cu_d^{2+}, although the spectra of the cytochromes overlap considerably, and EPR spectroscopy detects $\underline{a}^{3+}$ (a low spin heme) and Cu_d. From this it would appear that EPR spectroscopy is the inferior tool for studying cytochrome $\underline{c}$ oxidase. This may be true for the fully reduced form, where there are no EPR signals, and debatable for the oxidized form, but it is certainly not true for states of intermediate oxidation where spectrophotometry did not indicate the existence of forms with properties other than those of the oxidized or reduced ones.* That such states can be expected to exist is obvious from inspection of our schematic representation, since we can write 14 different intermediate states where one, two, or three electrons are randomly distributed among the four components of the oxidase. We do, of course, not expect such a random distribution, but expect some order in the distribution of electrons as they accumulate or leave the enzyme. We will now examine the evidence we have for the existence of specific intermediates. Our information is mainly derived from titrations, i.e., equilibrium situations, and partly from rapid reductions and oxidations with the shortest observation times of approximately 10 msec. We have no information on events which may occur more rapidly. Electron distribution for which we have no evidence may transiently exist under such conditions. According to the changes observed in the EPR spectra, the first electrons entering the active center of cytochrome $\underline{c}$ oxidase on titration are accepted by cytochrome $\underline{a}$ (low spin heme) and an EPR undetectable component (14), possibly Cu_u. We may represent the forms produced as follows:

$$
\begin{array}{cc cc cc}
\underline{a}^{2+} & \underline{a}_3^{3+} & \underline{a}^{3+} & \underline{a}_3^{3+} & \underline{a}^{2+} & \underline{a}_3^{3+} \\
Cu_d^{2+} & Cu_u^{2+} & Cu_d^{2+} & Cu_u^{+} & Cu_d^{2+} & Cu_u^{+} \\
\text{(C)} & & \text{(D)} & & \text{(E)} &
\end{array}
$$

These, or some of these intermediate forms, have a new property, namely they now show an EPR signal of cytochrome $\underline{a}_3^{3+}$ at g = 6, which is undetectable in the oxidized state (A), i.e., the oxidized form of cytochrome $\underline{a}^{3+}$ appears on reduction as a typical high spin ferric heme (17). The fact that this signal was absent in form (A) indicates that in this form an interaction occurs between $\underline{a}_3^{3+}$ and another paramagnetic component, possibly Cu_u, whose lack of detectability

*Since the "oxygenated" form is most likely an alternative, fully oxidized species, it is omitted from this consideration.

could be explained by the very same argument. We have previously considered antiferromagnetic interaction between
these components as one of the most likely possibilities (14,
18). If, as we cannot at present exclude, multiples of the
minimal functional unit have to be considered, interaction
between 2 $\underline{a}_3^{3+}$ may also explain the absence of an EPR signal
for $\underline{a}_3^{3+}$. For the sake of simplicity, we will restrict ourselves to interactions which might occur within individual
minimal functional units.

If copper-heme interaction is responsible for the lack
of an EPR signal for $\underline{a}_3^{3+}$ in form (A), we would expect $\underline{a}_3^{3+}$ to
become detectable, when Cu_u is reduced, that is, forms (D)
and (E) will be those that are distinguished by a high spin
ferric signal at g = 6.

As additional electrons are accepted, Cu_d and $\underline{a}_3^{3+}$ are
also reduced, as indicated by the disappearance of their EPR
spectra, i.e., the following forms will appear:

$$\begin{array}{cc} \underline{a}^{2+} \quad \underline{a}_3^{3+} & \underline{a}^{2+} \quad \underline{a}_3^{2+} \\ Cu_d^+ \quad Cu_u^+ \; ; & Cu_d^+ \quad Cu_u^+ \\ \quad (F) & \quad (G) \end{array}$$

and finally (B), the fully reduced form.

The principal difference between observation after equilibration and rapid observation appears to be that within msec
Cu_d is found reduced together with $\underline{a}$. The forms discussed
above (C-G) may thus partly be reached via different states,
such as for instance

$$\begin{array}{cc} \underline{a}^{3+} \quad \underline{a}_3^{3+} & \underline{a}^{2+} \quad \underline{a}_3^{3+} \\ Cu_d^+ \quad Cu_u^{2+} & Cu_d^+ \quad Cu_u^{2+} \\ \quad (H) & \quad (I) \end{array}$$

The EPR signal of $\underline{a}_3^{3+}$ does, however, arise without lag.

On reoxidation with the natural oxidant O_2, we now en-
counter another interesting situation. If we assume, in
keeping with rather generally accepted ideas (3-5), that the
reduced form of $\underline{a}_3$ is the component that interacts with O_2,
we realize that, if there is any reality in our formalism,
forms such as (C), (D), (E), and (F), which have $\underline{a}_3$ still in
the ferric state, will be refractory to oxidation, at least
to rapid reoxidation, on the time scale of the normal cytochrome $\underline{c}$ oxidase reaction. This, indeed, appears to be the
case. We have found that, in agreement with Gibson and

Greenwoods experiments (19), cytochrome $\underline{c}$ oxidase, when fully reduced by an excess of dithionite, is completely re-oxidized in 10 msec, that is faster than the resolution of our rapid freezing technique allows us to follow. No inter-mediate forms are seen under these conditions. If an excess of reductant is present, such forms do appear within seconds by secondary reduction. When we start, however, with partly reduced cytochrome $\underline{c}$ oxidase, in which forms such as (C)-(F) are present, produced by less than stochiometric amounts of dithionite, reoxidation is slow and extends to seconds. We think that reoxidation eventually occurs through slow intra- or intermolecular oxidation-reductions which may not neces-sarily involve oxygen directly.

We see one important aspect in this behavior of cyto-chrome $\underline{c}$ oxidase: the functional unit is built for a four electron oxidant; it is only under conditions when four electron oxidation by O_2 is possible that a reaction rate is observed characteristic of an enzyme and particularly of this notoriously efficient one. The interaction with another component, possibly Cu_u, makes $\underline{a}_3^{3+}$ unavailable for reduction, at least by the first 2 electrons that enter the functional unit. Only when this interaction has been broken, i.e., when the EPR signal of $\underline{a}_3^{3+}$ has appeared, can $\underline{a}_3^{2+}$ be formed by the addition of electrons. The oxygen receptor site, $\underline{a}_3^{2+}$, is thus available only when all components are reduced, that is when 4 electrons have been accepted by the unit. This mech-anism would appear to provide an electron gate, which is set for the rapid passage of four, and only four, electrons.*

When we followed the course of the slow reoxidation of partly reduced cytochrome $\underline{c}$ oxidase by oxygen, we observed that the signal of the high spin ferric heme $\underline{a}_3^{3+}$ disappeared only after Cu_d and $\underline{a}$ were reoxidized. It thus appears that during such reoxidation, forms such as (D), (E) or (K) are intermediates:

$$\underline{a}^{3+} \quad \underline{a}_3^{3+}$$
$$Cu_d^+ \quad Cu_u^+$$
$$(K)$$

The observation that forms (D), (E), or (K), in which Cu_u is cuprous and $\underline{a}_3$ is therefore uncoupled from the inter-

*This implies only that the transfer of 4 electrons is very rapid and complete within a time range into which the rapid reaction techniques thus far applied do not extend.

action, are not readily oxidized by oxygen, and the thought that the uncoupled a_3^{3+} might have an optical manifestation, differentiating it from the EPR undetectable coupled form, aroused our interest in the so-called oxygenated form of cytochrome c oxidase, which had been reported to appear on exposure to oxygen of cytochrome oxidase reduced by a number of reductants.

In most instances, dithionite was used and an excess of reductant was present together with oxygen when the oxygenated species was produced. A number of the conditions necessary for the production of this species were those which also gave rise to the species showing the rather persistent high spin ferric heme signal (g = 6) on the reoxidation of partly reduced cytochrome c oxidase. We thus considered the possibility that a mixture of forms such as (D), (E), and (K)--in this order of prevalence--might constitute what is thought to be the "oxygenated" form. Supporting this idea was our consistent finding of strong EPR signals of the uncoupled a_3^{3+} in samples of the oxygenated form which had been prepared in the classical manner, namely by shaking mixtures of oxidase and an excess of reducing agent with air. Before the EPR signals were recorded, the optical spectra were examined and after freezing and EPR spectroscopy, the reflectance spectra were recorded to ensure that no changes had taken place in the sample. The light absorption spectra we observed for what we considered to be the "oxygenated" form, were of the type published by several investigators, with a Soret maximum at 428-430 mμ (Figure 2).

According to these ideas, the "oxygenated" form should have been a mixture of partly reduced species with the common characteristic of uncoupled a_3^{3+} and not a single species, as postulated by others. The lowermost spectrum of Figure 2 also shows clearly that a simple mixture of oxidized and reduced cytochrome c oxidase does not show the characteristics of the "oxygenated" form.*

Indeed, the variability of the light absorption spectra published in the past for the "oxygenated" form (see Figure 3) supported the idea that this form is not a single species but a mixture of species (23-30). Even spectra published from

---

*As pointed out previously (14), oxidized and reduced cytochrome c oxidase do not exchange electrons readily.

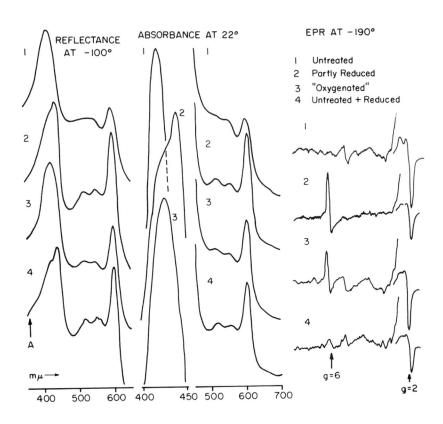

Figure 2.

Figure 2. EPR and light absorption spectra of cytochrome c
oxidase after different treatments, resulting in different
mixtures of species: (1) untreated "oxidized"; (2) partly
reduced anaerobically with dithionite; (3) a solution similar
to that of (2) briefly exposed to air, i.e., "oxygenated";
(4) a mixture of equal amounts of (1) with fully reduced enzyme. The enzyme was prepared as described in Figure 1. For
(2), 0.9 moles of dithionite per mole of heme were added anaerobically (cf. 20) to 0.50 ml of enzyme; for (3) air was admitted
(0.2 ml to a total enclosed volume of 15 ml) to a solution similar
to that of (2) and the solution was gently shaken for 20 sec;
for (4), 0.25 ml aliquots of enzyme were kept in separate
compartments of one tube (cf, 21) for 2.5 hr, one untreated
and one reduced with 0.5 mg of potassium ascorbate in the
presence of 1 µM cytochrome c monomer. The aliquots were then
mixed and the solution was frozen within 2 min. Light absorption was measured at 22° for (1) to (3). With (2), 20
min were allowed after adding dithionite and this sample was
frozen after an additional 10 min. (3) was frozen within 5
min from admitting air. EPR and reflectance spectra were
then recorded after these other measurements, and subsequent
thawing. Since under these conditions slow reduction continues, no representative spectrum from the UV region could
be obtained. The figure shows, from left to right, reflectance, absorbance and EPR spectra as marked. The abscissa
represents wavelengths in mµ and magnetic field strength,
respectively; the ordinate, represents apparent absorbance as
measured by diffuse reflectance spectroscopy (cf, 22), absorbance as measured at 22°, and for the EPR spectra first
derivative of the EPR absorption. The curves are staggered
for convenient comparision. For reflectance, the absorbance
values preprinted on the chart apply directly but must be
divided by 2 (0-0.5 O.D. slidewire). For absorbance at 22°
between 450-600 mµ, the preprinted values apply directly,
but the expanded UV portion is merely shown for comparison of
shapes, not absolute intensities. The ordinate for the EPR
spectra is in arbitrary units. The conditions of EPR spectroscopy were: microwave power, 45 mwatt; modulation amplitude, 12 gauss; scanning rate, 1000 gauss/min; and time constant, 1 sec. For accurate evaluation of the original re-
cords, which are shown here in reproduction, corrections
(<20%)for differences in concentration have to be applied.
Spectra for a fully reduced sample of the enzyme are not
shown. For such a sample at 22°, a $\Delta_{A605-620\ m\mu}$ of 0.5 for
the α-peak and no significant EPR signals were found.

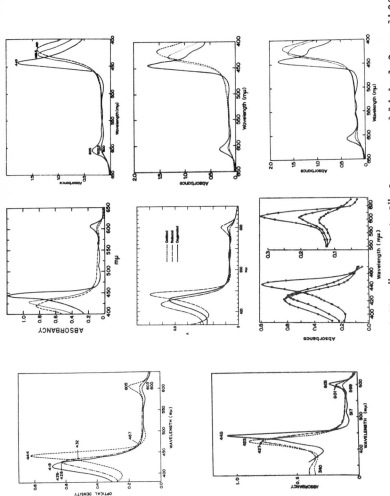

Figure 3. Light absorption spectra of the "oxygenated" form as published from different laboratories at various times. The spectra are taken, with permission, from the following references (counting from top to bottom in each column starting on the left side): 29, 27; 30, 26, 28; and 24, 23, 23.

the same laboratory at different times differ considerably among each other (23-25,27-29), and no agreement exists on the absorptivities for the α and Soret peaks and isosbestic points.

Nevertheless, a number of observations, which have just come to my attention during my journey to this meeting and in the corridors between sessions, are not in agreement with such an interpretation of the nature of what we may call the true "oxygenated" form. There is no doubt that "oxygenated" forms prepared in the classical way and in several instances, for which spectra appear in Figure 3 (e.g., Figure 3, bottom row), must have been a mixture of species containing forms specified above and exhibiting the g = 6 signal. Examples of spectra obtained in our work from such mixtures are collected in Figure 4. We may call such mixtures with spectra similar to the "oxygenated" form the "pseudo-oxygenated" form. Although the introduction of this term may seem to confuse the issue even more, it is true that the very appearance of such mixtures has, in the past as well as in our own experimental work, indeed led to confusion. The separation of the reducing agent from cytochrome c̲ oxidase, which is now possible through gel filtration (31), does, however, produce a relatively stable material, with a Soret peak at 426-427 mμ which does not show a g = 6 signal by EPR. We have not been able to detect differences between the EPR spectra of this form and the oxidized form (A) of cytochrome c̲ oxidase which exceeded differences observed between the spectra of a number of samples of the oxidized form by itself prepared by the same or by different methods. We hesitate, therefore, at this time to consider what differences we did see as significant. The EPR spectra were investigated at temperatures down to 10°K. Since, according to this, the signals exhibited by the oxidized form, viz̲. those of Cu_d and a̲, appear to be the same in shape and intensity as those of the species with the typical light absorption spectrum and Soret peak at 426-427 mμ, the latter seems to be an alternative oxidized form. This is also suggested by results obtained very recently by van Gelder* and by Williams, Lemberg and Cutler (31), namely that the "oxygenated" form is at the same oxidation state as the oxidized form. Such determinations of the oxidation state were, of course, possible only when the "oxygenated"

*B.V. van Gelder, personal communication, and E.C. Slater, this Symposium, p.

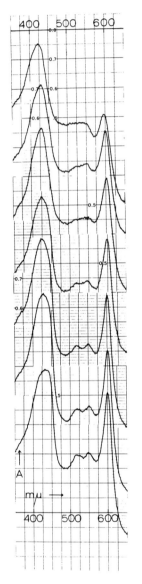

Figure 4. Reflectance spectra, recorded at 173°K, of cytochrome c oxidase in the oxidized state (top curve) and at various states of reduction (lower curves), illustrating the gradual shift of the "absorption" peak from 424 mμ to longer wavelengths. Cytochrome c oxidase, approximately 50 μM, was prepared as described for Figure 1 and reduced anaerobically to various degrees (50-80%) and then briefly exposed to air as described for Figure 2, curves 2 and 3. The oxidation state is thus determined by the degree of initial reduction and subsequent exposure to oxygen. All samples showed significant signals at g = 6. The sample showing the spectrum of the third curve comes closest to what is called the "pseudo-oxygenated" form in the text.

form could be separated from excess reducing agent and the results obtained, if correct, exclude the possibility that the "oxygenated" form consists of a mixture of partly reduced forms such as (D), (E), and (K) above, which should be titrated with 2 or 3 electron equivalents per unit instead of four, as required by the oxidized form (A).

There is also spectrophotometric information which indicates that the true "oxygenated" form is indeed a single species. Some of the more recently obtained light absorption spectra hint at this (25,26,30), but CD spectra of the "oxygenated" form in particular leave little doubt that in a pure "oxygenated" form very few, if any, other species can be present.* Such evidence, by itself, does, of course, still not rule out that this species might not be form (D) above or even a mixture of (D) and (K), as such a form of forms, with an uncoupled $\underline{a}_3^{3+}$ (EPR signal at g = 6) and $\underline{a}$ in the ferric state, would be expected to have a spectrum different from (A) or (B).

A third line of evidence indicates that the characteristics of reduction of the "oxygenated" form are different from those of the oxidized (31).** Again, this could only be ascertained after excess reducing agent could be separated from the "oxygenated" form.

All this evidence taken together has convinced us that the "oxygenated" form must be a variant of form (A).

We have been wondering whether the "oxygenated" form may not arise when forms such as (D), (E), and (K), which have the uncoupled $\underline{a}_3^{3+}$ (EPR signal at g = 6), are slowly reoxidized--as we reported above--as opposed to the rapid reoxidation, which completely reduced cytochrome oxidase undergoes and which should presumably lead to the oxidized form (A). We, therefore, compared the product of reoxidation of cytochrome $\underline{c}$ oxidase which had been reduced with an excess of dithionite and passed through a column of Sephadex G-25 in the presence of air--a situation where the enzyme would be exposed to reductant and oxygen simultaneously for a while-- and the product of oxidation of reduced samples which had been freed of dithionite on an anaerobic column of G-25. The enzyme emerged from this column completely reduced and,

*B.V. van Gelder, personal communication, and E.C. Slater, this Symposium, p.

**Y.P. Meyer, personal communication and this Symposium, p.

when shaken with air, promptly formed the "oxygenated" species with a Soret peak at 426–428 mμ. These experiments would then rule out the possibility that reducing agent or its products and oxygen have to be present simultaneously in order to generate the "oxygenated" form and that slow reoxidation via species with uncoupled $\underline{a}_3^{3+}$ may be necessary for its formation.

Wharton and Gibson have made a study of the reoxidation of reduced cytochrome $\underline{c}$ oxidase and came to the conclusion that the oxidized form is the initial product of oxidation within the first few msec. We have therefore applied the rapid freezing technique followed by EPR spectroscopic and optical (low temperature reflectance) observation. The product obtained at a nominal reaction time of 6 msec, which together with the effective quenching time may have to be corrected upward to $\sim$ 10 msec, had a Soret peak at 425–426 mμ. According to our experience, the peak position measured at room temperature by absorption spectrophotometry does not differ by more than 0.5 mμ from that measured at 173°K by the reflectance technique. It thus becomes more and more difficult to judge what form one is, in fact, dealing with, particularly if we consider that the enzyme as isolated, which one generally considers to be the oxidized form, very often has its Soret absorption at 422–424 mμ. The preparation used in our rapid reaction studies, for instance, had a Soret maximum at 423 mμ. When this preparation was left at 0° for one day, the Soret peak moved to 420 mμ; occasionally we have seen it move to 418 mμ in such instances. According to Lemberg's interpretation, preparations such as those we mentioned, namely with a Soret peak at 422–424 mμ, are, or at least contain, "oxygenated" form. There is no evidence why this should be so and there is no evidence that such preparations are enzymatically less active.

The question now obviously arises: Which form of the oxidase is the active oxidized form?

We feel at this moment that, before the identity of the form in which the enzyme is isolated is settled, it will be impossible to come to decisive conclusions as to what the various forms or mixtures of forms represent that arise under various conditions and have their Soret peaks at 425, 426, 427 or 428 mμ.

Lemberg has suggested (6,31) that in the "oxygenated" form, $\underline{a}_3$ is present as $\underline{a}_3^{4+}$ and one of the coppers (Cu_u ?) in the cuprous form. This would correspond to an overall oxidation state identical to that with $\underline{a}_3^{3+}$ and Cu^{2+}, i.e., (A). Since EPR at $\sim 10°K$ detects neither $\underline{a}_3$ nor Cu_u in the oxidized or "oxygenated" state, it cannot make any contribution to this question. We would, however, prefer not to invoke a fencing off, so to speak, of one electron within the cytochrome $\underline{c}$ oxidase unit so that Fe^{4+} and Cu^+ could coexist in a relatively stable state.

Lemberg has also suggested (32) that the appearance of the high spin ferric EPR signal at g = 6 on reduction of our preparations (7,8,14) indicates reduction of Fe^{4+} to Fe^{3+} (high spin), since our preparations are supposedly in the "oxygenated" Fe^{4+} form to start with. However, the appearance of the signal at g = 6 on partial reduction has been observed in all preparations of different origin which we investigated whether the Soret peak was initially at 418 or 424 mμ. If Lemberg were correct, we should have seen a strong dependence on the initial position of the Soret band. Also, maximal appearance of the high spin form is seen only after addition of 2 electrons out of the total 4-5 needed to reduce the oxidase unit completely, whereas the ideal oxidized form should be at the same oxidation level as the "oxygenated" form and should, therefore, show $\underline{a}_3$ in its ferric state before addition of any electrons. This has never been observed.

Although, obviously, we now know no better what the "oxygenated" form is than our colleagues and predecessors in this field, maybe we have a few new hints as to what it is not. Let us summarize what we think we have learned from the experiments (excluding those published previously) discussed in this contribution.

1. Partly reduced cytochrome $\underline{c}$ oxidase is only slowly (0.1-10 sec) reoxidized by O_2, whereas completely reduced enzyme is reoxidized within < 10 msec, as previously known (19). This we attribute to the fact that $\underline{a}_3^{3+}$ is reduced after the other components and partly reduced cytochrome $\underline{c}$ oxidase units will, therefore, have ferric $\underline{a}_3$ and thus no receptor for O_2.

2. Mixtures of partly reduced and partly reoxidized forms of cytochrome oxidase have a number of properties of the so-called "oxygenated" form, including light absorption spectra, as they have often been shown for this form. The Soret peak is generally found at 428 mμ.

3. Nevertheless, there is evidence, which we have confirmed, that there is a true apparently single species, at the oxidation state of the oxidized enzyme, which has the properties of the "oxygenated" form and a Soret peak at 426–428 mμ. There is, however, no compelling evidence that this form incorporates the oxygen molecule as part of its structure.

4. The "oxygenated" form with Soret peak at 426–428 mμ is formed on rapid (> 10 msec) or slow reoxidation of partly reduced cytochrome c oxidase and of completely reduced oxidase in the presence of dithionite. When completely reduced oxidase is separated from dithionite, rapid reoxidation leads to a species with a Soret peak at 425–426 mμ whereas slow reoxidation produces a species with Soret peak at 427–428 mμ.

5. Since the enzyme, as isolated, may have a Soret peak anywhere between 418 and 424 mμ it becomes necessary to define the state of the "oxidized" enzyme as isolated, before the nature of the other forms mentioned under 3 and 4 can be understood.

6. EPR spectroscopy at temperatures between 10 and 100°K has thus far not shown significant differences between the oxidized and "oxygenated" forms of cytochrome c oxidase. This indicates that the difference between the two forms is probably not to be found in the a and EPR detectable copper components, namely those which are seen in the spectrum of the oxidized form. If it is related to the a_3 or undetectable copper (neither of which is detected in the oxidized form), it is not associated with a change in state of these components that would make them detectable by EPR in the specified temperature range.

Nevertheless, since no significant change is observed with EPR detectable components, it seems more likely that the difference between the two forms lies in some change in a_3 or the undetectable copper or both.

Acknowledgments

We are indebted to Miss Sabine Schweidler for assistance with the preparations. Meat byproducts were obtained from Oscar Mayer & Co., Madison, Wisconsin.

References

1. Yonetani, T. In The Enzymes (P.D. Boyer, H. Hardy, and K. Myrbäck, eds.), 2nd ed., Academic Press, New York, 1963, p. 41.

2. Symposium: Haematin Enzymes (J.E. Falk, R. Lemberg, and R.K. Morton, eds.), Proc. Symp. Intern. Union Biochem., Australian Acad. Sci., Canberra, 1959, Pergamon Press, London, 1961.

3. Oxidases and Related Redox Systems (T.E. King, H.S. Mason, and M. Morrison, eds.), J. Wiley and Sons, New York, 1965.

4. Hemes and Hemoproteins (B. Chance, R. Estabrook, and T. Yonetani, eds.), Academic Press, New York, 1966.

5. Structure and Function of Cytochromes (K. Okunuki, M.D. Kamen, and I. Sekuzu, eds.), University of Tokyo Press, Tokyo, 1968.

6. Lemberg, M.R. Phys. Rev., 49, 48 (1969).

7. van Gelder, B.V., W.H. Orme-Johnson, R.E. Hansen, and H. Beinert. Proc. Nat. Acad. Sci. U.S., 58, 1073 (1967).

8. Beinert, H., B.F. van Gelder, and R.E. Hansen. In Structure and Function of Cytochromes (K. Okunuki, M.D. Kamen, and I. Sekuzu, eds.), University of Tokyo Press, Tokyo, 1968, p. 141.

9. Sun, F.F., K.S. Prezbindowski, F.L. Crane, and E.E. Jacobs. Biochim. Biophys. Acta, 153, 804 (1968).

10. Beinert, H., D.E. Griffiths, D.C. Wharton, and R.H. Sands. J. Biol. Chem., 237, 2337 (1962).

11. Beinert, H., and G. Palmer. J. Biol. Chem., 239, 1221 (1964).

12. Beinert, H., and G. Palmer. In Oxidases and Related Redox Systems (T.E. King, H.S. Mason, and M. Morrison, eds.), John Wiley & Sons, Inc., New York, 1965, p. 567.

13. Beinert, H. In Biochemistry of Copper (J. Piesach, P. Aisen, and W.E. Blumberg, eds.), Academic Press, Inc., New York, 1966, p. 213.

14. van Gelder, B.F., and H. Beinert. Biochim. Biophys. Acta, 189, 1 (1969).

591

15. van Gelder, B.F. Biochim. Biophys. Acta, 118, 36 (1966).

16. Taylor, J.S., J.S. Leigh, and M. Cohn. Proc. Nat. Acad. Sci. U.S., 64, 219 (1969).

17. Griffiths, J.S. Proc. Roy. Soc. (London) A., 235, 23 (1956).

18. Orme-Johnson, W.H., and H. Beinert. Ann. New York Acad. Sci., 158, 336 (1969).

19. Gibson, Q.H., and C. Greenwood. J. Biol. Chem., 240, 2 2694 (1965).

20. Tsibris, J.C.M., R.L. Tsai, I.C. Gunsalus, W.H. Orme-Johnson, R.E. Hansen, and H. Beinert. Proc. Nat. Acad. Sci. U.S., 59, 959 (1968).

21. Rajagopalan, K.V., P. Handler, G. Palmer, and H. Beinert. J. Biol. Chem., 243, 3784 (1968).

22. Palmer, G., and H. Beinert. Anal. Biochem., 8, 95 (1964).

23. Lemberg, R., and G.E. Mansley. Biochim. Biophys. Acta, 118, 19 (1966).

24. Lemberg, R., and J.T. Stanbury. Biochim. Biophys. Acta, 143, 37 (1967).

25. Gilmour, M.V., R. Lemberg, and B. Chance. Biochim. Biophys. Acta, 172, 37 (1969).

26. Wainio, W.W. In Oxidases and Related Redox Systems (T.E. King, H.S. Mason, and M. Morrison, eds.), J. Wiley & Sons, New York, 1965, p. 622.

27. Sekuzu, I., H. Mizushima, and K. Okunuki. Biochim. Biophys. Acta, 85, 516 (1964).

28. Okunuki, K. In Comprehensive Biochemistry (M. Florkin and E.H. Stotz, eds.), Elsevier Publ. Co., Amsterdam, 1966, p. 232.

29. Orii, Y., and K. Okunuki. J. Biochem. (Tokyo), 53, 489 (1963).

30. Wharton, D.C., and Q.H. Gibson. J. Biol. Chem., 243, 702 (1968).

31. Williams, C.R., R. Lemberg, and M.E. Cutler. Canad. J. Biochem., 46, 1371 (1968).

32. Lemberg, R. In Structure and Function of Cytochromes (K. Okunuki, M.D. Kamer, and I. Sekuzu, eds.), University of Tokyo Press, Tokyo 1968, p. 151.

AZIDE AS A MONITOR OF ENERGY COUPLING IN CYTOCHROME OXIDASE*

D.F. Wilson

Johnson Research Foundation, School of Medicine
University of Pennsylvania, Philadelphia, Pennsylvania 19104

Cytochrome oxidase is a fascinating hemoprotein and its reaction with inhibitory ligands has been the principal method for studying its mechanism of action. Keilin (1) used the difference in reactivity toward cyanide and later carbon monoxide as the basis for the distinction between cytochromes a and a_3 and these differences were then generalized to other ligands for which spectrally distinguishable compounds could not be observed such as azide, sulfide and hydroxylamine. There now seems no doubt that sulfide reacts with oxidized cytochrome a_3 but azide and hydroxylamine are less certain at least as far as the site of inhibition in mitochondria is concerned. In intact, phosphorylating mitochondria the azide inhibition is difficult to explain in terms of a simple binding to oxidized cytochrome a_3 for three reasons:

1. The inhibitor constant for azide is more than 12 times greater in the presence of saturating amounts of chemical uncouplers than in State 3 with excess ADP and inorganic phosphate, that is, phosphorylating conditions (2,3,4,5). No comparable effect is seen for cyanide inhibition.

2. In the azide inhibited steady state the spectrum of reduced cytochrome a is shifted a new mμ to the blue and the Soret band is more split than for normal reduced cytochrome a (2,6,7). Again no shift is observed for cyanide, $S^=$ or CO inhibition (6,7).

3. Azide is kinetically uncompetitive with respect to substrate as shown by parallel lines on Lineweaver-Burke double reciprocal plots while cyanide is noncompetitive (3,5, 8,9).

At this time I would like to present some preliminary

*Supported by USPHS 12202.

evidence that the azide binding site is strongly influenced by the coupling of electron transport to the energy conservation reactions.

Figure 1 is a room temperature spectrum of the absorbance changes in rat liver mitochondria on going from the azide inhibited aerobic steady state to anaerobic as measured with a dual wavelength instrument. The Soret region changes are dominated by cytochrome a_3 but in the alpha region we see the characteristic shift of the a alpha band from 602 mμ in the azide inhibited state to 605 mμ on anaerobiosis. The difference gives a peak at 600 mμ and a trough at 608 mμ.

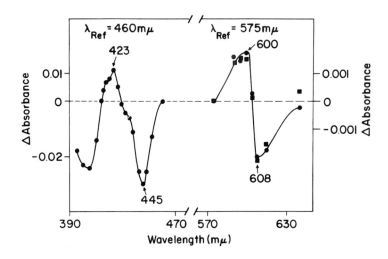

Figure 1. The wavelength dependence of the spectral change induced by the addition of azide to anaerobic mitochondria in the presence of ADP and P_i. The effective bandwidths of the spectrophotometer beams were between 0.8 and 1.5 mμ. The mitochondria were suspended at 3.3 mg/ml in 0.22 M mannitol, 0.5 M sucrose, 15 mM Tris and 10 mM P_i medium, pH 7.2. Succinate (6.7 mM) in the prsence of 3 μM rotenone was used as substrate. ●, absorbance change on addition of azide. ▪ , the opposite sign change on subsequent addition of 50 μM dicumarol. Reproduced from Biochim. Biophys. Acta, 131, 433 (1967), Figure 3.

In the present study we have monitored the kinetics of this transition by measuring at 608 mμ minus 600 mμ with a dual wavelength spectrophotometer.

As shown in Figure 2, if the mitochondria are first put in the azide inhibited steady state by sequential addition of succinate, phosphate, ADP, and azide, then an excess of cyanide (here 13 mM) added to stop respiration we see a first order transition with a half-time of approximately 8 seconds. If chemical uncouplers are added prior to the cyanide the rate is increased at least 4-fold.

In phosphorylating mitochondria the half-time in ·the absence of uncoupler is dependent on the concentration of added cyanide only up to about 4 mM and then is cyanide concentration independent at least up to 15 mM cyanide (Figure 3). The observed half-time of the cyanide independent region (here approximately 8 seconds) is the same as is observed if the oxygen consumption is stopped with excess sodium sulfide or by dithionite. This gives us reasonable confidence that the reaction is a property of the azide binding site and not the method used to stop respiration.

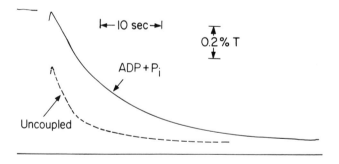

Figure 2. The cyanide induced transition from cytochrome a_{600} to cytochrome a_{605}. Rat liver mitochondria were suspended in a mannitol, sucrose, Tris (30 mM), phosphate (10 mM) medium pH 7.2 at 3 mg protein/ml. Succinate (6.7 mM), ADP (660 μM) or 0.5 μM 5-Cl-3-t-butyl,2'-Cl,4'-NO$_2$-salicylanilide and NaN$_3$ (1.33 mM). The transition was initiated by adding 13 mM KCN and monitored with a dual wavelength spectrophotometer at 600 mμ minus 608 mμ.

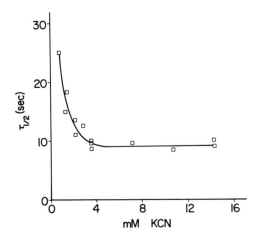

Figure 3. The cyanide concentration dependence of the cytochrome a_{600} to cytochrome a_{605} transition. Rat liver mitochondria were suspended at 6 mg protein/ml in MST-P_i (30 mM Tris and 10 mM P_i) pH 7.3. Rotenone (2 µM), succinate (7 mM), ADP (600 µM) and KN_3 (2.86 mM) were added and then the designated KCN concentration added. The transition was measured at 600–608 mµ with a dual wavelength spectrophotometer.

The azide concentration dependence (Figure 4) was determined using a cyanide concentration of 13 mM to stop the respiration at a series of different azide concentrations. On the left is the axis for the observed half-times and on the right is the observed extent of the transition. If we first look at the extent of the change in transmission we see a typical binding curve with an apparent dissociation constant of approximately 130 µM, in excellent agreement with the binding constant measured previously by direct titration (3,10). The half-time does not change with the extent of the transition. Since this dissociation constant is the same as the kinetically determined inhibitor constant, the half-times are independent of the steady state rate of electron transfer prior to adding the blocking agent. The rate of cytochrome a_{600} to a_{605} transition is therefore a property of the azide bound sites and not of a site in equilibrium with the azide. These are the properties expected if the transition is controlled by the rate of azide leaving. Dr. Wohlrab and I have therefore measured the "on" velocity constant for azide in order to see if the calculated azide dis-

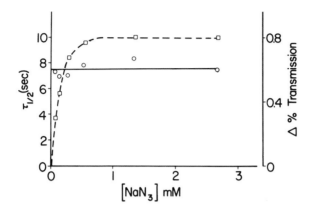

Figure 4. The azide concentration dependence of the extent (☐) and rate (o) of the cytochrome $\underline{a}_{600}$ to cytochrome $\underline{a}_{605}$ transition. The rat liver mitochondria were suspended in MST-P_i (30 mM Tris and 10 mM P_i) at a final concentration of 0.4 M cytochrome $\underline{a}$. Rotenone (2 μM), succinate (12 mM), ADP (600 μM) and the designated NaN$_3$ amount were added and then 13 mM KCN. The transition was measured at 600-608 mμ.

sociation constant is in reasonable agreement with the experimentally determined values.

When azide is mixed with mitochondria in the aerobic steady state we see a pseudo first order reduction of cytochrome $\underline{a}$ (Figure 5) which corresponds to azide binding and the transition to the azide inhibited steady state. This pseudo first order rate constant is directly proportional to the concentration of added azide, yielding a calculated second order constant which is strongly pH dependent (Table I). When it is recalculated for hydrazoic acid the rate constant is pH independent within experimental error suggesting that hydrazoic acid is the principal reactant and that the proper rate constant is approximately 1×10^6 M^{-1}sec^{-1}.

We can now make a balance sheet (Table II) and see if the observed value for the dissociation constant is consistent with the measured on and off velocity constants. The "on" velocity constant at pH 7.3 is approximately 4.5×10^3 M^{-1}sec^{-1}

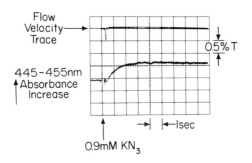

Figure 5. The rate of reaction of azide with mitochondria in the aerobic steady state. Rat liver mitochondria were suspended at 3 mg protein/ml in pH 7.3 MST-P$_i$. 20 mM succinate and 0.2 µM 5-Cl,3-t-butyl,2'-Cl,4'-NO$_2$-salicylanilide were added and then 0.9 mM KN$_3$ rapidly added (8 msec) in a stopped flow apparatus. The sample chamber light path was 0.6 cm.

TABLE I
The "On" Velocity Constant for the Azide Reaction
With Mitochondrial Cytochrome Oxidase

Reactant	Calculated Rate Constants ($M^{-1}sec^{-1}$)		
	pH 7.8	pH 7.3	pH 6.8
N$_3^-$	8×10^2	4.5×10^3	1.5×10^4
*HN$_3$	1×10^6	2×10^6	2×10^6

The rat liver mitochondria were suspended at 27 mg protein per ml in a MSET-P$_i$ medium containing 17 µM TTFB, 8 mM succinate and 1.2 mM ADP. The reduction of cytochrome a was measured at 445-455 mµ after mixing with an equal amount of aerobic MSET-P$_i$-azide solution in a stopped flow apparatus.

*Calculated assuming a pK of 4.7.

TABLE II

A Comparison of the Measured and Calculated Dissociation
Constants for the Azide-Cytochrome Oxidase Complex
in Uncoupled Mitochondria

$k_1 (N_3^-)$	k_{-1}	k_d (calculated)	k_d (measured)
$4.5 \times 10^3 M^{-1} sec^{-1}$	$0.5 \ sec^{-1}$	$110 \ \mu M$	$100-200 \ \mu M$

$$\text{cyt ox} + N_3^- \xrightleftharpoons[k_{-1}]{k_1} \text{cyt ox-}N_3$$

calculated for the azide ion while the "off" velocity constant is approximately $0.5 \ sec^{-1}$ for uncoupled mitochondria.
Their ratio yields a calculated dissociation constant of 110
μM, certainly within the 100 to 200 μM range obtained by titrating the cytochrome a reduction as shown in Figure 4.

If this cytochrome a_{600} to cytochrome a_{605} reaction is
indeed a measure of the "off" velocity for azide then it is
necessary to postulate that uncoupling causes a modification
of the cytochrome oxidase, possibly a structural rearrangement, which increases this rate by four-fold. Such an increase in the azide "off" velocity does not solve the question
of the release of azide inhibition by uncouplers because the
azide dissociation constant does not increase (2-4) suggesting that the "on" velocity is also increased as measured by
the steady state reduction of cytochrome a. The cytochrome
a_{600} to cytochrome a_{605} transition is therefore one more
measure of the control exercised by energy coupling reactions
on the mechanism of electron transport and the properties of
the electron transport carriers of cytochrome oxidase.

References

1. Keilin, D. and Hartree, E.F. Proc. Roy. Soc. London
    Ser. B, 127, 167 (1939).

2. Wilson, D.F. and Chance, B. Biochem. Biophys. Res.
    Commun., 23, 75 (1966).

3. Wilson, D.F. and Chance, B. Biochim. Biophys. Acta,
    131, 421 (1967).

4. Nicholls, P. and Kimelberg, H.K. Biochim. Biophys.
    Acta, 162, 11 (1968).

5. Palmierig, F. and Klingenberg, M. Europ. J. Biochem., 1, 439 (1967).

6. Wilson, D.F. and Gilmour, M.V. Biochim. Biophys. Acta, 143, 52 (1967).

7. Gilmour, M.V., Wilson, D.F. and Lemberg, R. Biochim. Biophys. Acta, 143, 487 (1967).

8. Minaert, K. Biochim. Biophys. Acta, 54, 26 (1961).

9. Winzler, R.J. J. Cell. Comp. Physiol., 21, 229 (1943).

10. Wilson, D.F. Biochim. Biophys. Acta, 131, 431 (1967).

SOME PROPERTIES OF "OXYGENATED" CYTOCHROME $\underline{aa}_3$

R.H. Tiesjema, A.O. Muijsers, B.F. Van Gelder,
M.F.Y. Blokzijl, and E.C. Slater

Laboratory of Biochemistry, B.C.P. Jansen Institute
University of Amsterdam, Amsterdam, The Netherlands

"Oxygenated" cytochrome $\underline{aa}_3$, characterized by a Soret peak at 428 nm (cf. ferrocytochrome $\underline{aa}_3$, 443-44 nm; ferrocytochrome $\underline{aa}_3$, 418-424 nm) and a more intense α-band, is formed by passing oxygen through a solution of cytochrome $\underline{aa}_3$ reduced by $Na_2S_2O_4$ (1) or other agents (2), or by adding H_2O_2 to ferricytochrome $\underline{aa}_3$ (3).

"Oxygenated" cytochrome $\underline{aa}_3$ is unstable and is spontaneously converted to ferricytochrome $\underline{aa}_3$ in a first-order reaction (Figure 1). THe first-order constant lies between 4 and 6 x 10^{-4} sec^{-1}. A value of 6 x 10^{-4} sec^{-1} was found when the conversion of the "oxygenated" cytochrome $\underline{aa}_3$ into ferricytochrome $\underline{aa}_3$ was measured anaerobically. Thus the rate of conversion is independent of the amount of oxygen present.

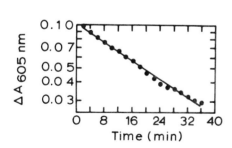

Figure 1. Decomposition of "oxygenated" cytochrome $\underline{aa}_3$ as measured by decline of A_{605nm}. 0.05 M potassium phosphate, 0.03 M sodium cholate, 58.3 μM heme $\underline{a}$, pH 7.2, at 22°. Light path, 1 cm. ΔA_{605nm} represents the difference between A_{605nm} at time $\underline{t}$ and final value.

The "oxygenated" form can also be made by passing the dithionite-reduced enzyme through a Sephadex G-25 column equilibrated with 3% cholate and 50 mM phosphate, pH 7.3 (4).

The same method may also be used after reduction of the enzyme under anaerobic conditions with an excess of NADH and a catalytic amount of phenazine metholsulfate. The first-order rate constants for the conversion of "oxygenated" cytochrome aa_3 into the ferri form were 7.5×10^{-4} and 15×10^{-4} sec^{-1}, respectively, for the "oxygenated" enzyme made in this way.

The EPR spectrum for the "oxygenated" form of cytochrome aa_3, prepared in the classical way by reduction of the enzyme with $Na_2S_2O_4$ followed by passing O_2 through the solution, is shown in Figure 2, trace A. In agreement with Beinert (personal communication), a large signal at g = 6 and the usual signals belonging to the oxidized enzyme are found. However,

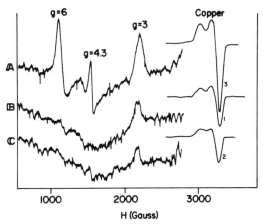

Figure 2. The EPR spectra of the "oxygenated" form of cytochrome aa_3. A. The "oxygenated" form was prepared by reducing the enzyme (0.48 mM in heme) anaerobically with a small excess of $Na_2S_2O_4$ followed by passing oxygen through the solution. B,C. The "oxygenated" form was prepared by reducing the enzyme (0.48 mM in heme) anaerobically with a small excess of $Na_2S_2O_4$ (b) or NADH and phenazine methosulfate (C) followed by passing the enzyme through a Sephadex G-25 column equilibrated with 3% cholate and 50 mM phosphate buffer, pH 7.3. The enzyme was about 4X diluted after passing the column. The conditions of EPR spectroscopy were: microwave power, 200 mW; modulation amplitude, 12 gauss; microwave frequency, 9180 Gc; scanning rate, 625 gauss min^{-1}; time constant, 0.3 sec; temperature, 79°K. The gain for the high-field part of the spectrum was 32X less than that for the low-field part. The gain in Trace A was one-half as large as that in Traces B and C.

when the "oxygenated" form was prepared by passing the enzyme
reduced by dithionite, or NADH and catalytic amounts of phenazine metholsulfate, through a G-25 column, no g = 6 signal was
observed (Figure 2, traces B and C). In all three cases, the
maximum of the Soret peak of the "oxygenated" form was found
at 427 nm. It may be concluded that the g = 6 signal is not
an essential feature of the "oxygenated" form. It may be due
to the presence of oxidation products of dithionite. This
phenomenon is still under investigation.

The oxidation state of "oxygenated" cytochrome aa_3 was
determined by applying van Gelder's titration (5) with NADH
in the presence of phenazine metholsulfate. The reduction
with an excess of NADH is biphasic, as shown in Figure 3, a
rapid increase at A_{605nm} being followed by a much slower
increase. Under these conditions, ferricytochrome aa_3 is
completely reduced within 1 to 2 minutes.

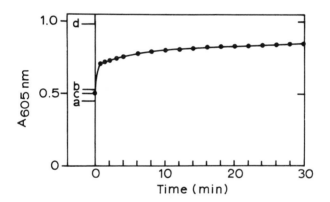

Figure 3. Reduction of "oxygenated" cytochrome aa_3 at 605 nm
with excess NADH in the presence of phenazine metholsulfate
and absence of oxygen. The "oxygenated" enzyme was prepared
by reduction with a small excess of $Na_2S_2O_4$, followed by passing O_2 through the solution. 0.05 M potassium phosphate, 4 mM
Tris sulfate, 0.03 M sodium cholate, 47.5 μM heme a, 125 μM
NADH, 0.7 μM phenazine metholsulfate, pH 7.2. Light path, 1
cm. Absorbances corrected for evaporation and dilution. The
four levels at the left-hand side give the A_{605nm} values of
a, ferricytochrome aa_3; b, "oxygenated" cytochrome aa_3; c,
value at beginning of titration (14 min after preparation of
"oxygenated" form); d, ferrocytochrome aa_3 (reduced with
$Na_2S_2O_4$). Temperature, 22°.

During the evacuation procedure, 34% of the "oxygenated" cytochrome aa_3 was converted into the ferri form. The absorbance expected after reduction of this amount of the ferri form is 0.686. The absorbance reached in the rapid phase is 0.696. The slow phase is a first-order reaction (Figure 4) with a first-order constant of 4.5×10^{-4} sec^{-1}, which is within the limits of the constant for the conversion of the "oxygenated" form to the ferri form. On the basis of this constant, it may be calculated that one would expect a conversion of 32% of the "oxygenated" form into the ferri form during the evacuation procedure (cf. 34% found).

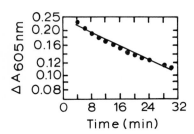

Figure 4. Slow phase of the course of reduction of "oxygenated" cytochrome aa_3 at 605 nm, carried out as in Figure 3. 41.6 µM heme a. A_{605nm} is the difference between A_{605nm} at time t and final value.

Thus, the fast phase in Figure 3 is due to reduction of ferricytochrome aa_3 formed from the "oxygenated" enzyme during the conversion of the remaining "oxygenated" form to the ferri form, followed by reduction of the latter.

A straight line is obtained by plotting A_{340nm} against A_{640nm}, in Figure 5. By correcting for the increase of absorbance at 340 nm of cytochrome aa_3 on reduction, the extinction coefficient in the difference spectrum at 605 nm (reduced-minus-"oxygenated") was calculated to be 10.3 cm^2 per mole heme a, assuming that each heme a molecule is reduced by 2 equivalents (5). The directly measured extinction coefficient was the same. Thus the titration of "oxygenated" cytochrome aa_3 is described by the scheme

$$\text{"oxygenated" } aa_3 \longrightarrow \text{ferri } aa_3 \xrightarrow{4e^-} \text{ferro } aa_3$$

and it is very likely that "oxygenated" cytochrome aa_3 is in the same oxidation state as the ferri form.

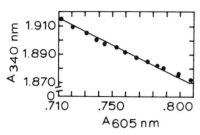

Figure 5. Relationship between A_{340nm} and A_{605nm} during reduction of "oxygenated" cytochrome aa_3 carried out as in Figure 3. 41.6 µM heme a.

The results support the suggestion made by Wharton and Gibson (6) and Williams et al. (4) that "oxygenated" cytochrome aa_3 is another conformational form of ferricytochrome aa_3.

References

1. Sekuzu, I., S. Takemori, T. Yonetani and K. Okunuki. J. Biochem., Tokyo, 46, 43 (1959).

2. Minnaert, K. Biochim. Biophys. Acta, 35, 282 (1959).

3. Greenwood, C. Biochem. J., 86, 535 (1963).

4. Williams, G.R., R. Lemberg and M.E. Cutler. Canad. J. Biochem., 46, 1371 (1968).

5. Beinert, H. Personal communication.

6. Van Gelder, B.F. Biochim. Biophys. Acta, 73, 663 (1963).

7. Wharton, D.C. and Q.H. Gibson. J. Biol. Chem., 243, 702 (1968).

Brand: I would like to indicate that it is also possible to make sulfonyl chloride derivatives of the N-acrylaminonaphthalene sulfonates, although this cannot be done with the 1,8 derivatives. It is possible that you might see larger fluorescence changes than those you observed with the DNS conjugates. Dr. Turner has prepared the sulfonyl chloride of 1,7 ANS.

Turner: I would point out that DNS, when it is coupled to amino acids, actually has fluorescence properties very similar to free ANS. Therefore, I think that Dr. Azzi's remarks about the interpretation of the fluorescence of DNS conjugated to protein are entirely in order.

Nicholls: I have a question about ligand binding to cytochrome oxidase relevant to the papers of Dr. Beinert and Dr. Wilson. We can consider three forms of the oxidase: $\underline{a}^{3+}\ \underline{a}_3^{3+}$, $\underline{a}^{2+}\ \underline{a}_3^{3+}$, and $\underline{a}^{2+}\ \underline{a}_3^{2+}$. Then according to the classical convention, azide and cyanide can bind to the first two forms, but presumably not the last one. However, we know from Dr. Beinert's EPR studies that in $\underline{a}^{3+}\ \underline{a}_3^{3+}$, and $\underline{a}_3$ signal is not seen and the corresponding a_3-azide signal is not seen. Both signals appear only in what is presumably the $\underline{a}^{2+}\ \underline{a}_3^{3+}$ complex; that is cytochrome $\underline{a}_3$ can see cytochrome $\underline{a}$ in the reduced form. Conversely, the spectroscopic observations of Dr. Wilson show that in this complex cytochrome $\underline{a}$ can see cytochrome $\underline{a}_3$ in the oxidized form, because the position of the absorption band of reduced cytochrome $\underline{a}$ differs according to whether a_3 is azide bound or free. The question is, therefore, is there any convincing evidence that azide, or indeed cyanide, binds to any form but the half-reduced form? It is possible that cytochrome $\underline{a}_3$ combines with azide only when cytochrome $\underline{a}$ is in the ferrous state.

Beinert: I am afraid I don't have the answer because, by EPR, we do not see azide bound to $\underline{a}_3$ in the fully oxidized form of the enzyme. We can go to the partly reduced state $\underline{a}^{2+}\ \underline{a}_3^{3+}$ and then the azide will bind to $\underline{a}_3^{3+}$, but when we then reoxidize this completely, the a_3^{3+}-azide signal disappears. So we apparently cannot keep a_3^{3+} tied up, at least as far as we can detect by EPR, when we go back to the completely oxidized form and similarly it seems that azide does not go on to a_3^{3+} in the first place in the fully oxidized form.

Wilson: Of course, you have a couple of problems with the azide. I agree that it won't bind to the reduced form, and you can't tell if it is bound in the oxidized form. Moreover, the kinetics in the intact system are that of an un competitive inhibitor with respect to the succinate or with respect to the TMPD, for whatever that means. In answer to Dr. Nicholls' question, I think Dr. Yonetani's data (1) would show that you probably can put cyanide on the oxidized cytochrome a_3, but it is a very slow reaction. He reported that if you reduced cytochrome oxidase and then add cyanide, you get a very rapid equilibrium with cyanide. If you add cyanide to the oxidized form, you have to add higher concentrations and let it go longer but it will react.

Gilmour: I have a couple of very different questions. First, I would like to ask Dr. Caughey if he still thinks that the 830 mμ band of cytochrome oxidase might be due to heme A? I wonder if his finding of copper with heme a would change his mind on this?

Caughey: We have not given careful consideration to this point with the copper-heme A complexes. However, the uncertainties which apply to the intact oxidase may also apply to this complex. Iron(III) porphyrins (including hemin A), reduced hemes, and copper complexes can exhibit absorption bands in the region of the 830 mμ band. Therefore, the so-called copper band of the oxidase could arise from hemin or copper components. Until one knows the nature of effects of interactions between heme and copper moieties upon the spectrum of the individual components in this region, the relative contributions of hemins and coppers to the intensity of the 830 mμ band will remain in doubt.

Gilmour: Dr. Beinert had shown variations in the spectra of the oxygenated compound from different workers. I think I can agree, and I think this is troublesome, but it is also true that if you take the ratio of the fully oxidized peak to the fully reduced peak of different people's preparations,

you will see some variations in that too. It doesn't help
very much. At least this is what we found when comparing
Lemberg's and Wanio's preparations. I also don't quite know
what to do with the intermediate that Dr. Beinert was talking
about, and I don't quite know enough about EPR interpretation. Since you can use reducing equivalents to take the
oxygenated back to the ferric, I don't quite see how this
relates to the different forms that you gave. Dr. Lemberg
has often suggested that the oxygenated compound doesn't
have any oxygen on it, because when we tried to look for it,
we just didn't find it. We don't know whether it goes off
very rapidly when you add reducing equivalents or whether it
just isn't there to begin with. Also, I wonder if EPR can
detect any of the higher oxidation states which Dr. Lemberg
has suggested.

Caughey: Before this question is answered by Dr. Beinert,
perhaps we can decide whether we are still going to use the
term "oxygenated complex" or what we are going to call it.

Beinert: I would say that, as long as we can't identify a
definite species which represents the "oxygenated" form, I
would rather not call it anything at the moment.

How we explain that reducing equivalents may get the
so-called oxygenated form, which we obtain as a mixture of
intermediate species, back to the oxidized form? I didn't
have enough time to explain it in my talk. I said, the peculiar situation seems to be that molecular oxygen can get
completely reduced oxidase back to the oxidized oxidase very
fast, because a_3 is in the reduced form and receptive toward
oxygen. However, in these intermediate species (say, a_3^{3+} a^{3+}
Cu_u^+ Cu_d^{2+}) a_3 is uncoupled from the copper--or whatever it is
coupled to--and is in the oxidized form; since a_3 is not
reduced, the whole unit is not receptive to oxygen as an oxidant. All you have to do is either add reducing equivalents
to get the reduced a_3 and then the thing will easily go back
oxidized. Or, if you add one electron carriers (e.g., cytochrome c, ferri- or ferrocyanide), intermolecular, or possibly
also intramolecular, oxidation-reductions will ensue in the
course of which ferric oxidase is formed directly or again
via the completely reduced form plus oxygen. This is the way
we would explain it. We have done this with cytochrome c
and ferricyanide, and what we obtained as the "oxygenated"
form will go back easier to the oxidized form. We think
that you are getting stuck with species which have no

efficient receptor for oxygen because a_3 is uncoupled and in the ferric form.

Slater: I would like to make a comment on this point. I think the titration makes it very likely that the oxygenated compound is at the same oxidation state as the ferric. The difference is that it is not yet reduced by reduced phenazine methosulphate. It is first decomposed to the ferric, which is reduced by the reduced phenazine methionlphate. It is unlikely that oxidizing equivalents unreactive with reduced methsulphate escape from the oxygenated compound during its decomposition to the ferric.

Myer: Two points which I couldn't make during the earlier presentation and may have some impact upon the discussion are: We have clearly shown the CD spectrum of so-called oxygen complex which is distinct from that of the oxidized and the reduced preparation. The second is that upon addition of potassium ferricyanide, we could regenerate the CD spectrum of the oxidized oxidase. In other words, the ferricyanide did change the dichronic spectra of the oxygen complex to that of the native molecule. This does seem to imply that a certain portion of the oxidase in the oxygenated complex is in the reduced form, as has been mentioned by Dr. Beinert.

Beinert: I think I should provide an answer to Dr. Slater's comments and also to a recent paper by Williams, Lemberg, and Cutler (2), whose contents came to our attention only during this meeting. There are obviously discrepancies in our EPR findings and interpretations. Not only must the results obtained elsewhere be compatible with ours and our schemes, but also, what we found must be accomodated in the schemes that our colleagues put forth. I have originally tried to explain the "oxygenated" form with the knowledge that we have at present, that is on the basis of the forms of the enzyme we now know to exist (3,4). One of the reasons for trying this is that in the EPR spectrum there is no evidence that there is yet another species corresponding to the "oxygenated" one. This is, of course, negative evidence. Nevertheless, in the EPR spectrum, we saw so far nothing else of significance--other than what we can explain without invoking an additional "oxygenated" species--while in the same sample we saw the optical spectra of the "oxygenated form" as they have been published by others (Figure 3). In addition, we accounted for a large portion of the heme in the EPR spectra, where it appears as a (g = 3) and as the uncoupled a_3 (g = 6). Therefore, if there was yet another species around, it obviously must have been there in addition to the species which we can detect by their usual EPR signals; this again would say that it is a mixture of species with which we were dealing.

Since, according to van Gelder's findings, there can be an "oxygenated" form without a g = 6 signal, there is then apparently a separate entity with spectral characteristics of the "oxygenated" form; and according to what we saw there must be something like a pseudo-"oxygenated" species, which we observed and which others must also have had, something other than what Dr. Slater is looking at; this pseudo-species I visualize as a mixture of species showing optical spectral characteristics very similar to a true, separate "oxygenated" species (cf Figures 2-4); the latter would then be what Dr. Slater is dealing with.

References

1. Yonetani, T. and G.S. Ray. J. Biol. Chem., 240, 3392 (1965).

2. Williams, Lemberg, and Cutler. Can. J. Biochem., 46, 1371 (1968).

3. van Gelder, et al. Proc. Natl. Acad. Sci. U.S., 58, 1073 (1967).

4. van Gelder and Beinert. Biochim. Biophys. Acta, in press.

EPR AND NMR SPECTRA OF A HEME A - COPPER COMPLEX ISOLATED FROM BOVINE HEART MUSCLE[*]

Ronald A. Bayne, George A. Smythe[+] and Winslow S. Caughey[+]

Department of Chemistry
University of South Florida
Tampa, Florida 33620

On several occasions EPR spectra taken in the course of isolating heme A from bovine heart muscle following the procedure developed in our laboratory (1) have shown significant quantities of copper to be present, along with hemin A, through the chromatographic fractionation step. A typical EPR spectrum for such a copper-hemin A complex in chloroform at 103°K is shown in Fig. 1. The spectrum of the solid at this temperature was similar with signals ascribable to hemin iron at $\underline{g}$ values of ca. 6 and 4 and a typical copper(II) signal near $\underline{g}$ = 2. Copper was also detected by chemical analysis (2) in these preparations. Acid washing followed by chromatography readily removed copper.

EPR spectra of solid, copper-free hemin A preparations at 103°K, either as normally isolated (1) or as isolated by Dr. Sue McCoy in these laboratories using the acid-acetone extraction and subsequent procedures followed by Lynen and co-workers (3) (Fig. 2), showed strong $\underline{g}$ = 6 and weak $\underline{g}$ = 4 and 2 signals. Only trace amounts of copper were indicated in these preparations, a finding also confirmed by chemical analysis.

The NMR spectrum of the complex in pyridine-d_5 exhibited absorption in the 0.5 to 2.5δ region consistent with a long alkyl group (17 carbons) with unsaturation (three double bonds). These absorptions have been reported earlier for heme A preparations without copper (4). However, with copper present the spectrum for the pyridine-d_4 was strikingly different from the case with copper absent. The resonance due

[*]This work was supported by USPHS Grant HE-11807.
[+]Present address: Department of Chemistry, Arizona State University, Tempe, Arizona 85281.

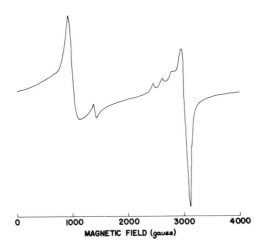

MAGNETIC FIELD (gauss)

Figure 1. EPR spectrum of hemin A-copper complex in chloroform. The measurement was carried out at 103°K using a Jeolco JES ME ESR spectrometer with a 100-KHz amplifier. Conditions were: modulation amplitude, 4 gauss; microwave power, 4 mW;
4 mW; frequency, 9.24 GHz; scanning speed, 1000 gauss per min.;
response, 0.03 sec.; gain, 7.1 x 10. Myoglobin fluoride and
$MnCl_2$ were used for calibration purposes.

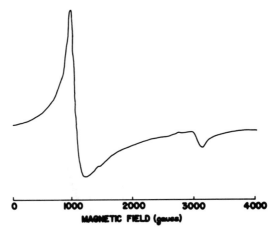

MAGNETIC FIELD (gauss)

Figure 2. EPR spectrum of acid-acetone purified hemin A.
Conditions of measurement were the same as described in Fig. 1.

to protons in the 2,6 positions of the pyridine at ca. 8.75δ were greatly broadened compared with those for the 4 and 3,5 protons. This broadening presumably resulted from solvent interaction with the copper(II) ion. Examination of the high-field and low-field regions, where resonances have been

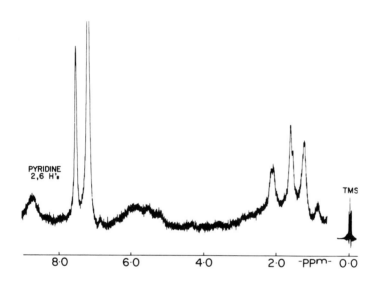

Figure 3. NMR spectrum of hemin A-copper complex in pyridine-d₅ obtained with a Jeolco JNM-4H-100 spectrometer operating at 100 MHz.

found in high-spin iron(III) porphyrins, revealed no resonances. However, the resonances observed, particularly within the 5 to 7δ region, were highly suggestive of a situation similar to that noted for dimeric hemins with an FeOFe linkage (5).

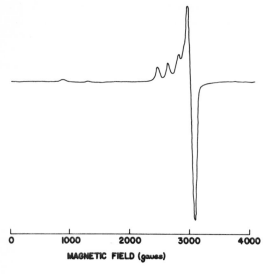

Figure 4. EPR
spectrum of hemin-
A copper complex
dissolved in pyri-
dine and incubated
at room temperature
overnight. The
conditions were the
same as in Fig. 1.

MAGNETIC FIELD (gauss)

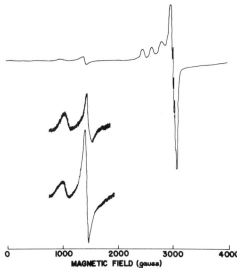

MAGNETIC FIELD (gauss)

Figure 5. EPR spectrum of the hemin A-copper complex in pyridine described in Fig. 4 after the addition of water. The
conditions of measurement of the complete spectrum were the
same as described in Fig. 1. The middle insert of the $g = 6$
and 4 region was run at a gain setting of 3.6 x 100 and a
response of 0.1 sec; bottom insert: same solution after
several days incubation in the pyridine-water with same experimental conditions.

The EPR spectrum of the copper-hemin A complex in pyridine (Fig. 4) showed only very weak signals in the g = 6 and g = 4 regions but an enhanced signal at g = 2, possibly due to combined copper(II) and FeOFe-like signals. Addition of water to the pyridine solution caused the signal in the g = 4 region to intensify (Fig. 5) giving a spectrum in the g = 6 and 4 regions not unlike that noted for a Yonetani "direct dialyzed" preparation (2) of cytochrome oxidase (Fig. 6), although the g = 2 region was significantly different.

The origin of the copper in these preparations is uncertain at present. Nevertheless it is possible that we are dealing with a heme A-copper complex native to cytochrome oxidase. Therefore, investigation of the properties, structures, and methods for isolation of the complexes from purified cytochrome oxidase, bovine heart muscle, and other tissues are continuing.

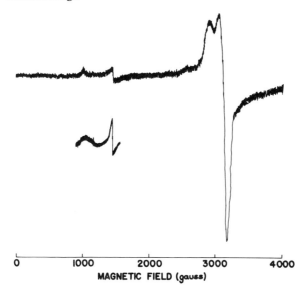

MAGNETIC FIELD (gauss)

Figure 6. EPR spectrum of cytochrome oxidase. The enzyme met all of the purity and activity criteria proposed by Yonetani (6). Conditions were: modulation amplitude, 4 gauss; microwave power, 8 mW; frequency, 9.24 GHz; scanning speed, 500 gauss per min.; response, 0.1 sec.; gain, 2.5 x 100; enzyme concentration, 0.4 mM. The bottom insert was run at 1000 gain setting; 1.0 sec. response; 200 gauss per min. scanning speed.

617

References

1. York, J. L., S. McCoy, D. N. Taylor and W. S. Caughey, J. Biol. Chem. 242, 908 (1967).

2. Yonetani, T., J. Biol. Chem. 236, 1680 (1961).

3. Grassl, M., G. Augsburg, U. Coy and F. Lynen, Biochem. J. 337, 35 (1963).

4. Caughey, W. S., J. L. York, S. McCoy and D. P. Hollis, in Hemes and Hemoproteins (B. Chance, R. W. Estabrook and T. Yonetani, eds.) Academic Press, New York, 1966, p. 25.

5. Smythe, G. A., R. A. Bayne and W. S. Caughey, this volume. p. 5

6. Yonetani, T., in Methods in Enzymology, Vol. X (R. W. Estabrook and M. E. Pullman, eds.) Academic Press, New York, 1967, p. 332.

SUBJECT INDEX

A

Absorption spectrum, 285, 344, 360,
461, 500, 547, 549
Acetylation, 87-89
Acrylamide gel electrophoresis, 133, 149
Acylation, 107, 110, 118, 129
Adair's formal equation, 343
ADP, 132, 135, 137, 141-143
ADP·Ribose
competitive inhibitor, 40-53
enhanced, 50
Aggregation-deaggregation systems, 420,
421
Alcohol, steroid derivatives, 25
Alcohol dehydrogenase, 22, 25-35, 37-46,
49-57
Aldol cleavage, 155-156, 162, 164
condensation, 155-156, 160
Aliphatic amino acid residue, 473
Alkylation, 129
"All or none" effect, 442, 444, 445
"All or none" reaction, 370
"Allohysteria," 112
gefilte fish, 133
Allosteric binding, 133
Allosteric enzymes, 132
Allosteric inhibition, 123
Allosteric regulation, 125
Allosteric transition, 65
Amino acid
analysis, 157
analyzer, 159
composition, 28, 71, 267, 278, 279, 281,
462, 497
residues, 468, 469, 481

sequences, 27, 33, 35, 85, 87, 88, 95-98,
203, 465, 488, 489, 540, 541
substitution, 236
Analog computer, 408
Analytical ultracentrifuge, 140, 155
ANS, 37-39, 59-61, 63-68, 161-163
Apo-CCP, 528
Apoenzyme, 5-7, 12, 13, 15, 16, 18,
22, 103, 109
Arrhenius plot, 206, 209, 212
Auramine O, 38-39, 43-46
Autocatalysis, 110

B

Bacterial catalases, 561, 568
Bacterial sulphite reductase, 231
Bacterial tryptophan pyrolase, 231
Binding
autocatalytic binding reaction, 373
carbon monoxide, 460, 461
coenzyme, 5
competitive, 38
constant, 106, 137, 310, 320
cyanide, 451, 452
four step binding model, 373
heavy metal, 76, 79
ligand, *see also* Ligand
of oxygen, 347, 350, 384
NAD-enzyme, 118
oxygen binding site, 346
of oxygen *to* hemoglobin, 371
rate constant of biomolecular oxygen,
367
TPNH, 123
xenon, 348

Hamiltonian, 232, 242, 248, 251, 254-257, 516
labeling, 49, 55, 241, 297, 459, 530
-orbit interaction, 248
polarization, 195, 468
Spinco Model E ultracentrifuge, 425
Statistical F-test, 352
Steady state, 119, 120
Stopped flow, 32, 104, 108, 110, 120, 124, 125, 127, 148, 150, 173, 407, 409, 432, 433, 598
apparatus, 598
Structural change, 21, 57
Structure
primary, 101
tertiary, 101, 175
Structure-spin state correlation, 321
Subunit interaction, absence of, 35
Superconducting magnet, 256

T

τ determinations, 150, 151
Temperature
cycling, 263, 264
jump, 148, 150, 153, 300, 549
relaxation method, 399
Tetramer-dimer
dissociation-association reaction, 368, 377, 378
equilibrium, 375, 377, 379
constant, 368
α-Thalassemia, 221
Thermal agitation, 253
Thermal denaturation, 354
Thermal equilibrium, 251, 253, 256
Thermal mixture, 202
Thermodynamic, 267, 452, 453
changes, 344-346, 359
data, 357
laws, 354
linkage, 309, 399
TNS, 38-40
Transacetylase reaction, 84
Transaldolase, 155-164

Transient kinetics, 108, 110
measurements, 119
period, 119-121
Transitions
"all or none," 114, 115
allosteric, 65
Tritium labeling, 157-161

U

Ultraviolet spectra, 134, 147, 152, 167, 286, 494, 537, 583
difference, 286

V

van Gelder's titration, 603
van't Hoff plot, 117, 207, 208, 344-346, 358
Varian EPR spectra spectrometer, 529
Varian HR-220 spectrometer, 466, 467
Varian X-band spectrometer, 561
"Verdo," 568
Visible spectrum, 167

W

Wave functions, 197

X

Xenon, 181, 183, 185, 363
-binding, 347, 348, 350, 352
flash, 335
hydroxide, 322, 324
X-ray, 11, 56, 71-73, 75, 76, 80, 98, 101, 144, 176, 181, 203, 482, 505, 509, 519, 558
crystallographic data, 495
crystallography, 528
diffraction, 346, 347, 391, 392
single crystal studies, 241
studies, 451, 465
of crystalline forms, 297